Edited by
Mandy C. Elschner, Sally J. Cutler,
Manfred Weidmann, and
Patrick Butaye

BSL3 and BSL4 Agents

Related Titles

Stulik, J., Toman, R., Butaye, P., Ulrich, R. G. (eds.)

BSL3 and BSL4 Agents

Proteomics, Glycomics, and Antigenicity

2011

ISBN: 978-3-527-32780-5

Katz, R., Zilinskas, R. A. (eds.)

Encyclopedia of Bioterrorism Defense

2011

ISBN: 978-0-470-50893-0

Bergman, N. H. (ed.)

Bacillus anthracis and Anthrax

2010

ISBN: 978-0-470-41011-0

Kostic, T., Butaye, P., Schrenzel, J. (eds.)

Detection of Highly Dangerous Pathogens

Microarray Methods for BSL3 and BSL4 Agents

2009

ISBN: 978-3-527-32275-6

Kwaik, Y. A., Metzger, D. W., Nano, F., Sjostedt, A., Titball, R. (eds.)

Francisella Tularensis

Biology, Pathogenicity, Epidemiology, and Biodefense

2007

ISBN: 978-1-57331-691-0

Torrence, P. F. (ed.)

Combating the Threat of Pandemic Influenza

Drug Discovery Approaches

2007

ISBN: 978-0-470-11879-5

Edited by Mandy C. Elschner, Sally J. Cutler, Manfred Weidmann, and Patrick Butaye

BSL3 and BSL4 Agents

Epidemiology, Microbiology, and Practical Guidelines

WILEY-BLACKWELL

The Editors

Dr. Mandy C. Elschner
Friedrich-Loeffler-Institut
Federal Research Institute for
Animal Health
Institute of Bacterial Infections
and Zoonoses
Naumburger Strasse 96a
07743 Jena
Germany

Dr. Sally J. Cutler
University of East London
School of Health, Sports and Bioscience
Water Lane, Stratford
London E15 4LZ
United Kingdom

Dr. Manfred Weidmann
University Medical Center Göttingen
Department of Virology
Kreuzbergring 57
37075 Göttingen
Germany

Prof. Patrick Butaye
Veterinary & Agrochemical Research Centre
(VAR – CODA – CERVA)
Department of Bacterial Diseases
Groeselenberg
99 B-1180 Brussels
Belgium

Cover
Display of *C. burnetii*-infected Cells in
the Scanning Electron Microscope
© Eye of Science, Reutlingen,
Germany

Library of Congress Card No.: applied for

British Library Cataloguing-in-Publication Data
A catalogue record for this book is available from the British Library.

Bibliographic information published by the Deutsche Nationalbibliothek
The Deutsche Nationalbibliothek lists this publication in the Deutsche Nationalbibliografie; detailed bibliographic data are available on the Internet at <http://dnb.d-nb.de>.

Wiley-Blackwell is an imprint of John Wiley & Sons, formed by the merger of Wiley's global Scientific, Technical, and Medical business with Blackwell Publishing.

Typesetting Laserwords Private Limited, Chennai, India
Printing and Binding betz-druck GmbH, Darmstadt
Cover Design Grafik-Design Schulz, Fußgönheim

Printed in the Federal Republic of Germany
Printed on acid-free paper

Print ISBN: 978-3-527-31715-8
ePDF ISBN: 978-3-527-64510-7
ePub ISBN: 978-3-527-64509-1
Mobi ISBN: 978-3-527-64508-4
oBook ISBN: 978-3-527-64511-4

Contents

Acknowledgment

This publication is supported by COST.

COST – the acronym for European Cooperation in Science and Technology – is the oldest and widest European intergovernmental network for cooperation in research. Established by the Ministerial Conference in November 1971, COST is presently used by the scientific communities of 35 European countries to cooperate in common research projects supported by national funds.

The funds provided by COST – less than 1% of the total value of the projects – support the COST cooperation networks (COST Actions) through which, with € 30 million/year, more than 30 000 European scientists are involved in research having a total value which exceeds € 2 billion/year. This is the financial worth of the European added value which COST achieves.

A "bottom up approach" (the initiative of launching a COST Action comes from the European scientists themselves), "à la carte participation" (only countries interested in the Action participate), "equality of access" (participation is open also to the scientific communities of countries not belonging to the European Union), and "flexible structure" (easy implementation and light management of the research initiatives) are the main characteristics of COST.

As a precursor of advanced multidisciplinary research COST has a very important role for the realization of the European Research Area (ERA) anticipating and complementing the activities of the "Framework Programmes", constituting a "bridge" toward the scientific communities of emerging countries, increasing the mobility of researchers across Europe and fostering the establishment of "Networks of Excellence" in many key scientific domains such as: Biomedicine and Molecular Biosciences, Food and Agriculture, Forests, their Products and Services, Materials, Physical and Nanosciences, Chemistry and Molecular Sciences and Technologies, Earth System Science and Environmental Management, Information and Communication Technologies, Transport and Urban Development, Individuals, Societies, Cultures, and Health. It covers basic and more applied research and also addresses issues of a pre-normative nature or of societal importance.

Web: *http://www.cost.esf.org*

Legal Notice by the COST Office

Preface

This is the third book written within the framework of the European project COST Action B28 which aims to increase knowledge on biosafety level (BSL)3 and BSL4 agents, to support the development of more accurate diagnostic assays, vaccines, and therapeutics, and to better understand the epidemiology of these highly pathogenic microorganisms that can potentially be used as biological weapons.

The research initiatives and interests of COST B28 partners are organized in five working packages/groups:

- **WG1:** Technology platform (including flow-cytometry and microarrays);
- **G2:** Antigenicity;
- **WG3:** Proteomics and glycomics;
- **WG4:** Genomics;
- **WG5:** Microbiology (bacteriology, virology, mycology).

The first book summarized knowledge on micro-array technology (WP1). The second book summarized knowledge on the proteomics, glycomics, and antigenicity of the BSL3/4 agents. This book provides a summary of knowledge on the organisms themselves.

The authors of the chapters are all involved in research concerning these agents and have been working with them extensively. Typically the authors also have access to BSL3/4 laboratories in which they can work with these agents and thus are familiar with the required precautions and legislation. Their expertise has also been employed for the assessment of outbreaks and understanding the epidemiological factors that facilitate their spread and subsequent control.

The book is divided into two major parts, separating the viruses and bacteria. Research, epidemiology therapeutics, and the management of these infections are indeed quite different. All agents can cause severe disease and outbreaks are monitored around the world. Some of the agents have been eliminated from certain parts of the world, such as *Bacillus anthracis* in parts of Western Europe, while others such as *Mycobacterium tuberculosis* are spreading at an ever increasing speed. Nevertheless, for most of the agents, there are regions where they remain endemic and thus the potential for reintroduction requires us to remain vigilant.

Few have actually been used for bioterroristic purposes such as *B. anthracis* in the United States in 2001, or in biowarfare such as *Burkholderia mallei*, used in

World War I to kill horses that were indispensable for warfare at the time. Others are well known for "natural" expansion due to a changed epidemiological situation as was the case for *Coxiella burnetii* in the Netherlands or Rift Valley Fever in Africa. To understand these differences and possibilities, this book offers thorough information and should be of great value for all those involved in research dealing with these highly pathogenic microorganisms, and for the non-expert who might encounter these pathogens and require the background and practical supporting information.

To achieve this latter objective, practical guidelines are included providing "at a glance" information for those handling these microorganisms.

Patrick Butaye
Sally J. Cutler
Mandy C. Elschner
Manfred Weidmann

List of Contributors

F. Xavier Abad
Centre de Recerca en Sanitat
Animal (CReSA)
UAB-IRTA
Campus de la Universitat
Autònoma de Barcelona
Bellaterra
08193 Barcelona
Spain

Massimo Amicosante
National Center of Infectious and
Parasitic Diseases
26 Yanko Sakazov Blvd
1504 Sofia
Bulgaria

and

University of Rome "Tor Vergata"
Department of Internal Medicine
Via Montpellier 1
I-00133 Rome
Italy

and

ProxAgen Ltd.
Sofia
Bulgaria

Markus Antwerpen
Bundeswehr Institute of
Microbiology
Neuherbergstrasse 11
80937 Munich
Germany

Elizabeta Bachiyska
National Center of Infectious and
Parasitic Diseases
26 Yanko Sakazov Blvd
1504 Sofia
Bulgaria

Sina Bavari
United States Army Medical
Research Institute of Infectious
Diseases
Fort Detrick
Frederick
Maryland 21702
USA

Steven B. Bradfute
United States Army Medical
Research Institute of Infectious
Diseases
Fort Detrick
Frederick
Maryland 21702
USA

Nadia Brankova
National Center of Infectious and
Parasitic Diseases
26 Yanko Sakazov Blvd
1504 Sofia
Bulgaria

Patrick Butaye
Veterinary & Agrochemical
Research Centre
(VAR – CODA – CERVA)
Department of Bacterial Diseases
Groeselenberg
99 B-1180 Brussels
Belgium

and

Faculty of Veterinary Medicine
Department of Pathology
Bacteriology
and Poultry Diseases
Ghent University
Salisburylaan 133
9820 Merelbeke
Belgium

Sally J. Cutler
University of East London
School of Health, Sports and
Bioscience
Water Lane, Stratford
London E15 4LZ
United Kingdom

Gerhard Dobler
Bundeswehr Institute of
Microbiology
Neuherbergstrasse 11
80937 Munich
Germany

Mandy C. Elschner
Friedrich-Loeffler-Institut
Federal Research Institute for
Animal Health
Institute of Bacterial Infections
and Zoonoses
Naumburger Strasse 96a
07743 Jena
Germany

Dimitrios Frangoulidis
Bundeswehr Institute of
Microbiology
Neuherbergstrasse 11
80937 Munich
Germany

Joachim Frey
University of Bern
Department of Infectious
Diseases and Pathobiology
Institute for Veterinary
Bacteriology
122 Laenggassstrasse
3012 Bern
Switzerland

Bruno Garin-Bastuji
French Agency for Food,
Occupational and Environmental
Health Safety (ANSES)
EU/OIE/FAO Reference
Laboratory for Brucellosis
Maisons-Alfort 94706
France

Nancy A. Gerloff
Institute of Immunology
Centre de Recherche Public de la
Santé/LNS
20A rue Auguste Lumière
Luxembourg 1950
Luxembourg

Marc Govaerts
Veterinary and Agrochemical
Research Centre
Groeselenberg 99
1190 Uccle
Belgium

Matthias Hanczaruk
Bundeswehr Institute of
Microbiology
Neuherbergstrasse 11
80937 Munich
Germany

Peter B. Jahrling
Integrated Research Facility at
Fort Detrick (IRF-Frederick)
National Institute of Allergy and
Infectious Diseases
National Institutes of Health
Fort Detrick
Frederick
Maryland 21702
USA

Anders Johansson
Umeå University
Department of Clinical
Microbiology
SE-901 87 Umeå
Sweden

Jonas Klingström
Swedish Institute for
Communicable Disease Control
Department of Preparedness
171 82 Solna
Sweden

and

Karolinska Institutet
Department of Microbiology,
Tumor and Cell Biology (MTC)
171 77 Stockholm
Sweden

Jens H. Kuhn
Integrated Research Facility at
Fort Detrick (IRF-Frederick)
National Institute of Allergy and
Infectious Diseases
National Institutes of Health
Fort Detrick
Frederick
Maryland 21702
USA

and

Tunnell Consulting Inc.
King of Prussia
Pennsylvania 19406
USA

Anne Laudisoit
University of Liverpool
Ecology, Evolution and Genomics
of Infectious Disease Research
Group
Biosciences
Crown Street
Liverpool L69 7ZB
UK

and

University of Antwerp
Evolutionary Biology
Groenenborgerlaan 171
Antwerp 2020
Belgium

Victoria Levterova
National Center of Infectious and
Parasitic Diseases
26 Yanko Sakazov Blvd
1504 Sofia
Bulgaria

Alice N. Maina
Institute of Tropical Medicine and
Infectious Diseases
PO Box 62000-00200
Nairobi
Kenya

Hermann Meyer
Bundeswehr Institute of
Microbiology
Neuherbergstrasse 11
80937 Munich
Germany

Ali Mirazimi
Swedish Institute for
Communicable Disease Control
171 82 Solna
Sweden

and

Karolinska Institutet
Department of Microbiology
171 77 Stockholm
Sweden

Claude P. Muller
Institute of Immunology
Centre de Recherche Public de la
Santé/LNS
20A rue Auguste Lumière
1950 Luxembourg
Luxembourg

Andreas Nitsche
Robert Koch-Institut
13353 Berlin
Nordufer 20
Germany

Plamen Padeshki
National Center of Infectious and
Parasitic Diseases
Department of Microbiology
26 Yanko Sakazov Blvd.
1504 Sofia
Bulgaria

Stefan Panaiotov
National Center of Infectious and
Parasitic Diseases
26 Yanko Sakazov Blvd
1504 Sofia
Bulgaria

Janusz T. Paweska
National Institute
for Communicable Diseases of the
National Health Laboratory Service
Special Pathogens Unit
Sandringham
Johannesburg
South Africa

Ariane Pietzka
Austrian Agency for Health and
Food Safety
Institute of Medical Microbiology
and Hygiene
Spargelfeldstrasse 191
1220 Vienna
Austria

Paola Pilo
University of Bern
Institute for Veterinary
Bacteriology
Department of Infections
Diseases and Pathobiology
Vetsuisse Faculty
Länggasstrasse 122
3012 Bern
Switzerland

Sheli R. Radoshitzky
United States Army Medical
Research Institute of Infections
Diseases
Fort Detrick
Frederick
Maryland 21702
USA

Werner Ruppitsch
Austrian Agency for Health and
Food Safety
Institute of Medical Microbiology
and Hygiene
Spargelfeldstrasse 191
1220 Vienna
Austria

Amy C. Shurtleff
United States Army Medical
Research Institute of Infectious
Diseases
Fort Detrick
Frederick
Maryland 21702
USA

Nigel Silman
Health Protection Agency
Microbiology Services
Porton Down
Salisbury SP4 0JG
UK

Anders Sjöstedt
Umeå University
Department of Clinical
Microbiology
SE-901 87 Umeå
Sweden

Chantal J. Snoeck
Institute of Immunology
Centre de Recherche Public de la
Santé/LNS
20A rue Auguste Lumière
Luxembourg 1950
Luxembourg

Lisa D. Sprague
Friedrich-Loeffler-Institut
Federal Research Institute for
Animal Health
Institute of Bacterial Infections
and Zoonoses
Naumburger Strasse 96a
07743 Jena
Germany

Stephanie Speck
Bundeswehr Institute of
Microbiology
Neuherbergstrasse 11
80937 Munich
Germany

Eva Spitalska
Institute of Virology
SAS, Dubravska Cesta 9
845 05 Bratislava
Slovak Republic

Anna Stoeger
Austrian Agency for Health and Food Safety
Institute of Medical Microbiology and Hygiene
Spargelfeldstrasse 191
1220 Vienna
Austria

Radu I. Tanasa
National Institute of Research and Development in Microbiology and Immunology "Cantacuzino"
Spl. Independentei 103
Sector 5
Bucharest 050096
Romania

Rudolf Toman
Laboratory for Diagnosis and Prevention of Rickettsial and Chlamydial Infections
Institute of Virology
Slovak Academy of Sciences
Dubravska cesta 9
845 05 Bratislava 45
Slovakia

Herbert Tomaso
Friedrich-Loeffler-Institut
Federal Research Institute for Animal Health
Institute of Bacterial Infections and Zoonoses
Naumburger Str. 96a
07743 Jena
Deutschland

Victoria Wahl-Jensen
Integrated Research Facility at Fort Detrick (IRF-Frederick)
National Institute of Allergy and Infectious Diseases
National Institutes of Health
Fort Detrick
Frederick
Maryland 21702
USA

and

Tunnell Consulting Inc.
King of Prussia
Pennsylvania 19406
USA

Pierre Wattiau
Veterinary and Agrochemical Research Centre
Groeselenberg 99
1180 Brussels
Belgium

Manfred Weidmann
University Medical Center Göttingen
Department of Virology
Kreuzbergring 57
37075 Göttingen
Germany

Michel S. Zygmunt
Institut National de la Recherche
Agronomique (INRA)
UMR1282 Infectiologie et Santé
Publique
F-37380 Nouzilly
France

and

Institut National de la Recherche
Agronomique (INRA)
Université François Rabelais de
Tours
UMR1282 Infectiologie et Santé
Publique
F-37000 Tours
France

Part A
Pathogens

BSL3 and BSL4 Agents: Epidemiology, Microbiology, and Practical Guidelines, First Edition.
Edited by Mandy C. Elschner, Sally J. Cutler, Manfred Weidmann, and Patrick Butaye.

Part I
Bacteria

BSL3 and BSL4 Agents: Epidemiology, Microbiology, and Practical Guidelines, First Edition.
Edited by Mandy C. Elschner, Sally J. Cutler, Manfred Weidmann, and Patrick Butaye.

1
Bacillus anthracis: Anthrax

Markus Antwerpen, Paola Pilo, Pierre Wattiau, Patrick Butaye, Joachim Frey, and Dimitrios Frangoulidis

1.1
Introduction

Anthrax is an acute infection, caused by the bacterium *Bacillus anthracis*. This zoonosis can be transmitted from grass-eating animals or their products to humans. However, it should be noted also that *B. anthracis* has all the characteristics of an environmentally adapted bacterium. Normally, anthrax occurs in 95% of all human reported cases as cutaneous anthrax, caused by bacteria penetrating the skin through wounds or micro fissures, but the bacterium can also manifest in the mouth and the intestinal tract. Infection of the lungs after inhalation of the spores is very rarely observed [1, 2].

B. anthracis is listed as a category A bioterrorism agent by the Centers for Disease Control and Prevention [3]. Besides the knowledge that *B. anthracis* has a history as a biological weapon in different national military programs, the general public became aware of this pathogen in 2001 when specially processed spore-filled letters were sent across the United States of America, killed five persons and infected up to 26. Since that time "white powder" became a synonym for the bioterroristic threat all over the world. In addition to this special aspect, anthrax is still an endemic disease in many countries and sporadically occurs all over the world [2]. Today, imported wool or hides from ruminants (e.g., goats, sheep, cows) could be contaminated with spores and occasionally lead to infections in industrial ("wool sorting factory") or private settings (Bongo drums) in countries where the disease is absent or infrequent [4, 5].

1.2
Characteristics of the Agent

B. anthracis is a nonmotile, Gram-positive, spore-forming aerobic rod, arranged in long chains within the tissue. Sporulation occurs in the presence of free oxygen and endospores develop in central positions that are considered as infectious particles.

BSL3 and BSL4 Agents: Epidemiology, Microbiology, and Practical Guidelines, First Edition.
Edited by Mandy C. Elschner, Sally J. Cutler, Manfred Weidmann, and Patrick Butaye.

They are extremely resistant to harsh environmental conditions, such as heat, dehydration, pH, desiccation, chemicals, irradiation, or ice. In this dormant state, they can survive up to decades or centuries in the environment without loss of virulence [1, 2]. Robert Koch discovered this agent and its sporulated form in 1877 and demonstrated for the first time his so-called postulates that establish a causal relationship between a pathogen and a disease [6].

1.3 Diagnosis

1.3.1 Phenotypical Identification

B. anthracis belongs to the *B. cereus* group, which comprises six closely related members: *B. anthracis, B. cereus, B. mycoides, B. pseudomycoides, B. thuringiensis,* and *B. weihenstephanensis.* These species are phylogenetically and phenotypically closely related and their phenotypical identification can be complex, especially because some isolates may display unusual biochemical and/or genetic properties that complicate their distinction. Furthermore, several of these species and also yet-undefined species of this group are normal contaminants in environmental samples such as soil or dust.

Phenotypical identification is the classical method of identification in routine laboratories. Some simple characteristics have been useful for distinguishing between different *Bacillus* species, but they have limitations particularly in distinguishing between *B. cereus* and *B. anthracis* [7]. The following sections focus on the different methods and highlight their advantages and disadvantages.

1.3.2 Growth Characteristics

B. anthracis is a Gram-positive aerobic spore-forming rod which grows well on all types of blood-supplemented agars. After cultivation at 37 °C overnight, *B. anthracis* shows the following morphological characteristics [2]:

- No hemolysis on blood agars;
- Spike-forming (see Figure 1.1);
- "Sticky" colonies (see Figure 1.2);
- Medusa head;
- While growing on Cereus-Ident agar (Heipha, Germany), it forms colonies that are silver-gray, silky-shiny, and polycyclic without any pigmentation [8]. As an alternative, trimethoprim-sulfamethoxazole-polymyxin-blood agar (TSPB agar) can be used, containing several antibiotics and resulting in a high selectivity against Gram-negative bacteria.

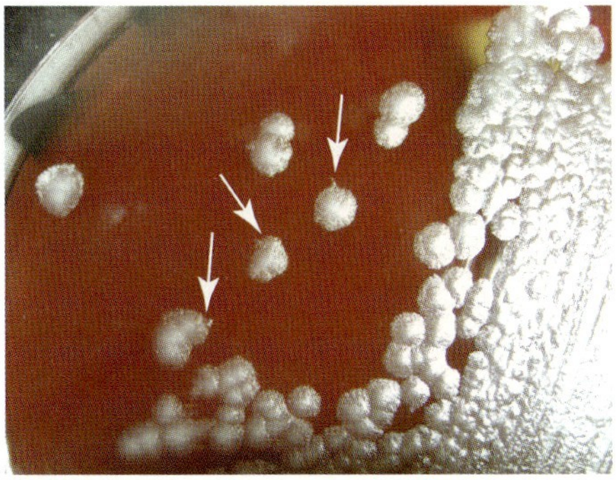

Figure 1.1 Spike-like projections at the colony edge from *Bacillus anthracis* (in addition no hemolysis on Columbia agar; Bundeswehr Institute of Microbiology, Munich, Germany).

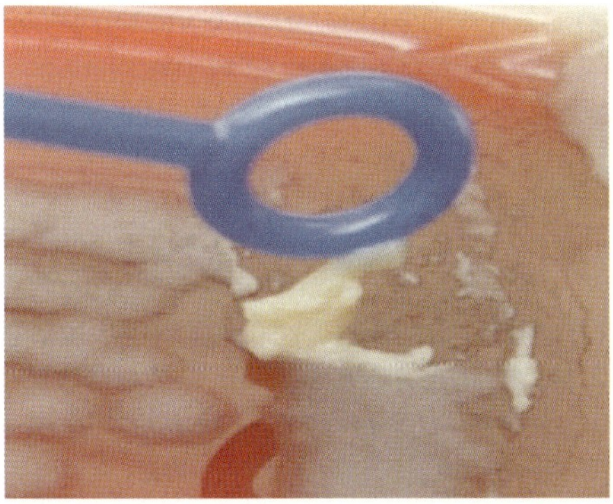

Figure 1.2 "Sticky colony" phenomenon of *B. anthracis* after manipulation with a loop (J. Frey, Bern, Switzerland).

Even if only some of these features are observed, the results should be considered as suspicious for *B. anthracis* and further analysis for differentiation should be mandatory.

1.3.3 Antibiotic Resistance

The so-called "string of pearls"/"Perlschnur" test is based on the fact that most of the strains isolated in nature are sensitive to penicillin (in contrast to *B. cereus* isolates) and show a special stress form of the bacteria when it is faced with sublethal doses of the antibiotic. This very simple test is still used today to get a first very easy confirmation of suspicious colonies.

The determination of the pattern of antibiotic resistance is of utmost interest, primarily for the assessment of a potential release of bacteria with a bioterroristic background [9], with genetically modified strains, and to detect rarely occurring natural resistances. To establish the antibiotic resistance profiles of strains, the minimum inhibition concentrations (MICs) have to be tested. They are fixed by the Clinical and Laboratory Standards Institute (CLSI). The CLSI publishes guidelines containing protocols and breakpoint values to evaluate antibiotic resistances profiles [10, 11].

1.3.4 Phage Testing and Biochemistry

Phage testing and biochemical tests are methods that are still in use today in veterinary diagnostic laboratories. Biochemical tests do not lead to a clear and valuable result for differentiating *B. anthracis* from *B. cereus*, while phage testing with bacteriophage gamma is more specific. The feature of sensitivity of *B. anthracis* to gamma phage is still used for identification although exceptions were described in this case, too. In former times veterinary laboratories used a mouse inoculation test for confirmation. When all phenotypic characteristics were present and the phage test was positive, and finally the mouse died within one day after intraperitonal inoculation, and bacillus-like bacteria were seen on colored blood samples, the isolated bacterium was seen as a virulent strain of *B. anthracis*.

1.3.5 Antigen Detection

Newer methods have been developed for the identification of *B. anthracis* from the presence of specific surface proteins. Such systems are encapsulated in lateral flow assays, commercialized for example by the companies Tetracore or Cleartest. Even for these assays, some isolates of *B. cereus* that result in a weak positive signal have been described. Based on their low sensitivity, it is recommended to use this kind of experiment only as a confirmation test of suspect colonies rather than for the detection of *B. anthracis* in environmental samples, since the concentration of bacteria may not be adequate in such samples. The advantage of such a test is the ease of specimen handling. Suspicious material is rubbed into a buffer and the suspension is dropped onto the cartridge. After 15 min of incubation, the results can be evaluated like a pregnancy test. If the test is used for checking questionable colonies (especially when they show no hemolysis on blood agar), it gives very specific results in the identification of *B. anthracis*.

1.3.6 Molecular Identification

1.3.6.1 Virulence Plasmid pXO1

The virulence plasmid pXO1 is said to be "*anthracis*-specific." The genes coding for "lethal factor" (LF; *lef*), "edema factor" (EF; *cya*), and "protective antigen" (PA; *pagA*) are normally present on this plasmid and are major virulence factors

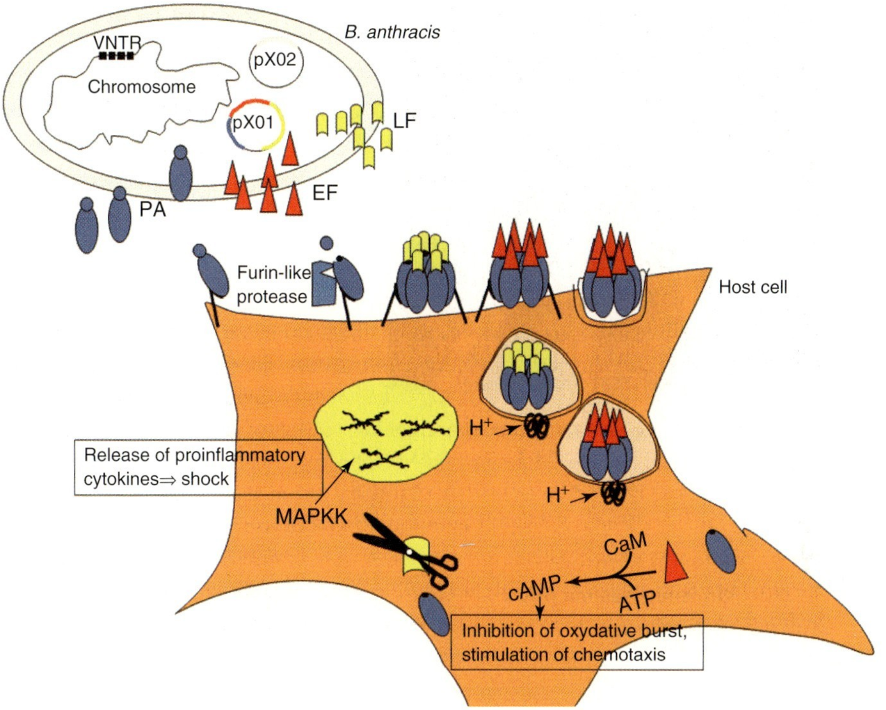

Figure 1.3 Molecular mechanism of virulence of *B. anthracis*. The bacterium *B. anthracis* (upper left), which is protected against phagocytosis by the capsule, a poly-D-glutamic acid polymer, (gray) encoded on plasmid pXO2 secretes an adhesion peptide or protective antigen (PA), the edema factor (EF), and the lethal factor (LF) whose genes are encoded on plasmid pXO1. PA binds to specific cell receptors of the host cell (lower right) where it is processed by a furin-like protease from the host, a process that will lead to multimer formation of PA and association of EF or LF. The PA::EF or PA::LF complexes are taken up by the host cell by endocytosis and are subsequently released in the cell at low pH, when the complexes dissociate and release active EF, a calmodulin-dependent adenylate cyclase, which leads to an inhibition of oxidative burst and an overstimulation of chemotaxis, and actives LF, a zinc-metalloprotease that cleaves mitogen activated protein kinase kinase (MAPKK), which leads to the overproduction and release of pro-inflammatory cytokines and resulting shock. (Courtesy of Elsevier and P. Pilo 2011, see Ref. 30).

of *B. anthracis* (see Figure 1.3). Due to this, most of the commercially available PCR diagnostic kits target sequences of these genes to detect *B. anthracis*. Recently, strains of *B. cereus* could be isolated from patients presenting the clinical symptoms of an "anthrax-like" disease. In these strains at least one of the above-mentioned genes could be observed. Therefore, this plasmid cannot be used as a specific marker for *B. anthracis*, thus complicating the definition and differentiation of species belonging to the *B. cereus* group [7, 12–16]. However, from a clinical point of view, it is not important which species is present or not if these virulence genes are present and the patient's symptoms are similar to those caused by *B. anthracis*.

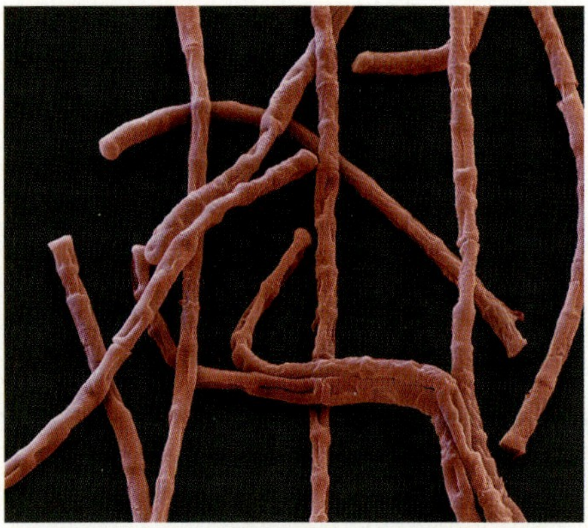

Figure 1.4 Scanning electron microscopic image of *B. anthracis* (6000×) with typical bamboo stick-like appearance of the rods (courtesy of Eye-of-Science, Reutlingen, Germany, 2008).

1.3.6.2 **Virulence Plasmid pXO2**

The virulence plasmid pXO2 contains the genes coding for the capsule-forming enzyme complex (see Figure 1.4). In contrast to *B. cereus*, *B. anthracis* forms a capsule composed of poly-gamma-D-glutamic-acid that is suspected to enable the survival of the bacteria during phagocytosis. The genes *capA*, *capB*, and *capC* are involved in capsule biosynthesis and are normally used as targets for detection by different PCR kits of *B. anthracis*. *B. cereus* does not carry this plasmid or only genetically related versions containing parts or variants [7, 12, 13, 16]. Therefore, PCR detection of both toxin genes on pXO1 and capsule genes on pXO2 is recommended for the identification of *B. anthracis* isolates.

1.3.7
Chromosome

A lot of specific targets for the reliable detection and identification of *B. anthracis* have been published. Most of them showed false-positive results during or even after the evaluation of a panel of *B. cereus* strains. The necessity of an unambiguous chromosomal target is due to the presence of so-called "*anthracis*-specific" genes in *B. cereus* and because the *rrs* (16S rRNA) gene sequencing does not provide the information needed. Even the other "classical" targets for species differentiation fail in this specific case and do not provide discriminatory power. Genes encoding the β subunit of the RNA polymerase (*rpoB*) and the subunit A of the DNA-gyrase (*gyrA*) do not display extensive DNA polymorphism between *B. cereus* and *B. anthracis*. The World Health Organization (WHO) recommends the use of the gene *saspB* as a reliable target and has published primers and probes for an assay based on Roche's Lightcycler technology [2]. A Taqman probe-based assay for the detection of another chromosomal target was published recently [17]. Both assays are up to date without any failure in the identification of *B. anthracis* and differentiation with *B. cereus*.

1.3.8
MLVA, SNR, and SNP Typing

In the case of outbreak investigations but also in forensic examinations of historical or unusual isolates, molecular typing is envisaged and the obtained information can be compared with data retrieved from previously isolated strains [18]. The results of such tests can be used to trace back the origin of the investigated isolates.

In variable number tandem repeats (VNTRs), multi-locus variable analysis (MLVA), or single nucleotide repeat (SNR) analysis, the gene loci comprising repeated motifs bordered by conserved DNA regions are investigated. The combined lengths of the examined loci are likely to differ from strain to strain. The length of the repeated units varies from one base in the case of SNR to several dozens (macro-satellites) in *B. anthracis*. In other organisms, such as *Enterococcus* species, repeat units may harbor up to 300 bp. When the length of the amplified fragment is determined and the amount of the repeated units calculated, the obtained data are comparable to databases similar to paternity testing in humans. Using this strategy, strains can be sourced down to the isolate level [5, 18–21].

Discrimination at cluster or group level is possible using the single nucleotide polymorphism (SNP) typing method. Using this tool, point mutations are investigated and compared to various established populations and, based on the obtained information, they are grouped into different clusters [5].

Such typing techniques can be performed in a reliable way only in specialized laboratories which are experienced in the generation of typing data and the interpretation of them. The use of standardized protocols is highly recommendable in order to correlate data from *B. anthracis* strains isolated worldwide and to trace back their origin [2, 16, 18, 22].

1.3.9
Serological Investigations

Due to the rapid clinical course of systemic anthrax, antibody detection is not a useful diagnostic of acute disease. To control titers after vaccination or to confirm an anthrax case suspicion *a posteriori*, specific antibodies directed against PA could be checked. A commercial enzyme-linked immunosorbent assay (ELISA) for that issue is available on the market (Virion/Serion, Würzburg, Germany). Questionable or very low titer results should be confirmed by blotting techniques [23]. Bead-based assays will be available in the future which could be powerful enough to overcome some problems in sensitivity and specificity [24].

1.4 Pathogenesis

1.4.1 Animals

Mainly ruminants are affected by this disease. The following animal species are categorized in order of susceptibility to the infection: ruminants (bovines, sheep, goat, camel, antelopes), horses, pigs, humans, carnivores (cats, dogs, lions, etc.), birds, amphibian, reptiles, and fish. The latter four in general are little susceptible [25].

In general herbivores get infected by spores from the ground penetrating through micro fissures in the mouth that arise by dryness or by overgrazing on hard and spiny range land. In the lesion, the spores germinate and multiply. After passing the local lymph barriers, sepsis and death can follow within a period of 6 h to 4 days. In general, animals are found suddenly dead, since the animals die within 1–2 h after the clinical symptoms become obvious. Marks of agony may present on the floor (typically signs of leg movements). Less frequent is the acute form, with death within 48 h after the start of clinical symptoms. Typically there is pre-terminal bleeding from the mouth, nose, and anus; and clotting of the blood may be impaired. This blood is highly contagious by germinated anthrax bacilli that convert rapidly to spores under the influence of oxygen. The contaminated skin and the ground around the animals cause the start of a cycle of infection [1, 2].

Birds seem to be not susceptible to the disease and may even be a vector in spreading the bacterium. Especially birds of prey, eating animals killed by an anthrax infection, may excrete the spores over a long distance [26].

1.4.2 Humans

Humans are less susceptible to anthrax than ruminants. The clinical appearance of anthrax in humans depends on the route of infection. In 95% of the cases, *B. anthracis* invades the deep layers of skin through wounds, micro fissures, or even abrasion sites. Through this route, cutaneous anthrax or pustula maligna develops [2]. The predominant locations are the extremities, face, thorax, and abdomen.

Spores in uncooked meat, penetrating the mucosa of the mouth or pharyngeal areas, are the main cause of oro-pharyngeal and intestinal anthrax. In this case, spores invade the lymph-associated barriers and provoke sepsis. This form of anthrax can be observed in families eating the same meal consisting of insufficiently cooked meat. Inhalational or pulmonary anthrax results from the inhalation of a sufficient amount of spores (about 10^4 for a non-immunocompromized host) that are transported to the "lymphandenous" tissue where they proliferate [2]. In former times, inhalational anthrax was a severe problem in wool, leather, and bone meal processing factories, where spores were present in high concentrations in the dust (hence the former name of woolsorter's disease). Today "biosafety management strategies" minimize the danger. In the case of respiratory infections, especially if more than one case occurs, a bioterroristic attack has to be ruled out.

Upon entry by the spores into the body, within a few hours the spores germinate and proliferate. Pathology is based on the release of three virulence factors (proteins): the PA that serves as an adhesion subunit for the two main protein toxins, the EF, and the LF that kills the surrounded tissue cells (Figure 1.3) [1, 2].

Immunity against *B. anthracis* can be acquired during a cutaneous anthrax infection but the duration of the protection is unclear. Survivors of systemic anthrax infections show a high antibody response, especially against PA and LF, comparable to a vaccination and will be protected for a long time.

Commercially available vaccines are based on the PA and generate moderate titers. Vaccination might be useful for persons at elevated risk to get in contact with *B. anthracis* and its spores, like laboratory personal, veterinarians, and workers in the above-mentioned industries. However, in most countries, no human anthrax vaccine is licensed. Since 1998, every soldier of the United States Army who is deployed to Iraq or Afghanistan is vaccinated [27]. The vaccination of cattle is still frequent in countries where the disease is enzootic. The strain Sterne is used worldwide as a live vaccine for ruminants. However, one should notice that vaccine strains have to be preserved cautiously because the emergence of non-immunogenic variants can happen as well as mixing with virulent strains from the field [28].

1.5 Clinical and Pathological Findings

One can distinguish the peracute, acute, and subacute forms, or otherwise also the skin form, the intestinal, and the pulmonary form. The incubation period ranges over 1–60 days, depending on the host and amount of bacteria incorporated.

In humans, cutaneous anthrax is the most common form of anthrax and is normally cured by antibiotic therapy without any complications. Typical lesions show a dark reddened papule, formed within 1–3 days with an aperture of only a few millimeters. During the next day, vesicles may appear at the border of this papule, grow, and merge. This cyst cracks within the next day and discharges a hemorrhagic exudate that is highly infectious. Under the skin, an ulcer is formed with a brown, later black dry ground with an aperture of 0.5–3.0 cm ("anthrax" originates from the Greek word for coal – describing the color of the skin lesions). At the beginning, this ulcer is painless, but later patients were said to feel a localized heat.

The edema leads to ablation of the skin, like after burning. Without any care, such confined existing pustules vanish within 3–6 weeks with little scarring. Therapy is nevertheless mandatory to prevent systemic spread from the initial infection site. Cutaneous anthrax does not exist in ruminants, and it is rarely seen in carnivores.

1.5.1
Oropharyngeal Anthrax

In humans this form of anthrax appears after the consumption of insufficiently cooked meat from infected animals. The lesion is visible in the stoma pharynx throat. Sometimes circumorale infections are also possible. Similar to the cutaneous form of anthrax, 1–3 days after infection papules appear, growing and transforming into ulcers, accompanied by heavy edemas and a swelling of the regional lymph nodes in soft tissues. Such a glottis or pharynx edema is a life-threatening complication that needs rapid intervention.

This form is also seen in carnivores eating infected or dead animals. In general these animals are not so susceptible to infection, but throat swelling may cause asphyxia. This is also seen in horses, where also death can occur due to excessive intestinal damage [29].

Oropharyngeal anthrax can also exist in a subacute or chronic form in pigs where swelling of the oropharynx causes dyspnea and dysphagia. Complete recovery is possible in these animals.

1.5.2
Abdominal or Intestinal Anthrax

In this case, an infectious dose of spores is ingested and can pass the gastric acid barrier. Germination takes place in the ileum, with a fast infiltration and expansion within the liver and spleen. The symptoms rapidly worsen, accompanied with body pain, vomiting, and bloody diarrhea. A hypovolemic shock with ascites caused by the toxin leads to death within 48 h.

In pigs, it can cause a severe enteritis, with a possibility for recovery. It should be noted that in this animal species intestinal anthrax is seldom accompanied with splenomegaly since septicemia is rare.

1.5.3
Inhalational or Pulmonary Anthrax

Inhalational anthrax is rare and, when a case appears, the possibility of a bioterroristic attack should be ruled out as soon as possible. However, rare deaths have recently been reported after the processing of Bongo drum animal hides. At the beginning, symptoms are fever, cough, pain, and adynamia. These non-specific disorders lead to acute mediastinitis, with shock and dyspnea. X-rays show an enlargement of the mediastinum and pleural effusion. The change to a systemic infection with a fulminant meningoencephalitis leads in most cases to death.

1.6 Epidemiology

Anthrax is an acute infection in ruminants, especially for cattle, buffalo, sheep, and goats, but also for horses, zebras, antelopes, and elephants [2]. A sporadic appearance can be observed in rabbits from infected fields and in pigs, dogs, and cats fed with infected meat. However, national programs for the control and eradication of anthrax led to the decrease of its natural incidence in animals and in humans worldwide and restricted the disease to animal-related professions. It is nearly absent in northern, middle, and western Europe. Due to stringent measures when cases are met, the discovery of antibiotics, and the development of an effective vaccine, anthrax is no longer considered as one of the most important causes for economical losses in livestock farming. When a case is encountered, the dead animal cannot be moved, and therefore the autopsy and sampling should happen on the spot. After autopsy, all contaminated material is burned on the spot and afterwards a disinfection of the place is performed using burnt lime. Other animals on the premises are checked every 6 h for any increase in body temperature. When there an increase is seen, intravenous injection with penicillin avoids the development of the disease and prevents the further spreading of spores.

Today, anthrax remains a problem, particularly in areas where surveillance systems, especially veterinary controls, are neglected or interrupted due to war activities or natural catastrophes. Also in areas with extended wildlife, control is difficult and has to rely mainly on the decontamination of sites where dead animals were found. Scavengers are problematic since they are little susceptible to the disease but may spread it. In these areas, the disease is still enzootic.

In Eurasia, anthrax was a severe problem in areas of the former Commonwealth of Independent States (CIS) and Turkey in the nineteenth up to the middle of the twentieth century, especially in livestock farming and the wool or leather processing industry. The development of veterinary controls, a ban on burying animal carcasses, the vaccination of animals, and the use of antibiotics led to the disappearance of anthrax. However sporadic cases of *B. anthracis*-infected animals are known from rural regions all over the world.

Due to a lack of precise data, WHO estimates the number of anthrax human cases to be less than 100 000 per year, which mostly present the cutaneous form. The appearance of this disease is mainly related to strong rainfall and manual agricultural activity (gardening). Human to human infections have not been observed [2].

1.7 Conclusion

Anthrax is still present worldwide sporadically or in some areas endemically. It is largely a ruminant disease in ruminants and for this reason its control should primarily start in animals. Strict sanitary measures are needed to handle it. Indeed, environmental contamination needs to be quite high to be able to infect an animal. Depending on the species, low (ruminants) or high concentrations of spores are necessary to cause the disease. Fortunately there is no direct host to host transmission. Antibiotic treatment is efficient unless the disease has reached the "point of no return," when too many toxins are in the body (this happens mostly when there is inhalation or ingestion of *B. anthracis* in humans). Nevertheless, this rarely occurs under natural conditions. Animal vaccines are available and are useful to control the disease in ruminants in endemic areas. This simplifies the management of outbreaks but the situation can be complicated in some regions because of political, economical, or natural problems. However, in the end, anthrax in wildlife may be difficult to control.

Laboratory diagnosis may sometimes be tricky because of unusual strains of *B. anthracis* and *B. cereus*. As described above, members of the *B. cereus* group are closely related and some characteristics are sometimes not specific. Besides, discussions are ongoing about *B. cereus* strains harboring *B. anthracis* plasmids, especially when both species show no more differences in the clinical outcome between them. Research is needed to understand the ecological and clinical status of these atypical strains.

Recently, a large amount of effort has been invested for the typing and subtyping of strains and isolates of *B. anthracis*. Worldwide data are becoming available to scientists to compare strains and to understand the epidemiology and ecology of *B. anthracis* [2, 16, 19, 22]. Indeed, due to its particular lifestyle, which passes through a sporulated form that can last for decades, this pathogen is highly monomorphic from a DNA sequence point of view.

Nowadays, one concern would come from specially prepared spores for intentional release. Anthrax is known to have been included in bioweapon programs, principally because of the stability and resistance of the spores. This requires trained personnel and validated protocols in order to identify rapidly and efficiently a criminal act. Another concern comes from the import (sometimes illegal) of non-disinfected animal hides and hairs, which can lead to human exposure in individuals living in countries where the disease is not endemic [4, 28]. Hence, anthrax merits medical alertness and awareness even in countries where the disease is not endemic in order to prevent fatal human infections.

References

1. Mock, M. and Fouet, A. (2001) Anthrax. *Annu. Rev. Microbiol.*, **55**, 647–671.
2. WHO (2008) *Guidelines for the Surveillance and Control of Anthrax in Humans and Animals*, 4th edn, World Health Organization, Geneva.
3. *www.bt.cdc.gov/agent/agentlist_category.asp.*
4. Anaraki, S., Addiman, S., Nixon, G., Krahé, D., Ghosh, R., Brooks, T., Lloyd, G., Spencer, R., Walsh, A., McCloskey, B., and Lightfoot, N. (2008) Investigations and control measures following a case of inhalation anthrax in East London in a drum maker and drummer. *Eur. Surveil.*, 13, 35–40.
5. Wattiau, P., Klee, S.R., Fretin, D., Van Hessche, M., Ménart, M., Franz, T., Chasseur, C., Butaye, P., and Imberechts, H. (2008) Occurrence and genetic diversity of Bacillus

anthracis strains isolated in an active wool-cleaning factory. *Appl. Environ. Microbiol.*, **74**, 4005–4011.

6. Münch, R. (2003) Robert Koch. *Microbes Infect.*, **5**, 69–74.
7. Kolstø, A.-B., Tourasse, N.J., and Økstad, O.A. (2009) What sets Bacillus anthracis apart from other Bacillus species? *Annu. Rev. Microbiol.*, **63**, 451–476.
8. Tomaso, H., Bartling, C., Al Dahouk, S., Hagen, R.M., Scholz, H.C., Beyer, W. *et al.* (2006) Growth characteristics of Bacillus anthracis compared to other Bacillus spp. on the selective nutrient media anthrax blood agar and cereus ident agar. *Syst. Appl. Microbiol.*, **29**(1), 24–28.
9. Bush, L., Malecki, J., Wiersma, S., Cahill, K., Fried, R., Grossman, M. *et al.* (2001) Update: investigation of bioterrorism-related anthrax and interim guidelines for exposure management and antimicrobial therapy. *MMWR Morb. Mortal. Wkly Rep.*, **50**, 909–919.
10. Clinical and Laboratory Standards Institute (2009) Methods for Dilution Antimicrobial Susceptibility Tests for Bacteria that Grow Aerobically, 8th edn, vol. 29(2), Approved Standard M07-A8, Clinical and Laboratory Standards Institute, Wayne.
11. Clinical and Laboratory Standards Institute (2009) Performance Standards for Antimicrobial Susceptibility Testing; Nineteenth Informational Supplement M100-S19, vol. 29(3), Clinical and Laboratory Standards Institute, Wayne.
12. Avashia, S., Riggins, W., Lindley, C., Hoffmaster, A., Drumgoole, R., Nekomoto, T. *et al.* (2007) Fatal pneumonia among metalworkers due to inhalation exposure to Bacillus cereus containing Bacillus anthracis toxin genes. *Emerg. Infect. Dis.*, **44**(3), 414–416.
13. Hoffmaster, A.R., Ravel, J., Rasko, D.A., Chapman, G.D., Chute, M.D., Marston, C.K. *et al.* (2004) Identification of anthrax toxin genes in a Bacillus cereus associated with an illness resembling inhalation anthrax. *Proc. Natl Acad. Sci. USA*, **101**(22), 8449–8454.
14. Klee, S.R., Ozel, M., Appel, B., Boesch, C., Ellerbrok, H., Jacob, D. *et al.* (2006) Characterization of Bacillus anthracis-like bacteria isolated from wild great apes from Cote d'Ivoire and Cameroon. *J. Bacteriol.*, **188**(15), 5333–5344.
15. Pannucci, J., Okinaka, R., Sabin, R., and Kuske, C. (2002) Bacillus anthracis pXO1 plasmid sequence conservation among closely related bacterial species. *J. Bacteriol.*, **184**, 134–141.
16. Simonson, T.S., Okinaka, R.T., Wang, B., Easterday, W.R., Huynh, L., U'Ren, J.M., Dukerich, M., Zanecki, S.R., Kenefic, L.J., Beaudry, J., Schupp, J.M., Pearson, T., Wagner, D.M., Hoffmaster, A., Ravel, J., and Keim, P. (2009) Bacillus anthracis in China and its relationship to worldwide lineages. *BMC Microbiol.*, **9**, 71.
17. Antwerpen, M., Zimmermann, P., Frangoulidis, D., Bewley, K., and Meyer, H. (2008) Real-time PCR system targeting a chromosomal marker specific for Bacillus anthracis. *Mol. Cell Probes*, **22**(5), 313–315.
18. Keim, P., Van Ert, M.N., Pearson, T., Vogler, A.J., Huynh, L.Y., and Wagner, D.M. (2004) Anthrax molecular epidemiology and forensics: using the appropriate marker for different evolutionary scales. *Infect. Genet. Evol.*, **4**, 205–213.
19. Keim, P., Price, L., Klevytska, A., Smith, K., Schupp, J., Okinaka, R. *et al.* (2000) Multiple-locus variable-number tandem repeat analysis reveals genetic relationships within Bacillus anthracis. *J. Bacteriol.*, **182**, 2928–2936.
20. Lista, F., Faggioni, G., Valjevac, S., Ciammaruconi, A., Vaissaire, J., le Doujet, C. *et al.* (2006) Genotyping of Bacillus anthracis strains based on automated capillary 25-loci multiple locus variable-number tandem repeats analysis. *BMC Microbiol.*, **6**, 33.
21. Stratilo, C., Lewis, C., Bryden, L., Mulvey, M., and Bader, D. (2006) Single-nucleotide repeat analysis for subtyping Bacillus anthracis isolates. *J. Clin. Microbiol.*, **44**, 777–782.
22. Pilo, P., Perreten, V., and Frey, J. (2008) Molecular epidemiology of Bacillus anthracis: determining the correct

origin. *Appl. Environ. Microbiol.*, **74**, 2928–2931.

23. Sirisanthana, T., Nelson, K.E., Ezzell, J.W., and Abshire, T.G. (1988) Serological studies of patients with cutaneous and oral-oropharyngeal anthrax from northern Thailand. *Am. J. Trop. Med. Hyg.*, **39**(6), 575–578.
24. Biagini, R.E., Sammons, D.L., Smith, J.P., MacKenzie, B.A., Striley, C.A., Semenova, V., Steward-Clark, E., Stamey, K., Freeman, A.E., Quinn, C.P., and Snawder, J.E. (2004) Comparison of a multiplexed fluorescent covalent microsphere immunoassay and an enzyme-linked immunosorbent assay for measurement of human immunoglobulin G antibodies to anthrax toxins. *Clin. Diagn. Lab. Immunol.*, **11**(1), 50–55.
25. Bathnagar, R. and Batra, S. (2001) Anthrax toxin. *Crit. Rev. Microbiol.*, **27**, 167–200.
26. Pil, A., Butaye, P., and Imberechts, H. (2003) Bacillus anthracis. *Vlaams Diergen. Tijds.*, **72**, 43–50.
27. Lange, J.L., Lesikar, S.E., Rubertone, M.V., and Brundage, J.F. (2003) Comprehensive systematic surveillance for adverse effects of anthrax vaccine adsorbed, US Armed Forces, 1998–2000. *Vaccine*, **21**(15), 1620–1628.
28. Wattiau, P., Govaerts, M., Frangoulidis, D., Fretin, D., Kissling, E., Van Hessche, M., China, B., Poncin, M., Pirenne, Y., and Hanquet, G. (2009) Immunological response of unvaccinated workers exposed to B. anthracis spores, Belgium. *Emerg. Infect. Dis.*, **15**(10), 1637–1640.
29. Radostis, O.P.M., Douglas, C.C., Blood, G.C., and Hinchliff, K.W. (1999) *Veterinary Medicine: A Textbook of the Diseases of Cattle Sheep Pigs Goats and Horses*, 9th edn, Saunders, Philadelphia, pp. 747–751.
30. Pilo, P. and Frey, J. (2011) Bacillus anthracis: molecular taxonomy, population genetics, phylogeny and patho-evolution. *Infect. Genet. Evol.*, **11**(6), 1218–1224.

2
Brucella Species: Brucellosis

Sally J. Cutler, Michel S. Zygmunt, and Bruno Garin-Bastuji

2.1
Introduction

Globally, brucellosis remains a major bacterial zoonosis, still devastating the productivity of livestock and affecting mankind, either directly through infection, or indirectly through a reduction of productivity among food-producing livestock. It is estimated that more than 500 000 new cases of human brucellosis occur each year; however this is likely to be an underestimate through the insidious nature of this infection [1]. Many nations have deployed eradication campaigns, with some countries such as the United Kingdom, France, and most other northern European countries successfully gaining brucellosis-free status. Even within these countries, comprehensive surveillance is an essential prerequisite to maintain this status, exemplified by recent introductions of *Brucella*-infected cattle to the United Kingdom, despite pre- and post-export control. Furthermore, in some southeastern European regions such as Albania, Bosnia-Herzegovina, Bulgaria, Croatia, Kosovo, and FYROM we are now seeing the re-emergence of brucellosis. Consequently, even if brucellosis-free status is achieved, we must avoid complacency.

Recent advances in both detection and typing methods, reinforced by the availability of whole genome sequencing have seen renewed efforts to elucidate the pathophysiological mechanisms employed by these microbes. Despite the homogeneous nature of the brucellae on a genetic basis, they display intriguing host specificity. Indeed, we are rapidly gaining new insights into the host microbial interactions of these organisms, but much remains to be resolved. Novel strains of *Brucella* have been found in marine mammals and wildlife. The implications of these extensions of brucellosis among diverse wildlife species remain to be fully resolved.

During the confines of this chapter, we will give an overview of current topics and remaining issues in the field of *Brucella* research.

BSL3 and BSL4 Agents: Epidemiology, Microbiology, and Practical Guidelines, First Edition.
Edited by Mandy C. Elschner, Sally J. Cutler, Manfred Weidmann, and Patrick Butaye.

2.2
Characteristics of the Agent

The etiological agent of brucellosis is a small, pleomorphic aerobic Gram-negative rod (see Figure 2.1), which was first isolated from human clinical cases of undulant fever by David Bruce (1887). Later, in 1904, members of the Mediterranean Fever Commission, under the leadership of David Bruce, identified the reservoir of infection in goats and subsequent transmission to man through the consumption of unpasteurized milk. In subsequent years, similar microbes were identified in cattle (*B. abortus*) and swine (*B. suis*).

This microbe belongs to the class of alpha-proteobacteria clustering along with *Bartonella*, *Ochrobactrum*, *Agrobacterium*, and *Rhizobium* as phylogenetic neighbors. The family *Brucellaceae* comprised of *Brucella*, *Mycoplana*, and *Ochrobactrum* has recently expanded to include *Pseudochrobactrum*, *Daeguia*, and *Crabtreella* [2]. The homogeneity within the species *O. intermedium* is of particular concern through the diagnostic challenges presented between this organism and *Brucella* species.

The genus was originally divided into species based on various biochemical capabilities, dye and phage susceptibilities, and host preferences (see Table 2.1). Some of these species in some cases were further divided into biovars (see Table 2.1). The validity of some biovars among *Brucella* species has been questioned, with increasing reports of atypical isolates. Following whole genomic hybridization studies, the high degree of homogeneity among the brucellae was noted, prompting the re-classification into a single species, *B. melitensis* [3]. Although justifiable according to phylogenetic criteria, this classification has proved unpopular, largely through distinct host susceptibilities and differences in host–pathogen interactions. The decision to revert to the original species designations was then taken [4]. Subsequently, isolates from marine mammals and the common vole were added to the list, giving 10 recognized species [5, 6], and more recently an isolate was recovered from a human breast implant [7]. The subject of *Brucella* taxonomy is still hotly debated, with studies based on 16s RNA gene sequences and those of *RecA* disclosing that the brucellae are a clade of *Ochrobactrum* (also accounting for

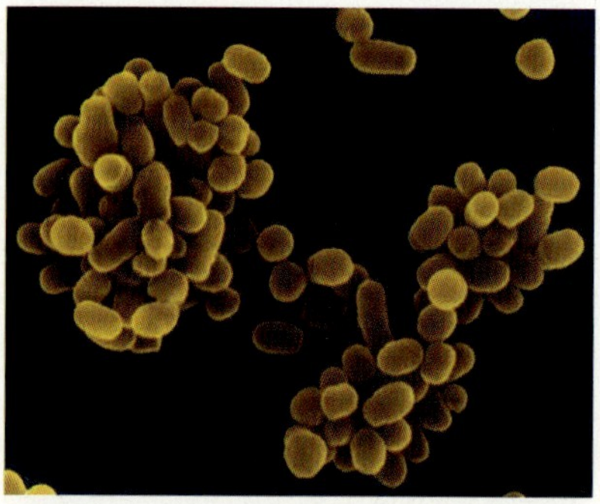

Figure 2.1 Scanning and transmission electron microscopy of *Brucella* species (Crown Copyright).

Table 2.1 Species, biovars, and host preferences among the brucellae.

***Brucella* species**	**Biovars**	**Reservoir host(s)**
B. melitensis	1–3	Sheep, goats, cattle
B. abortus	1–6, 7[a], 9	Cattle
B. suis	1, 3	Swine
	2	Swine, hare
	4	Reindeer
	5	Rodents
B. canis	1	Dogs
B. ovis	1	Sheep
B. neotomae	1	Rodents
B. pinnipedialis	Not determined	Otter, seal
B. ceti	Not determined	Dolphin, porpoise
B. microti	Not determined	European voles, foxes
B. inopinata[b]	Not determined	Not determined

[a]The reference strain of this biovar (63/75) represents a mixed culture. Strains conforming to the description of former biovar 7 were reported recently [10]. Biovar 8 was deemed null and void by the *Brucella* taxonomic subcommittee in 1978 [4].
[b]Formerly known as strain BO1 isolated from a prosthetic human breast implant [7, 11].

the cross-reactivity of *Ochrobactrum* in some *Brucella* molecular diagnostic assays) [8]. More recent analysis of 10 whole genomes gives yet another perspective upon *Brucella* phylogeny, with four distinct clades within the species, but a significant divergence from *Ochrobactrum* [9]. A major *Brucella* genome-sequencing project is currently underway with the analysis of representatives of each species and biovar. We must wait and watch to see if these more recent findings will impact upon the nomenclature of this genus.

The genomic complement of brucellae is approximately 3.29 Mb, typically split into two chromosomes with a large replicon of 2.11 Mb (chromosome 1) and a smaller replicon of 1.18 Mb (chromosome 2) [1]; however, *B. suis* biovar 3 possesses a single replicon of 3.1 Mb. Plasmids have not been reported within the brucellae, indeed, nor have any classical virulence-associated genes. It was believed that these organisms were devoid of flagellar genes; however, genomic sequencing revealed genes sufficient to encode for functional flagellar. It is believed that these are only transiently expressed *in vivo* and are thus likely to have a role in virulence of these bacteria [12, 13]. Certainly the surface lipopolysaccharide plays a crucial role in the virulence of the brucellae together with the expression of a type IV secretion system [14]. Microarray analysis of different *Brucella* species hybridized to *B. melitensis* 16M again showed limited diversity among species. This work identified differences between species suggestive of horizontal genetic acquisition, that appeared to be clustered into genomic islands; however, these differences alone may not be sufficient to account for the host specificity observed within the brucellae [15].

2.3
Diagnosis

Clinical diagnosis is discussed in Section 2.5; however, it has been suggested that human brucellosis should always be suspected in any sample until proven otherwise! This is a result of its ability to affect almost every major tissue within its host, resulting in a plethora of clinical signs often mimicking those of other conditions.

As a result of the myriad of clinical presentations that might be a consequence of brucellosis, laboratory confirmation is an absolute requirement. The gold standard remains cultivation of the organisms from clinical material. Though this might seem inconsequential, it is often fraught with difficulties. Greatest success from human patients is usually during acute infection where brucellae can be recovered from blood cultures or bone marrow. Interestingly, greatest success has been achieved with *B. melitensis*, with other species being more problematic to recover. Automated blood culture systems may not always produce reliable signals in response to the growth of *Brucella* and may require extended periods of incubation beyond those normally used [16]. When brucellosis is suspected, lysis centrifugation techniques have aided the recovery of brucellae [17, 18].

When testing animal samples, abortion material is usually successful (see Figure 2.2), in part through the associated large numbers of brucellae present. These can be detected with staining methods such as the Stamp stain (see Figure 2.3). In other animals, the choice of samples is more problematic. In swine, particular success has been achieved using tonsillar biopsies, while in other species lymph nodes have been particularly successful, with the head, genital (see Figures 2.4 and 2.5), and supra-mammary lymph nodes being the samples of choice. Brain tissue is also a potential site of persistence and particularly useful where there has been delay in sampling a fallen animal. Indeed, brain persistence has been demonstrated following experimental infection from eight years previously in an animal that transiently seroconverted for a year (personal observation; S.J.C.).

Brucellae will grow on a variety of the more nutritious microbiological cultivation media, including blood agar, *Brucella* agar (Oxoid), and serum dextrose agar; however, the best results from humans or marine mammals are generally obtained

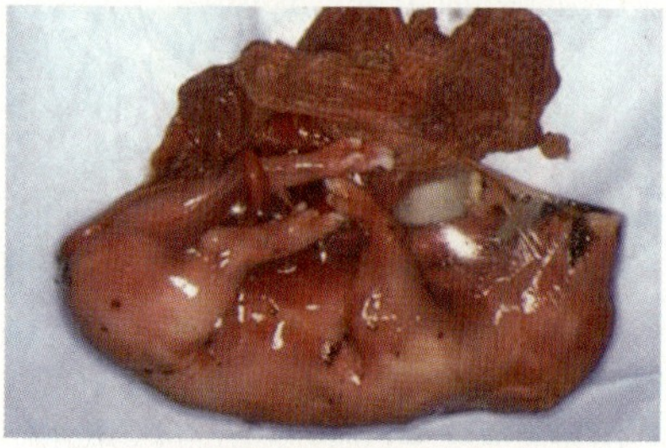

Figure 2.2 Aborted piglet (through infection with *B. suis* biovar 2). Picture courtesy of ANSES.

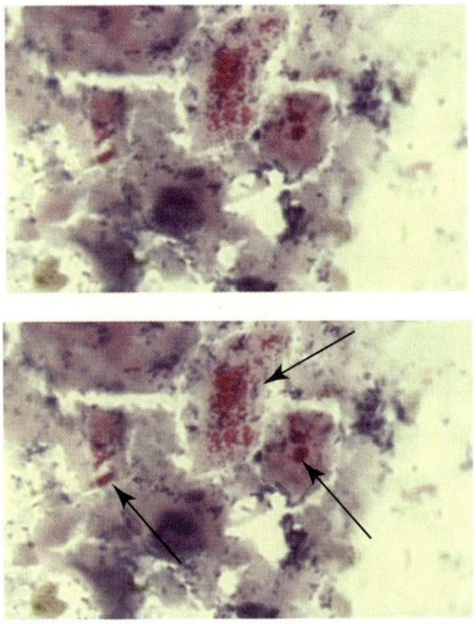

Figure 2.3 Abortion material infected with *B. melitensis* (Stamp's staining). Picture courtesy of ANSES.

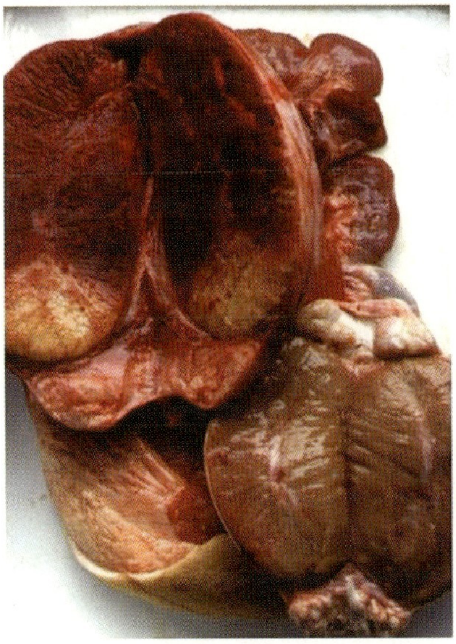

Figure 2.4 Orchitis in a boar (*B. suis* biovar 2). Picture courtesy of ANSES.

using a combination of liquid and solid media such as in a Castaneda bottle with prolonged incubation often in combination with a lysis centrifugation step [19].

When cultivated, the organisms look unremarkable (see Figure 2.6). Preliminary identification tests such as oxidase and urease tests should alert good microbiologists as to the possibility of *Brucella* and cultures should be removed immediately to

Figure 2.5 Orchitis in a ram (*B. melitensis*). Picture courtesy of ANSES.

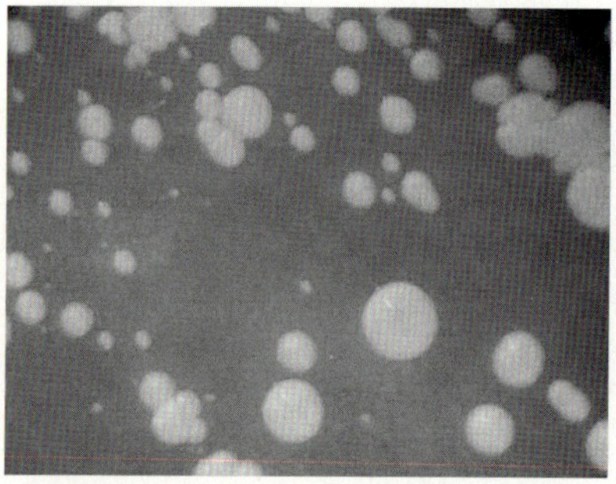

Figure 2.6 Growth of *Brucella* species on blood agar base showing characteristic morphology. Picture courtesy of ANSES.

suitable biosafety level (BSL)3 facilities. Once preliminary identification has been accomplished, a series of physiological tests, including phage and dye susceptibilities, requirements for CO_2, production of H_2S and urease, as well as agglutination with anti-A and -M monospecific sera, can be performed in order to biotype isolates [20]. This is a demanding highly specialized task best restricted to reference laboratories [20].

Serology has been the principal diagnostic method used for the detection of brucellosis for many years. Serological diagnostics have evolved over time from tests for agglutinins through to numerous ELISA formats and fluorescent antibody polarization assays [21]. Associated with this evolution is increased specificity and sensitivity; however, even this new generation of assays remain prone to detecting false-positive serologically reactive samples. These have been recently reviewed elsewhere [22]. The antigen offering superior sensitivity is lipopolysaccharide with its O-polysaccharide side chain of homopolymers of N-formyl-perosamine [N-formylated 4-amino, 4,6-dideoxyglucose] found either in α 1-2 linkages or α (1-2) linkages together with α (1-3). These variations result in the A and M serotype specificity of brucellae. The O-polysaccharide side chain shows distinct structural similarity with the N-formyl-perosamine homopolymer of N-formyl-perosamine of *Yersinia enterocolitica* O:9. This microbe is frequently found in association with swine, but also in other livestock species. It is likely that exposure to this microbe, or similar ones, may account for the serological "false alarms" encountered within brucellosis-free or almost free countries. Isolates with complete lipopolysaccharide including the O-side chain produce a smooth morphology when cultivated; however, this phenotype may revert to a rough variant associated with the loss of the O-side chain. The use of lipopolysaccharide antigens for serodiagnosis presents problems when naturally rough strains of *Brucella*, such as *B. canis* or *B. ovis* are the subject of investigation.

The search for improved diagnostic antigens has become something of a search for the "Holy Grail." Several different antigens have been assessed; however, the overall consensus of opinion is that they fail to offer diagnostic advantage over current serological assays [23].

More recent formats have become available for the diagnosis of human brucellosis, some enabling point of care testing [24–27], while others are believed to offer marginally superior diagnostic results [28].

Serodiagnosis is further complicated when used in countries where the disease is endemic/enzootic. Here significant proportions of the human population will naturally have elevated titers against brucellae. In animals, brucellosis eradication efforts utilizing live attenuated vaccine strains could result in positive titers in livestock recently vaccinated by the subcutaneous route.

2.3.1
Immunological Approaches

In order to overcome the problems outlined above, investigations have been undertaken to evaluate the use of other immunodiagnostic methods. As *Brucella* stimulates a strong Th1 response, the potential of using elevated interferon-gamma (IFN-γ) levels following specific stimulation has been explored. Early results look promising, with this approach clearly differentiating between brucellosis and *Yersinia* infections; however, no improvements in test sensitivity were offered [29]. Further limitations arise from the undulating levels observed with IFN-γ responses and their requirements for stimulation of blood cells within a tight time window [29, 30].

2.3.2
Polymerase Chain Reaction Assays

Many polymerase chain reaction (PCR) assays have been published for the detection of brucellae using various targets including the insertion sequence, IS*711*, and the outer membrane protein, BCSP31 [31]. A comparison of different PCR methods for diagnosis of brucellosis was recently undertaken, with the use of IS*711* giving marginally better sensitivity, while both this assay and BCSP31 were equivalent for their specificity [32]. Typically these will detect levels of 10 fg, which equates to three genomes. Many assays have also been modified for use on real-time thermocyclers, often accompanied by further increases in sensitivity. The limitation of this approach is the likelihood of sufficient bacteria present in blood or serum to enable detection. Host cells rapidly internalize brucellae, and the foci of infection are often found in lymph nodes, bone marrow, and various tissues, especially reproductive tissues in livestock. Application of PCR to these samples will be much more likely to detect *Brucella* DNA than the blood samples usually received.

Of increasing concern is the apparent persistence of DNA after therapy and clinical recovery. Several groups have now reported this, but clinical consequences remain to be elucidated [33, 34]. It has yet to be determined whether these individuals remain infected with pockets of viable organisms sequestered away in immunologically protected sites, or whether this is residual DNA following a successful cure.

2.4 Pathogenesis

Currently pathogenomic interrogation of sequenced genomes has met with enthusiasm, heralded as the likely way to dissect the basis for host specificity and virulence for these microbes; however, the biological basis for the observed traits remains elusive. The construction of microarrays based upon the genome of *B. melitensis* 16M and their comparison with the remaining five classical species of *Brucella* highlighted the remarkable similarity among this genus [15]. Only 217 of some 3198 ORFs were disclosed as missing or partial. Interestingly, these were clustered into nine genomic islands, suggesting the possibility of horizontal acquisition. Furthermore, possession of these islands with regard to virulence in different host species failed to reveal a significant correlation. Deletion studies of these islands revealed that all but one contributed to virulence, with mutation of genomic island 2 being associated with loss of smooth phenotype and having a role in either transportation or attachment of the O-side chain of smooth lipopolysaccharide [35]. Comparative analysis of whole genome sequences of wild-type and S19 vaccine strains of *B. abortus* again only showed limited differences among 45 ORFs, including the previously established deletions in erythritol utilization (*eryC*, *eryD*, and *eryF*) and a protein with homology with virulence factors of *Yersinia enterocolitica* and *Haemophilus influenzae*, but this gene is damaged in other fully virulent brucellae and is consequently unlikely to be responsible for the virulence of *B. abortus* [36]. Neither genomic nor limited proteomic investigations undertaken to date have to been able to unravel the complexities of the host–microbial interactions of the brucellae and their different vertebrate hosts.

The crucial nature of the initial interactions between the pathogen and host cells has been demonstrated. The mechanisms by which the brucellae interact with host phagocytic cells determine their fate. Here we see huge differences between the smooth and rough variants of brucellae. The smooth phenotype interacts with the lipid rafts on the host cell surface and, through "self-mediated" internalization, is able to avoid the degradative endocytic pathways. Though both are rapidly internalized, they utilize different mechanisms, with enormous consequences for the survival of *Brucella*. As a result of their atypical lipopolysaccharide, smooth *Brucella* have altered pathogen-associated molecular patterns (PAMPs) and associated reduced endotoxicity through their weak stimulation of toll-like receptors TLR2 and TLR4. This stealth-like property enables pathogenic *Brucella* to subvert normal cellular processes, establishing a unique intracellular replicative niche in which they persist with limited multiplication [37]. Conversely, rough strains are potent inducers of proinflammatory cytokines, including TNF-α interleukin-1 and interleukin-10. Our understanding of the complex interactions between the host cell and *Brucella* is still in its infancy; however, what is known is the pivotal role of the *vir*B system in ensuring correct internalization and biochemical communications with the host cell. Defects in genes of the *vir*B system characteristically result in attenuation.

2.5 Clinical and Pathological Findings

Human clinical disease is characterized by undulant fever, often accompanied by a myriad of nonspecific signs, including general malaise, abdominal cramps, headaches, insomnia, anorexia, arthralgia, and depression, to name but a few. Where focal lesions are apparent, other presentations will be predominant, including abscess formation particularly in the liver or spleen, arthritis, endocarditis, neurological disorders, orchitis, spondylitis, or spondylodiscitis, and less frequently respiratory signs. It is likely that the preferred intracellular location of this pathogen facilitates its hematogenous spread to varied sites within the host. However, in many cases, clinical signs are protean and often nonspecific. In some, the infection will show remarkable persistence over many years if untreated. A classic case of probable brucellosis was that described in Florence Nightingale. She suffered with undulant fevers, underwent personality changes and was bed-ridden with spondylitis for several years [38].

The two most infectious species for humans are *B. melitensis* and *B. suis* biovars 1 and 3, both with an infectious dose of approximately 10 organisms. *B. abortus* requires an estimated 100 organisms to infect humans. Although these species are the primary ones associated with human infection, cases have been encountered with various other species. Sporadic infection of humans has been recorded with *B. canis*, however this species is generally regarded as relatively non-infectious in humans. Interestingly, the recently described marine mammal isolates have been found to infect humans, initially established through a laboratory exposure [39], but

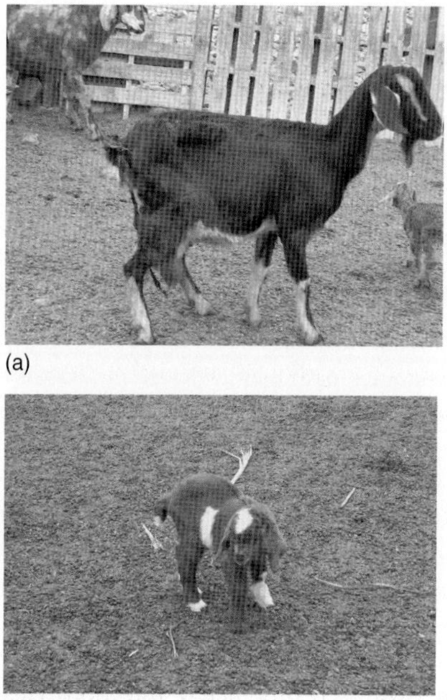

(a)

(b)

Figure 2.7 A probable case of brucellosis in a goat showing a retained placenta after delivery (a) and kid with spinal deformity (b).

then confirmed with naturally acquired infections [40, 41]. More recently, a *Brucella* species with the proposed name of *B. inopinata* was isolated from an infected breast implant [7, 11], highlighting the degree of clinical suspicion required.

In livestock, the disease typically manifests as reproductive failure, often through abortion or weak infected offspring (see Figures 2.2 and 2.7), but orchitis (Figures 2.4 and 2.5), epidedymitis and lameness can also occur, particularly in boars and small ruminants. Changing farming practices have introduced new susceptible species such as camels, deer, and water buffalo. Furthermore, alterations in husbandry practices upon animal welfare grounds have resulted in greater interaction between domestic livestock and potentially infected wildlife, leading to increased infection such as seen with free-ranging domestic pigs and wild boars. Furthermore, we now see far greater movement of livestock than previously, and facilitated by this is greater potential for rapid spread. It must be remembered that serological tests commonly used to screen livestock are not infallible. Furthermore, not all infected animals will give positive serological results. In consequence, any abortive episode among livestock should be investigated. In addition, tracking of animal movements is essential to rapidly reduce the spread of infection once detected.

Typically brucellosis surveillance of livestock is done through test and slaughter approaches; however, these are not appropriate in endemic regions or in developing countries where livestock have a major impact upon poverty and where veterinary services, laboratory infrastructure, and adequate budgets are not available.

2.6 Epidemiology, Molecular Typing, and Control Strategies

2.6.1 Epidemiology

The epidemiology of brucellosis is in a state of flux. Many countries in northern, central, and western Europe have successfully achieved Brucellosis-free status; however, these countries must remain vigilant to maintain this state. The infection has been more difficult to control in many other areas, particularly those where transhumance is largely practiced or with large wildlife reservoirs. Efforts are hindered by the regular migrations of tourists and even by religious festivals. Large human and animal population movements provide the potential for infections to arise. Furthermore, human socio-economical upheavals again provide a means for the spread of infection. Warfare or political instability typically results in a reduction of brucellosis control measures, usually met with a resurgence of infection. This is now the case in countries such as Iraq and several former USSR or Yugoslavian states.

Lack of education may also play a significant role in promoting infection. In areas of poverty, all protein (meat and milk) is used, even if derived from an animal that is/was diseased. Many traditional recipes utilizing dairy products still use unpasteurized ingredients. Education is urgently required to inform of the associated risks of brucellosis. In many endemic regions, vaccination coverage is sporadic, thus ineffective at controlling disease. Collectively, all of these factors influence the changing patterns of brucellosis epidemiology observed.

In those countries free from infection, tourists, either following vacations abroad, or immigrants returning home to visit relatives, can still regularly import brucellosis. In endemic countries, the incidence of brucellosis varies widely, in part to the factors outlined above. Human incidence rates of between 0.03 and more than 160 cases per 100 000 of the population have been recorded [42]. In many of these countries comprehensive livestock surveillance data is not available, thus human infection serves as a sentinel for infection in livestock.

2.6.2 Molecular Typing Methods

Classical biotyping has been the "gold standard" approach for distinguishing different isolates into their respective species and biovars. However, this requires highly skilled staff, large amounts of viable organism requiring containment level 3 facilities, and lengthy incubation periods. Given the reputation of brucellosis as the most frequently acquired laboratory infection [43], this approach is not to be undertaken lightly and is probably best reserved for reference facilities. To overcome these difficulties, various microbial typing methods, including OMP typing, ribotyping, amplified fragment length polymorphism (AFLP), pulsed-field gel electrophoresis (PFGE)–restriction fragment length polymorphism (RFLP), insertion sequence typing, to name but a few, have been applied to the *Brucella* with mixed success [44–46]. These approaches were able to successfully speciate isolates, but proved unable to offer differentiation down to biovar levels.

Many of these methods given above have a requirement for large amounts of starting materials, thus necessitating cultivation of the organism in sufficient quantities. Some advantage was offered by the AMOS PCR-based method for species identification of *B. abortus* biovars 1, 2, and 4, and *B. melitensis, B. ovis,* and *B. suis* biovar 1 through the use of a primer targeting the IS*711 Brucella*-specific insertion sequence and a down-stream species-specific site adjacent to another IS*711* site. This technique results in different sized amplicons indicating the identity of these species and biovars [47, 48]. Subsequent improvements and modifications could discriminate between vaccine strains [48, 49]; however, this assay was not fully inclusive for all brucellae, thus limiting its application. A more recent multiplex approach benefiting from genomic sequence data has been termed the "*bruce-ladder*" and can identify almost all species discriminated through differently sized amplicons [50]. This offers significant improvements but cannot reliably differentiate all strains of *B. canis* and the highly related *B. suis*.

More recently we have seen an expansion in typing approaches which utilize several loci and are thus inherently more robust by design. The first of these to be applied for typing brucellae was the hypervariable octameric oligonucleotide finger prints (Hoof Prints) scheme [51] that has proved to be a valuable method for trace-back studies when outbreaks occur. The method involved PCR amplification of eight variable loci identified from genome sequencing studies. However, the targets selected for use in this scheme were highly variable, thus most suited for epidemiological profiling and not necessarily for the identification of isolates through the possibility of homoplasy [52, 53]. Other schemes were quick to follow with PCR-based multi-locus variable number analysis (MLVA), also known as variable number tandem repeat typing (VNTR), being published by different groups [54–56]. Selection of more slowly evolving minisatellite repeats can reliably identify while the smaller more variable microsatellite repeats provide high-resolution typing. A publically available database can be found at *http://minisatellites.u-psud.fr/MLVAnet.*

An alternative highly discriminatory approach for the identification of sequence types of isolates can be achieved through a similarly multi-locus genomic approach, multi-locus sequence typing (MLST). This involves DNA sequence analysis of short regions of approximately 500 bp from housekeeping genes with conservational selective pressure and thus limited evolutionary diversity. Concatenation of this data enables the analysis of sequence types and analysis for phylogenetic inference. Analysis of 160 isolates representing all major species and biovars permitted the resolution of 27 distinct sequence types [57]. This analysis revealed distinct clusters among the established species of *B. abortus, B. melitensis, B. neotomae, B. ovis,* and marine mammal isolates, but also the greater heterogeneity among the biovars of

B. suis and the closely related *B. canis*. *B. suis* biovar 5 formed a unique unrelated cluster.

The MLST analysis has disclosed a series of single nucleotide polymorphisms (SNPs) that define particular clusters of brucellae. These can be incorporated into probe-based real-time assays, thus alleviating the requirement for sequence analysis. Two different assays have been published based upon SNPs using real-time platforms and minor groove binding probes, thus facilitating a high-throughput rapid testing [58, 59].

2.6.3 Control Strategies

As a zoonosis, brucellosis control is best aimed at the source. By reducing the burden of infection among livestock, this has the concomitant effect of reducing human infection. Eradication campaigns have been introduced in many countries with varying degrees of success. Typically, these are based on vaccination of susceptible hosts, surveillance to detect infected livestock and their subsequent slaughter, and movement restrictions where localized infection is present. The combination of these methods can lead to eventual freedom from disease; however, the likelihood of success is largely dependant on the compliance of many different groups (farmers, import/export, veterinary workers, public health workers, scientists, and government). Without this concerted, joined-up approach, the chances of success are very limited.

2.7 Conclusions

A new challenge is the growing incidence of *B. melitensis* infection in cattle, particularly in areas such as the Middle East, southeastern Europe, and central Asia. Vaccine efficacy is challenged by this new threat as cattle are routinely immunized with attenuated *B. abortus* strains (either S19 or RB51). The ability of these vaccines to protect against *B. melitensis* challenge in bovines remains largely untested. Also, the vaccination of swine, water buffalo, and camels remains an insufficiently explored issue.

A further challenge is posed by the threat of infection from wildlife reservoirs. This is a major challenge for the control of brucellosis worldwide. Particular problems have been documented with infected bison re-introducing brucellosis to cattle in the United States and with *B. suis* biovar 2-infected wild-boar or European hare posing significant infection risks to outdoor piggeries.

New potential threats are emerging, with significant infection levels being found among marine mammals. Infection levels of around 25% have been reported in samples submitted to northern and western European laboratories. Although experimental studies have demonstrated infection among terrestrial livestock [60], whether this poses a significant risk under natural circumstances remains to be established. Interestingly, human infection with these strains is possible and consequently, a genuine zoonotic threat exists [41].

Possibly the largest challenge is how to tackle the huge disease burden in developing countries. Here the consequences of brucellosis impact dramatically on both human health and on the livestock on which man depends. Diagnostic capability is sporadic and disease epidemiology largely unknown. We urgently require the introduction of a well thought-out vaccination strategy, pen-side diagnostic assays, education focused toward the spread of infection, and methods to efficiently control disease in livestock and reduce transmission to man.

References

1. Seleem, M.N., Boyle, S.M., and Sriranganathan, N. (2009) Brucellosis: a re-emerging zoonosis. *Vet. Microbiol.*, doi: 10.1016/j.vetmic.2009.06.021.
2. Whatmore, A.M. (2009) Current understanding of the genetic diversity of *Brucella*, an expanding genus of zoonotic pathogens. *Infect. Genet. Evol.*, **9**, 1168–1184.
3. Verger, J., Grimont, F., PAD, G., and Grayon, M. (1985) *Brucella*, a monospecific genus as shown by deoxyribonucleic acid hybridization. *Int. J. Syst. Bacteriol.*, **35**, 292–295.
4. Osterman, B. and Moriyon, I. (2006) International committee on systematics of prokaryotes; subcommittee on the taxonomy of *Brucella*: minutes of the meeting, 17 September 2003, Pamplona, Spain. *Int. J. Syst. Evol. Microbiol.*, **56**, 1173–1175.
5. Foster, G., Osterman, B.S., Godfroid, J., Jacques, I., and Cloeckaert, A. (2007) *Brucella ceti* sp. nov. and *Brucella pinnipedialis* sp. nov. For *Brucella* strains with cetaceans and seals as their preferred hosts. *Int. J. Syst. Evol. Microbiol.*, **57**, 2688–2693.
6. Scholz, H.C., Hubalek, Z., Sedlacek, I., Vergnaud, G., Tomaso, H., Al Dahouk, S., Melzer, F., Kampfer, P., Neubauer, H., Cloeckaert, A. *et al.* (2008) *Brucella microti* sp. nov., isolated from the common vole *Microtus arvalis*. *Int. J. Syst. Evol. Microbiol.*, **58**, 375–382.
7. Scholz, H.C., Nockler, K., Gollner, C., Bahn, P., Vergnaud, G., Tomaso, H., Al-Dahouk, S., Kampfer, P., Cloeckaert, A., Maquart, M. *et al.* (2010) *Brucella inopinata* sp. nov., isolated from a breast implant infection. *Int. J. Syst. Evol. Microbiol*, **60**, 801–808.
8. Scholz, H., Al Dahouk, S., Tomaso, H., Neubauer, H., Witte, A., Schloter, M., Kampfer, P., Falsen, E., Pfeffer, M., and Engel, M. (2008) Genetic diversity and phylogenetic relationships of bacteria belonging to the ochrobactrum-*Brucella* group by recA and 16S rRNA gene-based comparative sequence analysis. *Syst. Appl. Microbiol.*, **31**, 1–16.
9. Wattam, A.R., Williams, K.P., Snyder, E.E., Almeida, N.F. Jr., Shukla, M., Dickerman, A.W., Crasta, O.R., Kenyon, R., Lu, J., Shallom, J.M. *et al.* (2009) Analysis of ten *Brucella* genomes reveals evidence for horizontal gene transfer despite a preferred intracellular lifestyle. *J. Bacteriol.*, **191**, 3569–3579.
10. Allix, S., Le Carrou, G., Thiebaud, M., Albert, D., Perrett, L.L., Dawson, C.E., Groussaud, P., Stubberfield, E.J., Koylass, M., Whatmore, A.M. *et al.* (2008) *Brucella abortus* biovar 7: phenotypic and molecular evidence. Proceedings of the Brucellosis 2008 International Research Conference (Inc 61st Brucellosis Research Conference), Egham, London.
11. De, B.K., Stauffer, L., Koylass, M.S., Sharp, S.E., Gee, J.E., Helsel, L.O., Steigerwalt, A.G., Vega, R., Clark, T.A., Daneshvar, M.I. *et al.* (2008) Novel *Brucella* strain (BO1) associated with a prosthetic breast implant infection. *J. Clin. Microbiol.*, **46**, 43–49.

12. Fretin, D., Fauconnier, A., Kohler, S., Halling, S., Leonard, S., Nijskens, C., Ferooz, J., Lestrate, P., Delrue, R., Danese, I. *et al.* (2005) The sheathed flagellum of *Brucella melitensis* is involved in persistence in a murine model of infection. *Cell Microbiol.*, **7**, 687–698.
13. Zygmunt, M.S., Hagius, S., Walker, J., and Elzer, P. (2006) Identification of *Brucella melitensis* 16M genes required for bacterial survival in the caprine host. *Microbes Infect.*, **8**, 2849–2854.
14. Lapaque, N., Moriyon, I., Moreno, E., and Gorvel, J. (2005) *Brucella* lipopolysaccharide acts as a virulence factor. *Curr. Opin. Microbiol.*, **8**, 60–66.
15. Rajashekara, G., Glasner, J.D., Glover, D.A., and Splitter, G.A. (2004) Comparative whole-genome hybridization reveals genomic islands in *Brucella* species. *J. Bacteriol.*, **186**, 5040–5051.
16. Yagupsky, P. (1994) Detection of *Brucella melitensis* by bactec NR660 blood culture system. *J. Clin. Microbiol.*, **32**, 1899–1901.
17. Espinosa, B.J., Chacaltana, J., Mulder, M., Franco, M.P., Blazes, D.L., Gilman, R.H., Smits, H.L., and Hall, E.R. (2009) Comparison of culture techniques at different stages of brucellosis. *Am. J. Trop. Med. Hyg.*, **80**, 625–627.
18. Gaviria-Ruiz, M.M. and Cardona-Castro, N.M. (1995) Evaluation and comparison of different blood culture techniques for bacteriological isolation of *Salmonella typhi* and *Brucella abortus*. *J. Clin. Microbiol.*, **33**, 868–871.
19. Mantur, B.G. and Mangalgi, S.S. (2004) Evaluation of conventional Castaneda and lysis centrifugation blood culture techniques for diagnosis of human brucellosis. *J. Clin. Microbiol.*, **42**, 4327–4328.
20. Al Dahouk, S., Tomaso, H., Nockler, K., Neubauer, H., and Frangoulidis, D. (2003) Laboratory-based diagnosis of brucellosis-a review of the literature. Part I: techniques for direct detection and identification of *Brucella* spp. *Clin. Lab.*, **49**, 487–505.
21. Nielsen, K. and Gall, D. (2001) Fluorescence polarization assay for the diagnosis of brucellosis: a review. *J. Immunoassay Immunochem.*, **22**, 183–201.
22. Nielsen, K. (2002) Diagnosis of brucellosis by serology. *Vet. Microbiol.*, **90**, 447–459.
23. Letesson, J.J., Tibor, A., van Eynde, G., Wansard, V., Weynants, V., Denoel, P., and Saman, E. (1997) Humoral immune responses of *Brucella*-infected cattle, sheep, and goats to eight purified recombinant *Brucella* proteins in an indirect enzyme-linked immunosorbent assay. *Clin. Diagn. Lab. Immunol.*, **4**, 556–564.
24. Abdoel, T., Dias, I., Cardoso, R., and Smits, H. (2008) Simple and rapid field tests for brucellosis in livestock. *Vet. Microbiol.*, **130**, 312–319.
25. Abdoel, T. and Smits, H. (2007) Rapid latex agglutination test for the serodiagnosis of human brucellosis. *Diag. Microbiol. Infect. Dis.*, **57**, 123–128.
26. Clavijo, E., Diaz, R., Anguita, A., Garcia, A., Pinedo, A., and Smits, H.L. (2003) Comparison of a dipstick assay for detection of *Brucella*-specific immunoglobulin M antibodies with other tests for serodiagnosis of human brucellosis. *Clin. Diagn. Lab. Immunol.*, **10**, 612–615.
27. Smits, H.L., Basahi, M.A., Diaz, R., Marrodan, T., Douglas, J.T., Rocha, A., Veerman, J., Zheludkov, M.M., Witte, O.W., de Jong, J. *et al.* (1999) Development and evaluation of a rapid dipstick assay for serodiagnosis of acute human brucellosis. *J. Clin. Microbiol.*, **37**, 4179–4182.
28. Casanova, A., Ariza, J., Rubio, M., Masuet, C., and Diaz, R. (2009) Brucellacapt vs classical tests in the serological diagnosis and management of human brucellosis. *Clin. Vaccine Immunol.*, **16**, 844–851.
29. Thirlwall, R., Commander, N., Brew, S., Cutler, S., McGiven, J., and Stack, J. (2008) Improving the specificity of immunodiagnosis for porcine brucellosis. *Vet. Res. Commun.*, **32**, 209–213.
30. Jungersen, G., Sorensen, V., Giese, S., Stack, J., and Riber, U. (2006) Differentiation between serological responses to *Brucella suis* and *Yesinia enterocolitica* serotype O:9 after natural or experimental infection in pigs. *Epidemiol. Infect.*, **35**, 134.

31. Al Dahouk, S., Tomaso, H., Nockler, K., and Neubauer, H. (2004) The detection of *Brucella* spp. using PCR-ELISA and real-time PCR assays. *Clin. Lab.*, **50**, 387–394.
32. Bounaadja, L., Albert, D., Chenais, B., Henault, S., Zygmunt, M., Poliak, S., and Garin-Bastuji, B. (2009) Real-time PCR for identification of *Brucella* spp.: A comparative study of IS711, bcsp31 and per target genes. *Vet. Microbiol.*, **137**, 156–164.
33. Castano, M.J. and Solera, J. (2009) Chronic brucellosis and persistence of *Brucella melitensis* DNA. *J. Clin. Microbiol.*, **47**, 2084–2089.
34. Maas, K.S.J.S.M., Mendez, M., Zavaleta, M., Manrique, J., Franco, M.P., Mulder, M., Bonifacio, N., Casteneda, M.L., Chacaltana, J., Yagui, E. *et al.* (2007) Evaluation of brucellosis by PCR and persistence after treatment in patients returning to the hospital for follow-up. *Am. J. Trop. Med. Hyg.*, **76**, 698–702.
35. Rajashekara, G., Covert, J., Petersen, E., Eskra, L., and Splitter, G. (2008) Genomic island 2 of *Brucella melitensis* is a major virulence determinant: Functional analyses of genomic islands. *J. Bacteriol.*, **190**, 6243–6252.
36. Crasta, O., Folkerts, O., Fei, Z., Mane, S., Evans, C., Martino-Catt, S., Bricker, B., Yu, G., Du, L., and Sobral, B. (2008) Genome sequence of *Brucella abortus* vaccine strain S19 compared to virulent strains yields candidate virulence genes. *PLoS ONE*, **3**, e2193.
37. Kohler, S., Michaux-Charachon, S., Porte, F., Ramuz, M., and Liautard, J. (2003) What is the nature of the replicative niche of a stealthy bug named *Brucella*? *Trends Microbiol.*, **11**, 215–219.
38. Young, D.A. (1995) Florence nightingale's fever. *Br. Med. J.*, **311**, 1697–1700.
39. Brew, S.D., Perrett, L.L., Stack, J.A., MacMillan, A.P., and Staunton, N.J. (1999) Human exposure to *Brucella* recovered from a sea mammal. *Vet. Rec.*, **144**, 483.
40. McDonald, W.L., Jamaludin, R., Mackereth, G., Hansen, M., Humphrey, S., Short, P., Taylor, T., Swingler, J., Dawson, C.E., Whatmore, A.M. *et al.* (2006) Characterization of a *Brucella* sp. strain as a marine-mammal type despite isolation from a patient with spinal osteomyelitis in new zealand. *J. Clin. Microbiol.*, **44**, 4363–4370.
41. Sohn, A.H., Probert, W.S., Glaser, C.A., Gupta, N., Bollen, A.W., Wong, J.D., Grace, E.M., and McDonald, W.C. (2003) Human neurobrucellosis with intracerebral granuloma caused by a marine mammal *Brucella* spp. *Emerg. Infect. Dis.*, **9**, 485–488.
42. Pappas, G., Papadimitriou, P., Akritidis, N., Christou, L., and Tsianos, E. (2006) The new global map of human brucellosis. *Lancet Infect. Dis.*, **6**, 91–99.
43. Ergonul, O., Celikbas, A., Tezeren, D., Guvener, E., and Dokuzoguz, B. (2004) Analysis of risk factors for laboratory-acquired *Brucella* infections. *J. Hosp. Infect.*, **56**, 223–227.
44. Cloeckaert, A., Verger, J., Grayon, M., Paquet, J., Garin-Bastuji, B., Foster, G., and Godfroid, J. (2001) Classification of *Brucella* spp. isolated from marine mammals by DNA polymorphism at the omp2 locus. *Microbes Infect.*, **3**, 729–738.
45. Jensen, A.E., Cheville, N.F., Thoen, C.O., MacMillan, A.P., and Miller, W.G. (1999) Genomic fingerprinting and development of a dendrogram for *Brucella* spp. isolated from seals, porpoises, and dolphins. *J. Vet. Diagn. Invest.*, **11**, 152–157.
46. Tcherneva, E., Rijpens, N., Jersek, B., and Herman, L.M. (2000) Differentiation of *Brucella* species by random amplified polymorphic DNA analysis. *J. Appl. Microbiol.*, **88**, 69–80.
47. Bricker, B.J. and Halling, S.M. (1994) Differentiation of *Brucella abortus* bv. 1, 2, and 4, *Brucella melitensis, Brucella* ovis, and *Brucella suis* bv. 1 by PCR. *J. Clin. Microbiol.*, **32**, 2660–2666.
48. Bricker, B.J. and Halling, S.M. (1995) Enhancement of the *Brucella* AMOS PCR assay for differentiation of *Brucella abortus* vaccine strains S19 and RB51. *J. Clin. Microbiol.*, **33**, 1640–1642.
49. Ewalt, D.R. and Bricker, B.J. (2000) Validation of the abbreviated *Brucella* AMOS PCR as a rapid screening method for

differentiation of *Brucella abortus* field strain isolates and the vaccine strains, S19 and RB51. *J. Clin. Microbiol.*, **38**, 3085–3086.

50. Lopez-Goni, I., Garcia-Yoldi, D., Marin, C.M., de Miguel, M.J., Munoz, P.M., Blasco, J.M., Jacques, I., Grayon, M., Cloeckaert, A., Ferreira, A.C. *et al.* (2008) Evaluation of a multiplex PCR assay (bruce-ladder) for molecular typing of all *Brucella* species, including the vaccine strains. *J. Clin. Microbiol.*, **46**, 3484–3487.

51. Bricker, B.J., Ewalt, D.R., and Halling, S.M. (2003) *Brucella* 'hoof-prints': strain typing by multi-locus analysis of variable number tandem repeats (VNTRs). *BMC Microbiol.*, **3**, 15.

52. Bricker, B. and Ewalt, D. (2005) Evaluation of the hoof-print assay for typing *Brucella abortus* strains isolated from cattle in the United States: Results with four performance criteria. *BMC Microbiol.*, **5**, 37.

53. Bricker, B. and Ewalt, D. (2006) Hoof prints: *Brucella* strain typing by PCR amplification of multilocus tandem-repeat polymorphisms. *Methods Mol. Biol.*, **345**, 141–173.

54. Lc Flcche, P., Jacques, I., Grayon, M., Al Dahouk, S., Bouchon, P., Denoeud, F., Nockler, K., Neubauer, H., Guilloteau, L., and Vergnaud, G. (2006) Evaluation and selection of tandem repeat loci for a *Brucella* MLVA typing assay. *BMC Microbiol.*, **6**, 9.

55. Tiller, R.V., De, B.K., Boshra, M., Huynh, L.Y., Van Ert, M.N., Wagner, D.M., Klena, J., Mohsen, T.S., El-Shafie, S.S., Keim, P. *et al.* (2009) Comparison of two multiple locus variable number tandem repeat (VNTR) analysis (MLVA) methods for molecular strain typing human *Brucella melitensis* isolates from the Middle East. *J. Clin. Microbiol.*, **47**, 2226–2231.

56. Whatmore, A.M., Shankster, S.J., Perrett, L.L., Murphy, T.J., Brew, S.D., Thirlwall, R.E., Cutler, S.J., and MacMillan, A.P. (2006) Identification and characterization of variable-number tandem-repeat markers for typing of *Brucella* spp. *J. Clin. Microbiol.*, **44**, 1982–1993.

57. Whatmore, A.M., Perrett, L.L., and MacMillan, A.P. (2007) Characterisation of the genetic diversity of *Brucella* by multilocus sequencing. *BMC Microbiol.*, **7**, 34.

58. Foster, J.T., Okinaka, R.T., Svensson, R., Shaw, K., De, B.K., Robison, R.A., Probert, W.S., Kenefic, L.J., Brown, W.D., and Keim, P. (2008) Real-time PCR assays of single-nucleotide polymorphisms defining the major *Brucella* clades. *J. Clin. Microbiol.*, **46**, 296–301.

59. Gopaul, K., Koylass, M., Smith, C., and Whatmore, A. (2008) Rapid identification of *Brucella* isolates to the species level by real-time PCR based single nucleotide polymorphism (SNP) analysis. *BMC Microbiol.*, **8**, 86.

60. Rhyan, J.C., Gidlewski, T., Ewalt, D.R., Hennager, S.G., Lambourne, D.M., and Olsen, S.C. (2001) Seroconversion and abortion in cattle experimentally infected with *Brucella* sp. isolated from a pacific harbor seal (*Phoca vitulina richardsi*). *J. Vet. Diagn. Invest.*, **13**, 379–382.

3
Burkholderia mallei: Glanders

Lisa D. Sprague and Mandy C. Elschner

3.1
Introduction

Glanders, a rare but highly contagious and frequently fatal zoonotic disease of solipeds, including horses, mules, and donkeys, is caused by the bacterium *Burkholderia mallei*. The disease is characterized by ulcerating granulomatous lesions of the skin and mucous membranes. Fever, myalgia, headache, fatigue, diarrhea, and weight loss are typical symptoms of infection. The organism frequently spreads via lymph channels to the regional lymph nodes, often causing lymphangitis and lymphadenitis en route. A particular cutaneous form of glanders, also known as *farcy*, can be observed in horses. It manifests itself with nodular abscesses, also known as *farcy buds* which can ulcerate, and regional cutaneous enlarged and indurated lymphatic pathways (farcy pipes), secreting a glanders-typical yellow-green gelatinous pus (farcy oil).

Natural infection has been described in numerous animal species such as dogs, cats, carnivorous predators, and small ruminants, that is, goats, living in the vicinity of infected equids, or with access to contaminated carcasses. Camels are also susceptible to infection, usually in association with human disease.

Glanders has been eradicated in most countries, but is still found in parts of Africa, the Middle East, South America, and Eastern Europe. Outbreaks of the disease have to be notified to the World Organisation of Animal Health (OIE). *B. mallei* has gained increased attention as a possible warfare agent in the biological weapons programs of several countries.

3.2
Characteristics of the Agent

Burkholderia mallei is a small pleomorphic, Gram-negative, aerobic, bipolar staining, and amotile bacillus [1, 2], with an exopolysaccharide capsule [3]. A typical *B. mallei* colony on glycerol agar is whitish to cream colored, smooth surfaced, moist, and

BSL3 and BSL4 Agents: Epidemiology, Microbiology, and Practical Guidelines, First Edition.
Edited by Mandy C. Elschner, Sally J. Cutler, Manfred Weidmann, and Patrick Butaye.

viscid with a tendency to deliquesce. Older colonies are shiny yellow or brown. No hemolysis can be observed on blood agar [4, 5]. In broth cultures, the appearance of a filamentous form is reported [6].

The tenacity of the agent is low. *B. mallei* is sensitive to sunlight, heat, and desiccation. Bacteria-containing pus or exudates usually become sterile within a few days. Urine inactivates the agent within 40 min, stomach fluid within 30 min. However, the agent can remain viable for months in a dark and humid environment.

B. mallei is susceptible to various disinfectants. It can be inactivated by sodium hypochlorite at 500 ppm chlorine, 1% sodium hydroxide, 2% formalin, 1% potassium permanganate, and benzalkonium chloride (1 : 2000). Phenol and Lysol are ineffective [4, 7]. However, current sensitivity testing of the validity of disinfectants is missing.

3.3 Diagnosis

The internationally recognized and validated methods for diagnosing glanders in animals are published by the OIE in the Manual of Diagnostic Tests and Vaccines for Terrestrial Animals and updated regularly [8]. *B. mallei* is transmissible to humans by direct contact with sick animals or contaminated/infectious material. Any manipulation with potentially infected/contaminated material must take place in a laboratory that meets the requirements for Containment Group 3 pathogens.

3.3.1 Cultural Identification

The isolation of *B. mallei* from clinical samples is difficult, due to the low number of bacteria in the tissues. Previous antibiotic treatment of affected animals also reduces the amount of bacteria in samples. The bacteria grow well in ordinary culture media containing 3–4% glycerol. Colonies appear within 48–72 h but care must be taken to avoid the overgrowth of plates with contaminants like *Pseudomonadaceae* [9]. Other contaminating bacteria such as *Proteus* spp. can be inhibited by using a *B. mallei*-selective agar as described by Xie *et al.* [10]. The *B. pseudomallei*-selective Ashdown agar [11] is not suitable for *B. mallei* isolation. In order to avoid these problems, *B. mallei* can be isolated from samples by means of experimental infection of guinea pigs. Since the sensitivity of this method is approximately 20%, at least five animals have to be infected per sample [12]. After an intra peritoneal injection of *B. mallei*-containing material, the animals show the Strauss reaction, that is, severe peritonitis and orchitis, and the bacterium can be isolated from the infected testes. No commercially available test is suitable for the biochemical analysis of *B. mallei*.

3.3.2 Molecular Based Methods

Due to the high genetic similarity between *B. mallei* and *B. pseudomallei* (DNA-DNA homology >90%, 16S RNA homology 100%), most described assays only identify the *B. mallei/B. pseudomallei* complex based on the 16S rRNA and the *fliC* gene [13–19]. Other methods targeting the *flip* gene [20, 21], the metalloprotease gene [22], the type three secretion system [23], and the $bimA_{ma}$ gene [24] can be used to confirm the isolate to be *B. mallei*. A newly established microarray-based method allows the fast and accurate identification and differentiation of *B. mallei* and *B. pseudomallei* using at least four independent genetic markers including the16S rRNA gene, *fliC* – and *motB* gene [25].

3.3.3 Antigen Detection

B. mallei and *B. pseudomallei* are antigenetically closely related, as has been shown by several serological studies using polyclonal [26] and monoclonal antibodies (MAbs) [27, 28]. The use of MAbs specific for *B. pseudomallei* exopolysaccharide has revealed that some *B. mallei* strains display the immunoreactive component, thus demonstrating the antigenic relations between *B mallei* and *B. pseudomallei*; but it was also possible to determine the existence of distinctive types of *B. mallei* strains [27]. However, some *B. mallei* strains do not react with LPS specific MAbs, thus suggesting the existence of different subtypes of *B. mallei* with lipopolysaccharides (LPSs) lacking the entire *O*-polysaccharide moieties [28, 29]. These studies illustrate the difficulty to identify an overall *B. mallei* specific antigen, which explains the unavailability of a commercial antigen detection kit for glanders. So far, only one MAb, namely "3D11" (Biotrend Chemikalien GmbH, Cologne, Germany) raised against the LPS of *B. mallei* is commercially available. MAbs can be applied to various techniques: Western blot analysis for detecting the LPS, and immunofluorescent staining of heat inactivated bacteria [28]. However, these techniques have not yet been validated for their applicability to clinical samples.

3.3.4 Serology

The eradication of glanders from western Europe, Australia, and North America was achieved by serodiagnosis using a complement fixation test (CFT), serum agglutination test, and malleinization, a test for hypersensitivity, in combination with a rigorous culling policy [6]. At present CFT, ELISA techniques, and the mallein test are the most accurate and reliable methods for diagnostic applications [8].

Within the framework of international trade with equids (EU Directive 90/426/EWG), equids originating from designated countries have to be tested by

CFT in order to obtain a health certificate. Although CFT has a sensitivity of at least 97% [26], the occurrence of a considerable number of false CFT positive tested animals has been reported [9, 30]. To overcome these problems and to avoid the resulting unnecessary trade restrictions there is a need for the development and evaluation of alternative or confirmatory tests. Recently, two new cELISA have been described. One uses mallein as antigen [31], the other uses a commercially available *B. mallei* specific MAb (3D11) and a simple carbohydrate preparation of *B. mallei* as antigen [32]. Both tests show comparable sensitivities (98.9 and 98.6%, respectively) and a specificity of 100%. Other tests used for diagnosing glanders are: Rose Bengal agglutination test, immunoblotting, indirect hemagglutination test, counter immunoelectrophoreses, and indirect fluorescent antibody test [33–37]. New approaches using microarray technology for the serodiagnosis of glanders and melioidosis are based on polysaccharide antigens [38].

3.4
Clinical and Pathological Findings in Humans

In humans, the manifestation of glanders can vary, depending on the route of entry and virulence of the agent. Various forms of infection have been described, including nasal, pulmonary, septicemic, disseminated, and chronic infection. Localized infection can remain a local process or disseminate throughout the organism, resulting in septicemia. Acute glanders has an incubation period of one to seven days and is characterized by bacterial septicemia. The prodromal stage is accompanied by anorexia, fever, rigors/chills, nausea, myalgia, malaise, and painful joints. Up to 80% of the patients show pulmonary symptoms. Pneumonia, pulmonary abscesses, and pleural effusions can develop. Sudden onset of severe pain and fever is followed by hematogenic systemic dissemination of bacteria. With disease progression, granuloma and abscesses form subcutaneously and in the muscles. Patients die within 7–30 days without adequate treatment. The prognosis for acute *B. mallei* septicemia is guarded regardless of treatment, with case fatality rates as high as 95% [12, 39–44]. Reports on subclinical cases identified during autopsy exist [45].

Chronic infections can occur and may remain latent in affected individuals with exacerbations and remissions over many years. Nodular lesions form in numerous organs, including the lungs, liver, spleen, muscles, and skin, concomitantly with lymphadenopathy and weight loss.

A cutaneous manifestation of glanders can develop within one to five days after the entry of the agent through skin lacerations or abrasions. Nodules, abscesses, and ulcers appear on the skin, generally along the lymphatic vessels of the limbs and the face. The nodules can caseate or become calcified and localized erythema can develop in the affected region; the regional lymph nodes are usually enlarged.

3.5
Clinical and Pathological Findings in Animals

Glanders is primarily a disease of solipeds but natural infection has also occurred in goats, sheep, domestic, and wild carnivores. Rodents such as guinea pigs, field mice, and hamsters are particularly susceptible to glanders, whereas cattle, pigs, poultry, birds, and rats are considered to be more resistant to infection [4, 46, 47]. Large cats, that is, lions, tigers, and leopards, and carnivores such as jackals, hyenas, wolves, dogs, and polar bears can succumb to infection after consumption of infectious horse meat. Camels and small ruminants can fall ill if they are kept in the vicinity of infected horses. Cattle, pigs, poultry, and birds are considered to be resistant to the agent [8, 48, 49].

The respiratory tract is the principal target in both acute and chronic infection. Acute infection, usually seen in mules and donkeys, is characterized by fulminating broncho-pneumonia with coughing and high fever (40 °C). The incubation period is short and death can occur within a few days to a couple of weeks. In horses, the course of disease is more insidious and debilitating, usually accompanied by chronic pneumonia. Chronic infections can develop over several weeks to months. Latently affected animals can be ill for some months and, although clinical recovery may be observed, the infection can persist for years.

Glanders in horses can manifest itself either as an infection of the upper respiratory tract, the lung, or the skin. Infection of the upper respiratory tract is characterized by multiple nodules and ulcers in the upper airway and nasal cavities, resulting in unilateral or bilateral hemorrhagic, mucopurulent nasal discharge. The mucous membranes of the nostrils swell and can ulcerate, resulting in the rupture of the nasal septum. The regional lymph nodes are enlarged, painful and may suppurate and drain. The structure and consistency of the submandibular lymph nodes becomes knotty and hard; adherence to the skin and mandible may be observed [45]. A striking histopathological finding in the nasal mucous membranes is severe thrombosis in the large vessels [50].

Equids with pulmonary disease develop fever (40 °C) and show signs of depression, anorexia, coughing, and general malaise. Once the disease progresses, dyspnea and rales can develop. At necropsy the lesions can be attributed to septicemia and a diffuse, severe catarrhal bronchopneumonia. Chronic disease leads to a more miliary form of pneumonia, with scattered, discrete, round, and firm nodules (0.5–1.0 cm) embedded in the lung tissue. On incision these nodules reveal a white center and a dark gelatinous periphery. Histologically, the center consists of pus and the surrounding is mostly made up of polymorphonuclear leukocytes. Older lesions may be encircled by a layer of epithelioid and giant cells surrounding the necrotic center. Frequently, the bronchial and mediastinal lymph nodes are affected; additionally, the upper respiratory tract is infected as well. The observed lesions are ulcers with irregular borders which form stellate scars once healed [50, 51].

Cutaneous glanders (farcy) can result after skin inoculation or as a secondary manifestation of the respiratory disease. Nodules (farcy buds) are found not only on the skin of the head, neck, thorax, and abdomen but also in the subcutaneous tissue and the lymph nodes. Occasionally, nodules have also been found in internal organs, including the liver, spleen, and testes. Lymphatic vessels draining affected lymph nodes become firm, enlarged, and indurated (farcy pipes) and severe swelling can be observed. This manifestation usually occurs in the limbs. If the lesions rupture, they secrete glanders typical yellow-green gelatinous pus (farcy oil). During the healing process crater-shaped ulcers develop, which have the tendency to bleed very easily [45, 50, 51].

3.6 Epidemiology

Glanders is an ancient disease, the earliest reports dating back to the scripts of Hippocrates (450 to 425 BC), Aristoteles (330 BC), and Vegetius (300 AC). An even earlier description of glanders can assumedly be found in the first book of Homer's Iliad [52]. These days, glanders is a disease rarely heard of in Western Europe and in North America. However, up to the beginning of the twentieth century, this was not the case and glanders occurred throughout the world. Especially during war times the prevalence of glanders was high, particularly in army horses, and losses were severe. The geographic expansion of the disease was facilitated by the movement of troops and their horses. Moreover, animal to man transmission was not uncommon, the outcome usually fatal [53].

Since *B. mallei* is an obligate parasite with a limited host range (solipeds) and given that tests for detecting carriers of the disease exist, considerable steps toward global eradication have been made. Western Europe, north America, and Australia are considered to be glanders-free. Endemic spots still exist in some Asian countries, such as Turkey, Iran, Iraq, Pakistan, India, Mongolia, and China [6]. A recent outbreak of glanders in Brazil in 2000 revealed a second geographically independent endemic region (OIE press release, 15 March 2000). Glanders can also be regularly detected in imported, quarantined horses in Dubai, UAE (OIE press release, 15 October 2004). The imported case of glanders into Germany in 2006 revealed the potential risk of introducing this notifiable disease to glanders-free areas [54].

Infected equids excrete *B. mallei* with all secretions. Transmission can occur horizontally via direct or indirect contact between infected animals or contaminated objects. Especially water and food troughs and harnesses contaminated with saliva or nasal secretions are a significant source of infection [50]. Spreading of disease by means of aerosol formation during inter-animal contact is considered to be secondary. Crowding, unsanitary, and stressful housing conditions and the rapid introduction of horses from an unknown origin dramatically increase the transmission of infection.

Human infection usually develops after exposure to sick animals, contaminated objects, infected tissues, or bacterial cultures. Mucous membranes, small wounds, and skin abrasions are the typical entry sites for the agent [12]. Laboratory-acquired infections usually take place during post-mortem examinations via cutaneous injuries or by means of inhalation of infectious aerosols during the routine handling of samples or cultures [12, 45, 55].

Glanders is a notifiable animal disease and must be reported to the OIE. Infected animals must be culled, due to the potential for zoonotic disease spread. No vaccines for man or animal exist and infection does not result in natural immunity.

3.7 Molecular Typing

The sequence of the genome of the *Burkholderia mallei* type strain ATCC 23344T was published back in 2004 [56] and succeeded by several sequences of other *B. mallei* isolates. Molecular techniques such as pulsed field gel electrophoresis [57], PCR-restriction fragment length polymorphism [58], and ribotyping [59] have been used for the further discrimination of isolates. But these techniques have a major drawback: they are in-house tests of specialized laboratories since extensive strain collections are required [8]. The further improvement of such typing methods is a precondition for epidemiological tracing.

3.8 Conclusions

Glanders is a rare but highly contagious and frequently fatal zoonotic disease of solipeds. The disease is characterized by ulcerating granulomatous lesions of the skin and mucous membranes. Glanders has been eradicated in most countries, but is still found in parts of Africa, the Middle East, South America, and eastern Europe. Outbreaks of the disease have to be notified to the OIE.

Cultural identification of *B. mallei* from clinical samples is difficult; however, serological assays and a newly established microarray-based method allow the fast and accurate identification and differentiation of *B. mallei*. At present CFT, ELISA techniques, and the mallein test are the most accurate and reliable methods for diagnostic applications.

References

1. Yabuuchi, E., Kosako, Y., Oyaizu, H., Yano, I., Hotta, H., Hashimoto, Y., Ezaki, T., and Arakawa, M. (1992) Proposal of *Burkholderia* gen. nov. and transfer of seven species of the genus *Pseudomonas*homology group II to the new genus, with the type species *Burkholderia cepacia* (Palleroni and Holmes 1981) comb. nov. *Microbiol. Immunol.*, **36**, 1251–1275.
2. Brindle, C.S. and Cowan, S.T. (1951) Flagellation and taxonomy of whitmore's bacillus. *J. Pathol. Bact.*, **63**, 571–575.
3. Popov, S.F., Kurilov, V.Y., and Iakovlev, A.T. (1995) *Pseudomonas pseudomallei* and *Pseudomonas mallei* – capsule

forming bacteria. *Z. Microbiol. Epidemiol. Immunobiol.*, **5**, 32–36.
4. Maier, H. (1981) *Handbuch der Bakteriellen Infektionen bei Tieren.* Jena, Gustav Fischer Verlag, pp. 111–153.
5. Bongert, J. (1927) *Bakeriologische Diagnostik der Tierseuchen*, 7th edn, Verlagsbuchhandlung Richard Schötz, Berlin, pp. 406–455.
6. Neubauer, H., Sprague, L.D., Zacharia, R., Tomaso, H., Al Dahouk, S., Wernery, R., Wernery, U., and Scholz, H.C. (2005) Serodiagnosis of *Burkholderia mallei* infections in horses: state-of-the-art and perspectives. *J. Vet. Med. B Infect. Dis. Vet. Public Health*, **52** (5), 201–205.
7. Whitwell, K. and Hickman, J. (1992) *Manual of Standards for Diagnostic Tests and Vaccines*, Office International des Epizooties, Paris, pp. 485–492.
8. Anonymous (2008) World Organisation of Animal Health: Manual of Diagnostic Tests and Vaccines for Terrestrial Animals, online version 2008.
9. Wernery, U., Kinne, J., and Morton, T. (2004) *Pictoral Guide to the Diagnosis of Equine Glanders. CVRL Brochure 2004*, Central Veterinary Research Laboratory, Dubai.
10. Xie, X., Xu, F., Xu, B., Duan, X., and Gong, R. (1980) A New Selective Medium for Isolation of Glanders Bacilli. Collected papers of veterinary research. *Control Inst. Vet. Biol., Minist. Agric., Peking, China (People's Rep. of)*, **6**, 83–90.
11. Ashdown, L.R. (1979) An improved screening technique for isolation of Pseudomonas pseudomallei from clinical specimens. *Pathology*, **11**, 293–297.
12. Neubauer, H., Finke, E.J., and Meyer, H. (1997) Human glanders. *Int. Rev. Armed Forces Med. Serv.*, **LXX**, 10/11/12, 258–265.
13. Bauernfeind, A., Roller, C., Meyer, D., Jungwirth, R., and Schneider, I. (1998) Molecular procedure for rapid detection of Burkholderia mallei and Burkholderia pseudomallei. *J. Clin. Microbiol.*, **36** (9), 2737–2741.
14. Sprague, L.D., Zysk, G., Hagen, R.M., Meyer, H., Ellis, J., Anuntagool, N. *et al.* (2002) A possible pitfall in the identification of Burkholderia mallei using molecular identification systems based on the sequence of the flagellin fliC gene. *FEMS Immunol. Med. Microbiol.*, **34** (3), 231–236.
15. Tkachenko, G.A., Antonov, V.A., Zamaraev, V.S., and Iliukhin, V.I. (2003) Identification of the causative agents of glanders and melioidosis by polymerase chain reaction. *Mol. Gen. Mikrobiol. Virusol.*, **3**, 18–22.
16. Liu, Y., Wang, D., Yap, E.H., Yap, E., and Lee, M.A. (2002) Identification of a novel repetitive DNA element and its use as a molecular marker for strain typing and discrimination of ara_ from ara*i* Burkholderia pseudomallei isolates. *J. Med. Microbiol.*, **51** (1), 76–82.
17. Sonthayanon, P., Krasao, P., Wuthiekanun, V., Panyim, S., and Tungpradabkul, S. (2002) A simple method to detect and differentiate Burkholderia pseudomallei and Burkholderia thailandensis using specific flagellin gene primers. *Mol. Cell. Probes*, **16** (3), 217–222.
18. Tomaso, H., Scholz, H.C., Al Dahouk, S., Pitt, T.L., Treu, T.M., and Neubauer, H. (2004) Development of 5′ nuclease real-time assays for rapid identification of the Burkholderia mallei/pseudomallei complex. *Diagn. Mol. Pathol.*, **13**, 247–253.
19. Lee, M.A., Wang, D., and Yap, E.H. (2005) Detection and differentiation of Burkholderia pseudomallei, Burkholderia mallei and Burkholderia thailandensis by multiplex PCR. *FEMS Immunol. Med. Microbiol.*, **43** (3), 413–417.
20. Scholz, H.C., Joseph, M., Tomaso, H., Al Dahouk, S., Witte, A., Kinne, J. *et al.* (2006) Detection of the reemerging agent Burkholderia mallei in a recent outbreak of glanders in the United Arab Emirates by a newly developed fliP-based polymerase chain reaction assay. *Diagn. Microbiol. Infect. Dis.*, **54** (4), 241–247 [Epub 2006 Feb 8].
21. Tomaso, H., Scholz, H.C., Al Dahouk, S., Eickhoff, M., Treu, T.M., Wernery, R. *et al.* (2006) Development of a 50-nuclease real-time PCR assay targeting fliP for the rapid identification of

Burkholderia mallei in clinical samples. *Clin. Chem.*, **52** (2), 307–310.

22. Neubauer, H., Sprague, L.D., Joseph, M., Tomaso, H., Al Dahouk, S., Witte, A. *et al.* (2007) Development and clinical evaluation of a PCR assay targeting the metalloprotease gene (mprA) of B. pseudomallei. *Zoonoses Public Health*, **54** (1), 44–50.
23. Novak, R.T., Glass, M.B., Gee, J.E., Gal, D., Mayo, M.J., Currie, B.J. *et al.* (2006) Development and evaluation of a real-time PCR assay targeting the type III secretion system of Burkholderia pseudomallei. *J. Clin. Microbiol.*, **44** (1), 85–90.
24. Ulrich, M.P., Norwood, D.A., Christensen, D.R., and Ulrich, R.L. (2006) Using real-time PCR to specifically detect Burkholderia mallei. *J. Med. Microbiol.*, **55** (Pt 5), 551–559.
25. Schmoock, G., Ehricht, R., Melzer, F., Rassbach, A., Scholz, H.C., Neubauer, H., Sachse, K., Mota, R.A., Saqib, M., and Elschner, M. (2009) DNA microarray-based detection and identification of Burkholderia mallei, Burkholderia pseudomallei and Burkholderia spp. *Mol. Cell. Probes*, **23**, 178–187.
26. Cravitz, L. and Miller, W.R. (1950) Immunologic studies with Malleomyces mallei and Malleomyces pseudomallei I: Serological relationships between M. mallei and M. pseudomallei. *J. Infect. Dis.*, **86**, 46–51.
27. Anuntagool, N. and Sirisinha, S. (2002) Antigenic relatedness between Burkholderia pseudomallei and Burkholderia mallei. *Microbiol. Immunol.*, **46** (3), 143–150.
28. Feng, S.H., Tsai, S., Rodriguez, J., Newsome, T., Emanuel, P., and Lo, S.C. (2006) Development of mouse hybridomas for production of monoclonal antibodies specific to Burkholderia mallei and Burkholderia pseudomallei. *Hybridoma (Larchmt)*, **25** (4), 193–201.
29. Burtnick, M.N., Brett, P.J., and Woods, D.E. (2002) Molecular and physical characterization of Burkholderia mallei O antigens. *J. Bacteriol.*, **184** (3), 849–852.
30. Wernery, U.R., Zacharia, R., Wernery, S., Joseph, and L., Valsini (2005) Ten years of freedom from notifiable equine diseases in the United Arab Emirates. Proceedings of the 15th International Conference of Racing Analysts and Veterinarians, Dubai, pp. 1–4.
31. Katz, J., Dewald, R., and Nicholson, J. (2000) Procedurally similar competitive immunoassay systems for the serodiagnosis of Babesia equi, Babesia caballi, Trypanosoma equiperdum, and Burkholderia mallei infection in horses. *J. Vet. Diagn. Invest.*, **12** (1), 46–50.
32. Sprague, L.D., Zachariah, R., Neubauer, H., Wernery, R., Joseph, M., Scholz, H.C., and Wernery, U. (2009) Prevalence-dependent use of serological tests for diagnosing glanders in horses. *BMC Vet. Res.*, **5**, 32.
33. Naureen, A., Saqib, M., Muhammad, G., Hussain, M.H., and Asi, M.N. (2007) Comparative evaluation of Rose Bengal plate agglutination test, mallein test, and some conventional serological tests for diagnosis of equine glanders. *J. Vet. Diagn. Invest.*, **19** (4), 362–367.
34. Katz, J.B., Chieves, L.P., Hennager, S.G., Nicholson, J.M., Fisher, T.A., and Byers, P.E. (1999) Serodiagnosis of equine piroplasmosis, dourine, and glanders using an arrayed immunoblotting method. *J. Vet. Diagn. Invest.*, **11**, 292–294.
35. Jana, A.M., Gupta, A.K., Pandya, G. *et al.* (1982) Rapid diagnosis of glanders in equines by counter-immuno-electrophoresis. *Ind. Vet. J.*, **59**, 5–9.
36. Ma, C.L., Fan, S.M., Wang, X. *et al.* (1977) Diagnosis of glanders in horses by the indirect fluorescent antibody (IFA) technique. *Chin. J. Vet. Sci. Technol.*, **9**, 3–5.
37. Zhang, W.D. and Lu, Z.B. (1983) Application of an indirect haemagglutination test for the diagnosis of glanders and melioidiosis. *Chin. J. Vet. Med.*, **9**, 8–9.
38. Parthasarathy, N., DeShazer, D., England, M., and Waag, D.M. (2006) Polysaccharide microarray technology for the detection of Burkholderia pseudomallei and Burkholderia mallei antibodies. *Diagn. Microbiol. Infect. Dis.*, **56** (3), 329–332.

39. Domma, K. (1953) Rotz bei mensch und Tier. *Wien. Tierärztl. Monat.*, **40**, 426–432.
40. Mc Gilvray, C.D. (1944) The transmission of glanders from horse to man. *Can. J. Publ. Health*, **35**, 268–275.
41. Mc Gilvray, C.D. (1949) The transmission of glanders from horse to man. *J. Am. Vet. Med. Assoc.*, **104**, 255–261.
42. von Brunn, A. (1919) Über die Ursachen und die Häufigkeit des Vorkommens des Rotzes beim Menschen, sowie über die Maßregeln zur Verhütung der Rotzübertragungen. *Vierteljahrschr. Gerichtl. Med.*, **3**, 134–161.
43. Bernstein, J.M. and Carling, E.R. (1909) Observations on human glanders. *Br. Med. J.*, **1**, 319–323.
44. Crohn, B.B. (1908) Notes on blood cultures in human glanders. *Proc. N.Y. Path. Soc.*, **8**, 105–110.
45. Dvorak, G.D. and Spickler, A.R. (2008) Glanders. *J. Am. Vet. Med. Assoc.*, **233** (4), 570–577.
46. Redfearn, M.S., Palleroni, N.J., and Stanier, R.Y. (1966) A comparative study of *Pseudomonas pseudomallei* and *Bacillus mallei*. *J. Gen. Microbiol.*, **43**, 293–313.
47. Dedié, K., Bockemühl, J., Kühn, H., Volkmer, K.J., and Weinke, T. (1993) *Bakterielle Zoonosen bei Tier und Menschen*, Ferdinand Enke Verlag, Stuttgart, pp. 159–168.
48. Minett, F.C. (1959) in *Infectious Diseases of Animals*, Diseases due to Bacteria, vol **1** (eds A.W. Stableforth and I.A. Galloway), Butterworths Scientific Publications, London, pp. 296–309.
49. Illner, P. (1980) *Infektionskrankheiten der Haustiere*, 2. Aufl, VEB Gustav Fischer Verlag, Jena, pp. 511–514.
50. Arun, S., Neubauer, H., Gurel, A., Ayyildiz, G., Kuscu, B., Yesildere, T., Meyer, H., and Hermanns, W. (1999) Equine glanders in Turkey. *Vet. Rec.*, **6**, 255–258.
51. Knight, H.D. (1972) in *Equine Medicine and Surgery* (eds E.J. Catcott and J.F. Smithcors), American Veterinary Press, Wheaton, Ill, pp. 108–111.
52. Urso, C. (1993) Hypothesis on the achean disease pathologica. *Pathologica*, **85** (1097), 441–443.
53. Wilkinson, L. (1981) Glanders: medicine and veterinary medicine in common pursuit of a contagious disease. *Med. Hist.*, **25**, 363–384.
54. Elschner, M.C., Klaus, C.U., Liebler-Tenorio, E., Schmoock, G., Wohlsein, P., Tinschmann, O., Lange, E., Kaden, V., Klopfleisch, R., Melzer, F., Rassbach, A., and Neubauer, H. (2009) *Burkholderia mallei* infection in a horse imported from Brazil. *Equine Vet. Educ.*, **21** (3), 147–150.
55. Srinivasan, A., Kraus, C.N., DeShazer, D., Becker, P.M., Dick, J.D., Spacek, L., Bartlett, J.G., Byrne, W.R., and Thomas, D.L. (2001) Glanders in a military research microbiologist. *N. Engl. J. Med.*, **26**, 256–258.
56. Nierman, W.C., Deshazer, D., Kim, H.S., Tettelin, H., Nelson, K.E., Feldblyum, T., Ulrich, R.L., Ronning, C.M., Brinkac, L.M., Daugherty, S.C., Davidsen, T.D., Deboy, R.T., Dimitrov, G., Dodson, R.J., Durkin, A.S., Gwinn, M.L., Haft, D.H., Khouri, H., Kolonay, J.F., Madupu, R., Mohammoud, Y., Nelson, W.C., Radune, D., Romero, C.M., Sarria, S., Selengut, J., Shamblin, C., Sullivan, S.A., White, O., Yu, Y., Zafar, N., Zhou, L., and Fraser, C.M. (2004) Structural flexibility in the *Burkholderia mallei* genome. *Proc. Natl. Acad. Sci. U.S.A.*, **101**, 14246–14251.
57. Chantratita, N., Vesaratchavest, M., Wuthiekanun, V., Tiyawisutsri, R., Ulziitogtokh, T., Akcay, E., Day, N.P., and Peacock, S.J. (2006) Pulsed-field gel electrophoresis as a discriminatory typing technique for the biothreat agent *Burkholderia mallei*. *Am. J. Trop. Med. Hyg.*, **74**, 345–347.
58. Tanpiboonsak, S., Paemanee, A., Bunyarataphan, S., and Tungpradabkul, S. (2004) PCR-RFLP based differentiation of *Burkholderia mallei* and *Burkholderia pseudomallei*. *Mol. Cell. Probes*, **18**, 97–101.
59. Harvey, S.P. and Minter, J.M. (2005) Ribotyping of *Burkholderia mallei* isolates. *FEMS Immunol. Med. Microbiol.*, **44**, 91–97.

4
Burkholderia pseudomallei: Melioidosis

Lisa D. Sprague and Mandy C. Elschner

4.1
Introduction

Melioidosis is an infectious disease of humans and animals caused by the bacterium *Burkholderia pseudomallei*. The agent has an extremely broad host range; among domestic animals, melioidosis is most commonly reported in sheep, goats, and swine. But sporadic cases or small outbreaks of the disease have been reported in a variety of animals, such as the monkey, gibbon, orang-utan, kangaroo, wallaby, deer, buffalo, cow, camel, llama, zebra, koala, dog, cat, horse, mule, parrot, rat, hamster, rabbit, guinea pig, ground squirrel, seal, dolphin, and crocodile [1]. The clinical picture can present itself in various forms: acute fulminate septicemia, local infection, subacute illness, chronic infection, and subclinical disease. The incubation period for naturally occurring infection in animals is not known.

Melioidosis occurs in tropical areas between latitudes 20°N and 20°S, predominantly in southeast Asia and northern Australia. Cases in animals have also been reported from China, Thailand, the Indian subcontinent, Iran, Saudi Arabia, United Arab Emirates, Chad, South Africa, Taiwan, Singapore, and Brazil. Sporadic outbreaks have occurred in France and the United Kingdom; moreover, the agent has been isolated from horses in Spain and from drinking water in Italy.

Melioidosis must be considered an emerging disease. Migration and animal transport around the world but also tourism increase the risk that the disease can leave its known endemic boundaries and establish itself elsewhere.

4.2
Characteristics of the Agent

Burkholderia pseudomallei is a small, motile, Gram-negative, bipolar staining, oxidase-positive rod with an exopolysaccharide capsule [2] which grows readily on routine culture media under aerobic conditions. It can also multiply under anaerobic conditions when nitrate or arginine is present [3]. Colonies have a distinctive putrid (sweet), earthy odor which used to be the hallmark for preliminary

BSL3 and BSL4 Agents: Epidemiology, Microbiology, and Practical Guidelines, First Edition.
Edited by Mandy C. Elschner, Sally J. Cutler, Manfred Weidmann, and Patrick Butaye.

identification. *B. pseudomallei* is also able to survive in the absence of nutrients in distilled water for several years [4]. Positive broths regularly display a tough, dry pellicle with a matt surface [5] and biofilms may be produced at the air–fluid interspace [6]. The presence of a type II and III secretion apparatus has been demonstrated by numerous research groups [7–11].

Burkholderia pseudomallei is a natural saprophyte with a high tenacity that can be isolated from soil and muddy water in endemic areas [12]. It can survive in water at room temperature for up to eight weeks, in muddy water for up to seven months and in soil in the laboratory for up to 30 months [13]. The bacterium is not particularly resistant to UV irradiation or sunlight [14]. Chlorine has only a bacteriostatic effect on the agent as bacteria have been recovered from water containing up to 1000 ppm free chlorine. However, it can effectively reduce the number of viable bacteria. In addition, lower pH reduces the effect of the disinfectant [15]. Current sensitivity testing of the validity of disinfectants is missing.

4.3 Diagnosis

Any manipulation with potentially contaminated material must take place in a laboratory that meets the requirements for Containment Group 3 pathogens. Samples for the detection of the agent can be drawn from blood, serum, liquor, urine, stool, bronchial lavage, or from pharynx, mucosa, or skin swabs; post mortem material from liver, spleen, heart blood, lung, and ulcers are also suitable. The detection of the agent in environmental samples is also possible.

4.3.1 Cultural Identification

For the isolation of *B. pseudomallei* from clinical samples standard media like blood agar, MacConkey agar, and commercial *B. cepacia* media can be used. Ashdown's selective agar (ASA) [16] facilitates isolation of the agent from respiratory tract samples and *B. pseudomallei* selective agar (BPSA) improves isolation of mucoid strains without reducing recovery of the classic wrinkled type [17, 18].

In general, the colonies of most isolates are initially smooth but show a typical wrinkled wheel spoke structure after 48 h. The isolation of environmental samples should be performed using an enrichment broth based on Ashdown's agar and an incubation period of 48 h at 42 °C [19]. Further characterization of suspicious colonies can be done by Gram-staining, motility testing, biochemical identification, and some antibiotic resistance tests. *B. pseudomallei* is oxidase-positive, but gentamicin- and polymixin-resistant. The commercially available API 20 NE (BioMerieux) and VITEK2 GN systems are only of limited use as they recognized only 87 and 78% of the tested *B. pseudomallei* isolates, respectively [20]. Biochemical tests for the distinction of *B. pseudomallei* from related *Burkholderia* species like *B. oklahomensis* or *B. mallei* are not reliable and it is advisable to confirm the identification of *B. pseudomallei* by further antigen detection and molecular based methods.

4.3.2 Antigen Detection

Several tests have been described for antigen detection: fluorescence based methods [21], agglutination using monoclonal antibodies [22], enzyme-linked immunoassays (ELISAs) detecting the exo-polysaccharide antigen [23], and detecting the *B. pseudomallei* specific 30-kDa protein [24] in blood cultures. Gas–liquid chromatographic analysis of bacterial fatty acid methyl esters identified 98% of 71 *B. pseudomallei* isolates correctly [22]. However, commercial tests are not available and antigen detection methods should be used in combination with cultural and molecular based results.

4.3.3 Molecular Based Methods

Numerous assays detecting specifically *B. pseudomallei* or in combination with *B. mallei* have been described. These assays are based on the 16S rRNA gene, *fli*C, the type three secretion system, or the antibiotic resistance gene P27 [25–35]. Based on specific differences in the sequence of the metalloprotease gene a sensitive PCR was developed and clinically evaluated in a sick camel [36]. Altukhova *et al.* [37] designed a PCR assay using flagellin-based primers for diagnosing experimental glanders and melioidosis in clinical samples.

A newly established microarray-based method using a minimum of four independent genetic markers including the 16S rRNA gene, *fli*C, and *mot*B allows the fast and accurate identification and differentiation of *B. mallei* from *B. pseudomallei* [38].

Most of the described methods, however, are only applicable to laboratory cultures; human clinical samples still need to be evaluated in additional studies.

4.3.4 Serology

The detection of specific antibodies is only possible in the late phase of the infection. Therefore, serological assays are of less importance for the diagnosis of acute melioidosis [19]. In general, however, the diagnostic utility of serological testing in endemic regions is doubtful [39] since antibodies (IgG, IgM) can persist for years after recovery. Moreover, due to the high level of subclinical and chronic infections in endemic areas the seroprevalence is high [19]. Several assays have been established but all are non-commercial in-house assays and should only be used by specialized laboratories.

The detection of a fourfold or greater rise in *B. pseudomallei* antibodies is a supportive, but not a definitive indication of disease. In non-endemic areas the detection of anti-*B. pseudomallei* antibodies in sera from patients suffering from melioidosis might be useful to confirm clinical cases or detect chronically infected persons [19].

The most frequently used serological test to help confirm exposure to the causative organism is the indirect hemagglutination assay (IHA). One has to bear in mind that despite culture-confirmed disease, patients often have a negative IHA result and occasionally fail to seroconvert in serial testing. Of special interest is the finding that the presence of bacteremia is significantly associated with a negative IHA and the coexistence of diabetes is associated with a positive IHA [40]. In general, seroconversion is unlikely to occur early enough to affect treatment choices during the admission phase of severe, acute infection [41].

4.4 Clinical and Pathological Findings in Humans

In humans infection occurs after inhalation or cutaneous uptake of the agent. The course of disease can range from asymptomatic infection to acute septicemia. Mild and subclinical infections combined with flue-like symptoms appear to be the most common form of melioidosis [42]. Subacute pneumonia and weight loss are the most prominent features of disease; however, any other organ system can be affected with abscess formation, granulomatous or necrotizing lesions [42, 43]. High fever (40 °C) can be accompanied by central nervous symptoms in the form of confusion or stupor; Icterus and diarrhea can contribute to the clinical presentation. Metastatic abscess formation can be observed in many organs, primarily in the lung, liver, spleen, urinary tract, subcutaneous tissue, and joints. Irrespective of treatment acute melioidosis is characterized by a high fatality rate (50%) [42, 44].

Chronic melioidosis can be mistaken for tuberculosis or fungal infection and can exacerbate after months or even years. Of particular note is that a number of predisposing factors exist that can lead to the exacerbation of melioidosis: diabetes mellitus, acute leukemia, bronchial carcinoma, renal disease, liver cirrhosis, systemic lupus erythematosus, drug abuse, alcoholism, administration of glucocorticoids, and pregnancy. Diabetes mellitus, in particular, seems to be a significant factor for the development of septicemia [42–44].

Melioidosis should be suspected in patients from endemic regions with acute febrile disease or symptoms of chronic infection. In tourists and refugees from these regions melioidosis should always be taken into consideration, even if the sojourn dates back months or even years.

4.5 Clinical and Pathological Findings in Animals

The clinical picture of melioidosis can present itself in various forms: acute fulminate septicemia, local infection, subacute illness, chronic infection, and subclinical disease. Infection can occur through inhalation, ingestion, or via skin wounds after contact with contaminated water or dust particles [45]. Acute melioidosis, especially in young animals, is frequently a fulminate septic infection with hematogenous dissemination, abscess formation, and high mortality. The incubation period for naturally occurring infection is not known. The characteristic feature of melioidosis is the formation of multiple abscesses in most organs, not only in the lung, spleen, and liver, but also in the subcutis and the involvement of regional lymph nodes according to the tributary areas [46].

Pigs and cattle are allegedly resistant to *B. pseudomallei* infection; clinical signs in pigs, cattle, and small ruminants include fever, anorexia with progressive emaciation, discharge from nose and eyes, coughing, dyspnea, and diarrhea [47]. Coughing, salivation, lameness, severe mastitis, abortion, and central nervous system disorders may appear in small ruminants.

In cattle and camels the disease tends to run a chronic course; in the latter, the lung seems to be the predilection site for *B. pseudomallei*. Camels are usually affected with a hacking cough which is later accompanied by purulent nasal discharge and respiratory insufficiency [1]. Cold-blooded animals and birds appear to be resistant to *B. pseudomallei*. Sporadic reports exist which describe infections in crocodiles, snakes, and fish. In birds the clinical signs are lethargy, anorexia, and diarrhea.

Horses, dogs, monkeys, marsupials, and marine mammals succumb to *B. pseudomallei* infection. In horses the disease can manifest itself in various forms: superacute cases with high fever, septicemia, limb edema, diarrhea, and death occurring within 24 h; acute cases with edema in the limbs, slight colic, and intestinal hypermotility. In general, however, the course is subacute to chronic, lasting from three weeks to three months. Further symptoms are emaciation, weakness, edema, and lymphangitis of the limbs, mild colic, diarrhea, pneumonia, cough, and nasal discharge. Dogs can develop acute, subacute, and chronic forms of melioidosis. The most significant clinical signs in acute cases are fever (40 °C), severe diarrhea, fulminant pneumonia, and septicemia. The subacute form can last from several days to several months and frequently starts as a skin lesion with the development of lymphangitis, lymphadenitis, and septicemia. Pulmonary involvement with subsequent septicemia may be observed. Necrosis and granulomatous inflammation of localized lesions describe the chronic form. In monkeys melioidosis presents itself with anorexia, wasting, listlessness, intermittent cough, nasal discharge, mild respiratory disease which can lead to acute, fulminating bronchopneumonia, and general weakness. Tissue swelling with formation of multiple abscesses is frequently observed. Paralysis and nerve damage can occasionally be seen [48]. In marine mammals (dolphin, whale, sea lion, and gray seal) the disease may take either an acute or a chronic run, but usually manifests itself as acute septicemia. Inappetence, anorexia, listlessness, and pyrexia can be observed a few days preceding death. The clinical picture in marsupials (wallabies, tree climbing kangaroo, and koala) has not been reported so far [49–51].

4.6 Epidemiology

Melioidosis occurs in tropical areas between latitudes 20°N and 20°S, predominantly in southeast Asia (Thailand, Laos, Vietnam), Singapore, and northern Australia. In endemic regions melioidosis is frequently encountered in the population of rural areas, especially in rice farming communities. Sporadic case reports from India, China, Taiwan, and both north- and South America exist. *B. pseudomallei* is a natural saprophyte with an extended host range readily isolated from soil and muddy water (rice paddies) in endemic areas [12]. Although melioidosis is a very rare imported disease in Europe, endemic spots seem to have established themselves in France, Italy, and Spain.

Human infection usually develops after inoculation of the agent in skin abrasions and small cuts, but also when swimming or working in contaminated water. Uptake via aerosol formation is also possible. Transmission can occur via contaminated objects without animal contact. Man to man transmission is a very rare incident. Laboratory acquired infections can occur by inhalation of infectious aerosols during routine handling of samples or cultures, inoculation or when handling infectious laboratory animals.

In animals infections resemble the clinical picture of glanders in horses. Host animals, however, are not horses but sheep, goats, and pigs, although the latter have been considered to be resistant to the agent. Sporadic cases have also been reported in wild and zoo animals.

4.6.1 Molecular Typing

The typing of *B. pseudomallei* strains by DNA fingerprinting assays has enhanced the understanding of epidemiology and surveillance of melioidosis during recent years. The use of these methods enables the distinction between natural and intentional outbreaks.

B. pseudomallei strains can be typed by means of multi locus variable number tandem repeat analysis (MLVA) [52–54] and by multi-locus sequence typing (MLST) [55]. These methods revealed a high genetic diversity between the different *B. pseudomallei* strains [56, 57]. A study carried out in Singapore revealed 72 genotypes in 102 analyzed *B. pseudomallei* isolates [58]. Harmonization of this methodology between laboratories and an internationally shared data base would greatly simplify the typing of any new isolate.

4.7 Conclusions

Melioidosis is an infectious disease of humans and animals. The clinical picture can present itself in various forms, ranging from acute fulminate septicemia, local infection, subacute illness, to chronic infection, and subclinical disease. Although the disease predominantly occurs in tropical areas between latitudes 20°N and 20°S, endemic spots seem to have established themselves in Europe, that is, France, Italy, and Spain.

Isolation and identification of the agent should be done in a laboratory that meets the requirements for Containment Group 3 pathogens. Isolation of the agent from clinical samples can be done on standard media. Commercially available tests based on biochemical analyses are only of limited use and serological assays should be used in combination with cultural and molecular based assays. Numerous PCR-based assays exist which can either detect specifically *B. pseudomallei* or in combination with *B. mallei*.

References

1. Sprague, L.D. and Neubauer, H. (2004) Melioidosis in animals: a review on epizootiology, diagnosis and clinical presentation. *J. Vet. Med. B. Infect. Dis. Vet. Public Health*, **51** (7), 305–320.
2. Gauthier, Y.P., Hagen, R.M., Brochier, G.S., Neubauer, H., Splettstößer, W.D., Finke, E.-J., and Vidal, D.R. (2001) Study on the pathophysiology of experimental *Burkholderia pseudomallei* infection in mice. *FEMS. Immunol. Med. Microbiol.*, **30**, 53–63.
3. Yabuuchi, E., Kosako, Y., Oyaizu, H., Yano, I., Hotta, H., Hashimoto, Y., Ezaki, T., and Arakawa, M. (1992) Proposal of *Burkholderia* gen. nov. and transfer of seven species of the genus *Pseudomonas* homology group II to the new genus, with the type species *Burkholderia cepacia* (Palleroni and Holmes 1981) comb. nov. *Microbiol. Immunol.*, **36**, 1251–1275.
4. Wuthiekanun, V., Smith, M.D., and White, N.J. (1995) Survival of *Burkholderia pseudomallei* in the absence of nutrients. *Trans. R. Soc. Trop. Med. Hyg.*, **89**, 491.
5. Valade, E., Thibault, F.M., Gauthier, Y.P., Palencia, M., Popoff, M.Y., and Vidal, D.R. (2004) The PmlI-PmlR quorum-sensing system in *Burkholderia pseudomallei* plays a key role in virulence and modulates production of the MprA protease. *J. Bacteriol.*, **186**, 2288–2294.
6. Currie, B.J.M. Mayo, N.M. Anstey, P. Donohoe, A. Haase, and D.J. Kemp (2001) A cluster of melioidosis cases from an endemic region is clonal and is linked to the water supply using molecular typing of *Burkholderia pseudomallei* isolates. *Am. J. Trop. Med. Hyg.* **65**, 177–179.
7. DeShazer, D., Brett, P.J., Burtnick, M.N., and Woods, D.E. (1999) Molecular characterization of genetic loci required for secretion of exoproducts in *Burkholderia pseudomallei*. *J. Bacteriol.*, **181**, 4661–4664.
8. Winstanley, C., Hales, B.A., and Hart, C.A. (1999) Evidence for the presence in *Burkholderia pseudomallei* of a type III secretion system-associated gene cluster. *J. Med. Microbiol.*, **48**, 649–656.
9. Attree, O. and Attree, I. (2001) A second type III secretion system in *Burkholderia pseudomallei*: who is the real culprit? *Microbiology*, **147**, 3197–3199.
10. Rainbow, L., Hart, C.A., and Winstanley, C. (2002) Distribution of type III secretion gene clusters in *Burkholderia pseudomallei, B. thailandensis* and *B. mallei*. *J. Med. Microbiol.*, **51**, 374–384.
11. Smith-Vaughan, H.C., Gal, D., Lawrie, P.M., Winstanley, C., Sriprakash, K.S., and Currie, B.J. (2003) Ubiquity of putative type III secretion genes among clinical and environmental *Burkholderia pseudomallei* isolates in Northern Australia. *J. Clin. Microbiol.*, **41**, 883–885.
12. Nachiangmai, N., Patamasucon, P., Tipayamonthein, B., Kongpon, A., and Nakaviroj, S. (1985) *Pseudomonas pseudomallei* in southern Thailand. *Southeast*

Asian J. Trop. Med. Public Health, **16**, 83–87.

13. Thomas, A.D. and Forbes-Faulkner, J.C. (1981) Persistence of *Pseudomonas pseudomallei* in soil. *Aust. Vet. J.*, **57**, 535.
14. Tong, S., Yang, S., Lu, Z., and He, W. (1996) Laboratory investigation of ecological factors influencing the environmental presence of *Burkholderia pseudomallei*. *Microbiol. Immunol.*, **40**, 451–453.
15. Howard, K. and Inglis, T. (2003) Novel selective medium for isolation of *Burkholderia pseudomallei*. *J. Clin. Microbiol.*, **41**, 3312–3316.
16. Ashdown, L.R. (1979) An improved screening technique for isolation of Pseudomonas pseudomallei from clinical specimens. *Pathology*, **11**, 293–297.
17. Howard, K. and Inglis, T.J. (2003) The effect of free chlorine on *Burkholderia pseudomallei* in potable water. *Water Res.*, **37**, 4425–4432.
18. Wuthiekanun, V., Dance, D.A., Wattanagoon, Y. *et al.* (1990) The use of selectivemedia for the isolation of *Pseudomonas pseudomallei* in clinical practice. *J. Med. Microbiol.*, **33**, 121–126.
19. Kekulé, A.S., AlDahouk, S., Bartling, C., Beyer, W., Bockemühl, J., Diller, R., Dobler, G., Essbauer, S., Finke, E.J., Fleischer, B., Frangoulidis, D., Hagen, R.M., Henning, K., Meyer, H., Neubauer, H., Oehme, A., Pauli, G., Pfeffer, M., Rakin, A., Schmitz, H., Splettstösser, W.D., Sprague, L.D., Tomaso, H., Wölfel, R., and Zimmermann, P. (2007) *Mikrobiologisch-Infektiologische Qualitätsstandards (MiQ), Hochpathogene Erreger – Biologische Kampfstoffe*, vol. **27**, Elsevier, München, pp. 84–93.
20. Deepak, R.N., Crawley, B., and Phang, E. (2008) *Burkholderia pseudomallei* identification: a comparison between the API 20NE and VITEK2GN systems. *Trans. R. Soc. Trop. Med. Hyg.*, **102** (Suppl. 1), 42–44.
21. Wuthiekanun, V., Desakorn, V., Wongsuvan, G., Cheng, A.C., Maharjan, B., Limmathurotsakul, D. *et al.* (2005) Rapid immunofluorescence microscopy for diagnosis of melioidosis. *Clin. Diagn. Lab. Immunol.*, **12**, 555–556.
22. Inglis, T.J.J., Merritt, A., Chidlow, G., Aravena-Roman, M., and Harnett, G. (2005) Comparison of diagnostic laboratory methods for identification of Burkholderia pseudomallei. *J. Clin. Microbiol.*, **43**, 2201–2206.
23. Anuntagool, P., Intachote, P., Naigowit, P., and Sirisinha, S. (1996) Rapid antigen detection assay for identification of *Burkholderia (Pseudomonas) pseudomallei* infection. *J. Clin. Microbiol.*, **34** (4), 975–976.
24. Pongsunk, S., Thirawattanasuk, N., Piyasangthong, N., and Ekpo, P. (1999) Rapid identification of Burkholderia pseudomallei in blood cultures by a monoclonal antibody assay. *J. Clin. Microbiol.*, **37** (11), 3662–3667.
25. Bauernfeind, A., Roller, C., Meyer, D., Jungwirth, R., and Schneider, I. (1998) Molecular procedure for rapid detection of Burkholderia mallei and Burkholderia pseudomallei. *J. Clin. Microbiol.*, **36** (9), 2737–2741.
26. Sprague, L.D., Zysk, G., Hagen, R.M., Meyer, H., Ellis, J., Anuntagool, N. *et al.* (2002) A possible pitfall in the identification of Burkholderia mallei using molecular identification systems based on the sequence of the flagellin fliC gene. *FEMS Immunol. Med. Microbiol.*, **34** (3), 231–236.
27. Tkachenko, G.A., Antonov, V.A., Zamaraev, V.S., and Iliukhin, V.I. (2003) Identification of the causative agents of glanders and melioidosis by polymerase chain reaction. *Mol. Gen. Mikrobiol. Virusol.*, **3**, 18–22.
28. Liu, Y., Wang, D., Yap, E.H., Yap, E., and Lee, M.A. (2002) Identification of a novel repetitive DNA element and its use as a molecular marker for strain typing and discrimination of ara_ from ara*i* Burkholderia pseudomallei isolates. *J. Med. Microbiol.*, **51** (1), 76–82.
29. Sonthayanon, P., Krasao, P., Wuthiekanun, V., Panyim, S., and Tungpradabkul, S. (2002) A simple method to detect and differentiate Burkholderia pseudomallei and

Burkholderia thailandensis using specific flagellin gene primers. *Mol. Cell. Probes*, **16** (3), 217–222.

30. Tomaso, H., Scholz, H.C., Al Dahouk, S., Pitt, T.L., Treu, T.M., and Neubauer, H. (2004) Development of 5' nuclease real-time assays for rapid identification of the *Burkholderia mallei/pseudomallei* complex. *Diagn. Mol. Pathol.*, **13**, 247–253.
31. Thibault, F.M., Valade, E., and Vidal, D.R. (2004) Identification and discrimination of Burkholderia pseudomallei, B. mallei, and B. thailandensis by real-time PCR targeting type III secretion system genes. *J. Clin. Microbiol.*, **42**, 5871–5874.
32. Lee, M.A., Wang, D., and Yap, E.H. (2005) Detection and differentiation of Burkholderia pseudomallei, Burkholderia mallei and Burkholderia thailandensis by multiplex PCR. *FEMS Immunol. Med. Microbiol.*, **43**, 413–417.
33. Gal, D., Mayo, M., Spencer, E. *et al.* (2006) Application of a polymerase chain reaction. *J. Clin. Microbiol.*, **44**, 85–90.
34. U'Ren, J.M., Van Ert, M.N., Schupp, J.M., Easterday, W.R., Simonson, T.S., Okinaka, R.T., Pearson, T., and Keim, P. (2005) Use of a real-time PCR TaqMan assay for rapid identification and differentiation of Burkholderia pseudomallei and Burkholderia mallei. *J. Clin. Microbiol.*, **43**, 5771–5774.
35. Novak, R.T., Glass, M.B., Gee, J.E., Gal, D., Mayo, M.J., Currie, B.J., and Wilkins, P.P. (2006) Development and evaluation of a real-time PCR assay targeting the type III secretion system of *Burkholderia pseudomallei. J. Clin. Microbiol.*, **44**, 85–90.
36. Neubauer, H., Sprague, L.D., Joseph, M., Tomaso, H., Al Dahouk, S., Witt, A., Kinne, J., Hensel, A., Wernery, R., Wernery, U., and Scholz, H.C. (2007) Development and clinical evaluation of a PCR assay targeting the metalloprotease gene (mprA) of B. pseudomallei. *Zoonoses Public Health*, **54**, 44–50.
37. Altukhova, V.V., Antonov, V.A., Tkachenko, G.A., Zinchenko, O.V., Zamaraev, V.S., Plekhanova, N.G., Iliukhin, V.I., and Trofimov, D.I. (2007) The use of polymerase chain reaction for detection of the agents of glanders and melioidosis using experimental infection. *Mol. Gen. Mikrobiol. Virusol.*, **3**, 22–27.
38. Schmoock, G., Ehricht, R., Melzer, F., Rassbach, A., Scholz, H.C., Neubauer, H., Sachse, K., Mota, R.A., Saqib, M., and Elschner, M. (2009) DNA microarray-based detection and identification of Burkholderia mallei, Burkholderia pseudomallei and Burkholderia spp. *Mol. Cell. Probes*, **23**, 178–187.
39. Peacock, S.J. (2006) Melioidosis. *Curr. Opin. Infect. Dis.*, **19**, 421–428.
40. Harris, P.N., Ketheesan, N., Owens, L., and Norton, R.E. (2009) Clinical features that affect indirect-hemagglutination-assay responses to *Burkholderia pseudomallei. Clin. Vaccine Immunol.*, **16**, 924–930.
41. Inglis, T.J.J., Rolim, D.B., and Rodriguez, J.L.N. (2006) Clinical guideline for diagnosis and management of melioidosis. *Rev. Inst. Med. Trop. Sao Paulo*, **48**, 1–4.
42. Dance, D.A.B. (1999) in *Infectious Diseases* (eds D. Amstrong and J. Cohen), Mosby, London, pp. 6.34.3–6.34.5.
43. Zysk, G., Splettstößer, W.D., Meyer, H., and Neubauer, H. (2000) A review of the diagnostic procedures in melioidosis with special respect on molecular and immunological techniques. *Clin. Lab.*, **46**, 119–130.
44. Parry, C.M. and Harries, A.D. (1998) in *Infectious Diseases of the Respiratory Tract* (ed. M.E. Ellis), Cambridge University Press, Cambridge, pp. 486–488.
45. Thomas, A.D., Spinks, G.A., D'Arcy, T.L., Norton, J.H., and Trueman, K.F. (1988) Evaluation of four serological tests for the diagnosis of caprine melioidosis. *Aust. Vet. J.*, **65**, 261–264.
46. Omar, A.R. (1963) Pathology of melioidosis in pigs, goats and a horse. *J. Comp. Pathol.*, **73**, 359–372.
47. Mayer, H. (1982) Zum Vorkommen von Pseudorotz (Melioidose) in Westeuropa. *Tierärztl. Umsch.*, **2**, 126–131.
48. Dance, D.A., King, C., Aucken, H., Knott, C.D., West, P.G., and Pitt, T.L. (1992) An outbreak of melioidosis in

imported primates in Britain. *Vet. Rec.*, **130**, 525–529.

49. Saroja, S. (1979) Melioidosis in wallabies. *Aust. Vet. J.*, **55**, 439–440.
50. Egerton, J.R. (1963) Melioidosis in a tree climbing kangaroo. *Aust. Vet. J.*, **39**, 243–244.
51. Ladds, P.W., Thomas, A.D., Speare, R., and Brown, A.S. (1990) Melioidosis in a koala. *Aust. Vet. J.*, **67**, 304–305.
52. Ramisse, V., Houssu, P., Hernandez, E., Denoeud, F., Hilaire, V., Lisanti, O., Ramisse, F., Cavallo, J.D., and Vergnaud, G. (2004) Variable number of tandem repeats in Salmonella enterica subsp. enterica for typing purposes. *J. Clin. Microbiol.*, **42**, 5722–5730.
53. U'Ren, J.M., Schupp, J.M., Pearson, T., Hornstra, H., Friedman, C.L., Smith, K.L., Daugherty, R.R., Rhoton, S.D., Leadem, B., Georgia, S., Cardon, M., Huynh, L.Y., DeShazer, D., Harvey, S.P., Robison, R., Gal, D., Mayo, M.J., Wagner, D., Currie, B.J., and Keim, P. (2007) Tandem repeat regions within the *Burkholderia pseudomallei* genome and their application for high resolution genotyping. *BMC Microbiol.*, **30** (7), 23.
54. Liu, Y., Loh, J.P., Aw, L.T., Yap, E.P., Lee, M.A., and Ooi, E.E. (2006) Rapid molecular typing of *Burkholderia pseudomallei*, isolated in an outbreak of melioidosis in Singapore in 2004, based on variable-number tandem repeats. *Trans. R. Soc. Trop. Med. Hyg.*, **100**, 687–692.
55. Godoy, D., Randle, G., Simpson, A.J., Aanensen, D.M., Pitt, T.L., Kinoshita, R., and Spratt, B.G. (2003) Multilocus sequence typing and evolutionary relationships among the causative agents of melioidosis and glanders, *Burkholderia pseudomallei* and *Burkholderia mallei*. *J. Clin. Microbiol.*, **41** (5), 2068–2079. [Erratum in: *J. Clin. Microbiol*, (2003), **41**, 4913].
56. Pearson, T., U'Ren, J.M., Schupp, J.M., Allan, G.J., Foster, P.G., Mayo, M.J., Gal, D., Choy, J.L., Daugherty, R.L., Kachur, S., Friedman, C.L., Leadem, B., Georgia, S., Hornstra, H., Vogler, A.J., Wagner, D.M., Keim, P., and Currie, B.J. (2007) VNTR analysis of selected outbreaks of *Burkholderia pseudomallei* in Australia. *Infect. Genet. Evol.*, **4**, 416–423.
57. U'ren, J.M., Hornstra, H., Pearson, T., Schupp, J.M., Leadem, B., Georgia, S., Sermswan, R.W., and Keim, P. (2007) Fine-scale genetic diversity among *Burkholderia pseudomallei* soil isolates in northeast Thailand. *Appl. Environ. Microbiol.*, **73**, 6678–6681.
58. Yen, M.W.S., Lisanti, O., Thibault, F., San, T.S., Kee, L.G., Hilaire, V., Jiali, L., Neubauer, H., Vergnaud, G., and Ramisse, V. (2009) Validation of ten new polymorphic tandem repeat loci and application to the MLVA typing of *Burkholderia pseudomallei* isolates collected in Singapore from 1988 to 2004. *J. Microbiol. Methods*, **77**, 297–301.

5
Coxiella burnetii: Q Fever

Matthias Hanczaruk, Sally J. Cutler, Rudolf Toman, and Dimitrios Frangoulidis

5.1
Introduction

Q fever is a zoonosis caused by the Gram-negative coccobacillus *Coxiella burnetii* belonging to the class Gammaproteobacteria. It has a worldwide distribution but is rarely reported, probably resulting from the diagnostic challenges presented by *C. burnetii* and a lack of requirement for notification of cases in many countries [1]. Despite this, it persists with a large reservoir among multiple species, regularly leading to smaller outbreaks.

Edward Holbrook Derrick firstly described this febrile illness in abattoir workers in Australia in 1937 and proposed the term "*Q fever*" (for "Query fever"). In the same year Frank Macfarlane Burnet and Mavis Freeman documented the successful isolation of *C. burnetii*. Independently in 1938, Gordon Davis and Herald Rea Cox working in the United States successfully cultivated this pathogen in embryonated eggs [2]. In order to honor both groups, this pathogen was given its name, *Coxiella burnetii*. The organism is a highly infectious agent, with experimental estimates suggesting an infectious dose of less than 10 organisms being capable of establishing an infection. Furthermore, they are highly resistant against both heat and desiccation, ubiquitous available, relatively easy to cultivate, and their aerosolic state is infectious over several kilometers. These characteristics resulted in the inclusion of *C. burnetii* among agents tested in the old biological weapon programs of the United States and the former Soviet Union.

Similarly, this was also the justification for its inclusion in the Centers for Disease Control and Prevention (CDC) list of potential bioterrorism agents.

Against this background, recent attention focusing upon this pathogen has highlighted our remarkable lack of understanding of its epidemiology, population diversity, and the basic biology of this organism. Huge improvements in our diagnostic capabilities have been made with introduction of molecular techniques, although serological methods still play a vital role in population screening. The availability of whole-genome sequence data has enabled the design and application of high-resolution molecular typing systems that are useful in both tracing the likely origins of infectious episodes and elucidating the epidemiology of these

BSL3 and BSL4 Agents: Epidemiology, Microbiology, and Practical Guidelines, First Edition.
Edited by Mandy C. Elschner, Sally J. Cutler, Manfred Weidmann, and Patrick Butaye.

microbes. Despite these recent scientific leaps and bounds in our knowledge, there is still much that the scientific community needs to learn about this intriguing "Query" organism.

5.2
Characteristics of the Agent

C. burnetii is an obligate intracellular, small Gram-negative, nonmotile, pleomorphic, coccobacillary bacterium (0.2–0.4 μm × 0.4–1 μm). Although it possesses a Gram-negative membrane, it cannot be stained using Gram techniques, although it can be visualized by the Giminez method.

Initially as a result of phenotypic similarities, the *Coxiella* were placed in the *Rickettsiales* order. More recent application of phylogenetic investigations, mainly based on 16S rRNA gene sequence analysis, resulted in the re-classification of the *Coxiella* genus within the *Legionellales* order. Within the *Proteobacteria* they build their own family (*Coxiellaceae*) in the *Legionellales* order. Since 2003 five whole genome sequencing projects of different *Coxiella* strains have been completed, with two further strains currently in assembly. The circular genome of the Nine Mile RSA 493 has a length of about 1.99 Mio. basepairs [3].

In its development cycle, *C. burnetii* generates both large cell variant (LCV) and small cell variants (SCVs). The latter is more environmentally stable and is the form normally picked up by macrophages during early infection, while the LCV is formed within phagolysosomes. Upon their liberation from the cell, a spore-like form (likened to an endospore) can arise, which shows a reinforced membrane structure pivotal in the enhanced resistance exhibited by this morphological form of *C. burnetii*. "Endospores" are metabolically inactive and thus remain stable in soil and dust over many years. Furthermore, they are likely to account for the prolonged survival of *C. burnetii* in diary products, meat and its products, in surface water over weeks or months; and they can be spread in dust or windborne aerosols over distances of several kilometers [4, 5]. These bacteria show some important physical characteristics, including: (i) stability against acids (up to pH 4.5), (ii) temperature (up to 70 °C for 15 min), (iii) UV light, pressure (up to 300 000 kPa), and (iv) ability to live in 10% saline for more than 6 months, and at 4 °C (possibly even more than 10 months). *C. burnetii* can however be killed by 1% lysol, 5% H_2O_2, 0.5% hypochlorite, 70% ethanol (for 30 min), 5% chloroform, formaldehyde, or pasteurization (at least 72 °C for 40 s) [6].

C. burnetii exists in two antigen phases, which are analogous to the smooth (phase I) and rough (phase II) lipopolysaccharide (LPS) forms seen among the *Enterobacteriaceae*. Bacteria in phase I can be observed during natural infections of humans and animals, whereas bacteria in phase II, which are mainly nonvirulent, develop after several passages in embryonated hen eggs, as a normally nonreciprocal antigenic mutant of phase I. Fluent transitions between both forms have been described.

5.3 Diagnosis

Suitable human samples for the direct detection of *C. burnetii* or for its isolation include EDTA or citrate-blood (for cultivation from leucocytes), bone marrow aspirates, sputum, urine, and other tissues such as heart valves. For serological diagnosis, two venous blood or serum samples at intervals of one to three weeks are required, especially in acute cases.

5.3.1 Direct Detection

In cardiac valves or placental tissues where organism numbers can be sufficiently high, either electron microscopy or light microscopy with staining according to Stamp, Giemsa, and Giminez methods can be used to visualize *C. burnetii*, which are seen under light microscopy as small, short red- or violet-colored, intracellular rods. Immunological techniques have been used to detect *C. burnetii* in a range of affected organs.

5.3.2 *C. burnetii* Cultivation

As a result of the highly infectious nature of *C. burnetii*, cultivation should not be attempted without adequate BSL 3 laboratories. Even with these, isolation is a difficult and very time-consuming procedure. Moreover, culture is not as sensitive as other methods such as detection of *Coxiella*-specific DNA. Viable cultures are however necessary for further scientific investigations, thus remain a research if not diagnostic priority. As *C. burnetii* is a strict intracellular bacterium, options for cultivation were previously restricted to the use of guinea pigs, mice, and embryonated eggs [7]. These have now been largely abandoned for safety reasons. Instead the less hazardous *in vitro* use of cell cultures such as human embryonic lung fibroblasts (HEL cells); embryonic epithelial kidney cells like the Buffalo green monkey (BGM) cells; Vero cells or L929 have become the mainstay for cultivation work [8]. Cell monolayers can be infected with several mammalian specimens including placenta, cardiac valves, abortion materials, but also blood, cerebrospinal fluid, bone marrow, liver material, or milk. The culture of *C. burnetii* may require several weeks prior to the appearance of intracellular vacuoles, the hallmark of successful infection (see Figures 5.1 and 5.2). A popular adaptation has become known as the "*shell vial*" technique, by which the sample suspension is centrifuged on a cell monolayer, which is grown on round cover slips enclosed within sealed tubes. This has been shown to be effective and sensitive, with the centrifugation step enhancing the attachment and penetration of the pathogens [9]. The use of triplicate cultures permits the removal of cultures at timely intervals for examination by staining. This is normally achieved using either monoclonal or polyclonal fluorescently labeled antibodies. The staining methods described

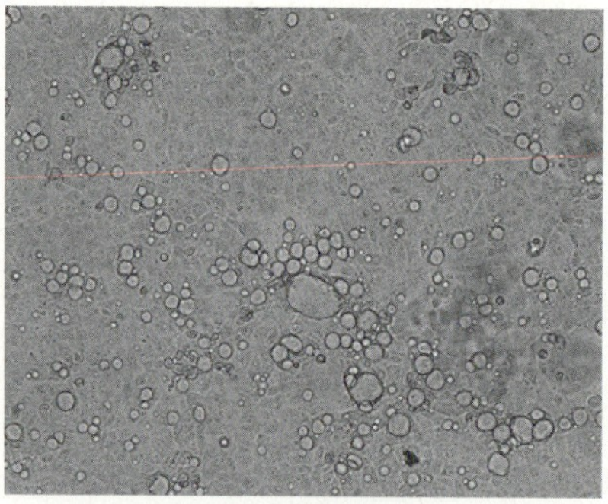

Figure 5.1 *C. burnetii*-infected BGM cells displaying the typical intracellular vacuoles (Bundeswehr Institute of Microbiology, Munich, Germany).

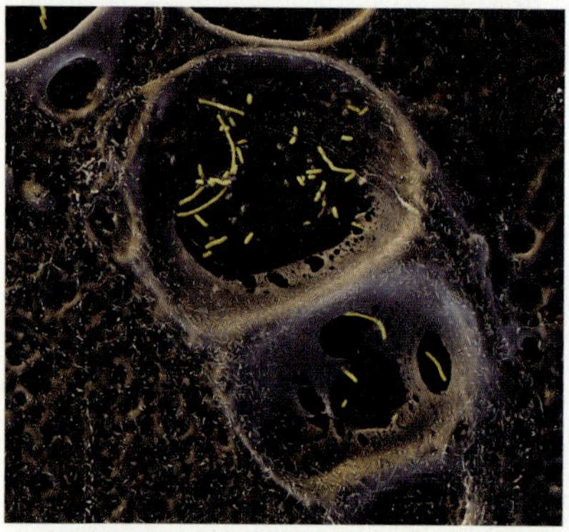

Figure 5.2 Scanning electron microscope display of *C. burnetii*-infected cells (© eye of science, Reutlingen, Germany).

above would suffice where fluorescent antibodies were not available. Although *C. burnetii* is an obligate intracellular pathogen that needs living cells for cultivation, Omsland *et al.* recently published the development of a complex nutrient medium that supported substantial growth of *C. burnetii* in a 2.5% oxygen environment under axenic (host cell-free) conditions [10]. The future will disclose whether this method has a potential role for cultivation in a diagnostic setting.

5.3.3
Detection of *C. burnetii* Specific DNA

An insertion sequence, IS1111, coding for a transposase, is found in multiple copies, sometimes up to 52 times in *C. burnetii* genomes. Consequently, it is often used as a preferential specific target for sensitive diagnostic PCRs. Furthermore, there are several commercial test kits available also utilizing this element, permitting detection in just a few hours by real-time PCR techniques [11–13]. Besides these

IS1111 gene sequences, other genes including *COM-1*, *SOP*, and *ICD* have been used as target sequences for PCR [14–16]. Recent diagnostic developments have focused upon the pre-analytic phase enabling high-throughput screening [17] and reducing manipulations by the use of ready to use mastermix and commercial kits. It is anticipated that increased molecular screening of suspected cases will accelerate the diagnosis of acute cases prior to the development of specific antibodies (see next section) [18, 19].

5.3.4 Serology

Serology remains the most frequently used diagnostic method for Q fever. In part this is often a result of time delays before clinicians or veterinarians consider the possibility of Q fever, and also it provides a cost-effective method compared with molecular diagnostics. Phase variation of *C. burnetii* has a direct impact on the serological diagnosis of Q fever. During acute Q fever, *C. burnetii* induces antibodies against phase II (protein antigens), while in the later stages of the disease, and especially in its chronic form often manifested as endocarditis, the high titers of antibodies are directed against phase I (LPS I antigen). Thus, the phase II antigen is used in diagnosing acute Q fever and the phase I antigen in diagnosing persistent/chronic infection with *C. burnetii*. Immunoreactive epitopes of the phase II antigen are easily accessible by the corresponding phase II antibodies because of the lack of any steric hindrance for the immunoreaction [20]. Identification of alternative specific and reliable immunodiagnostic proteins for diagnosing acute Q fever have yet to be determined. However, there has been progress in the elucidation of interactions of phase I antibodies with the LPS I antigen. A remarkable decrease in the serological activity of the O-polysaccharide antigen was observed when terminal virenose (Vir) and dihydrohydroxystreptose (Strep) were selectively removed from its chain [21]. At present, however, it is not known with certainty whether this immunoreaction involves both sugars in terminal positions or just with those Vir residues located in the O-polysaccharide backbone.

Either ELISA or indirect immunofluorescence (IF) tests using specific anti IgM, IgG, and IgA antibodies can be used to characterize the stage of infection. In the IF test, elevated IgM phase II ($>1:64$) and IgG II ($\geq 1:256$), or a fourfold increase in titer between two serum samples collected at intervals of two to six weeks confirms the diagnosis of acute Q fever [22].(Table 5.1).

While in chronic Q fever (persisting infection of six months; e.g., subacute endocarditis), the phase I titers are equal or higher than the phase II antibody response. In some chronic cases IgM antibodies can persist or yet in others they are absent.

On a cautionary note, it must be kept in mind, that IgM antibody titers can be falsely elevated such as in those cases with positive rheumatoid factor (auto-antibodies against gamma globulins) [24]. Furthermore, elevated IgG antibody titers may persist in some cases 7–10 years after an infection.

Table 5.1 Constellation for the serological diagnosis of Q fever with the IF test.

Phase	IgM		IgG		IgA	
	I	II	I	II	I	II
Acute	Negative/ positive < II	>1 : 64[a]	Negative/ positive < II	>1 : 256[a]	Positive/ negative	Positive/ negative
Chronic	Positive/ negative	Positive/ negative	>1 : 800[a]	Positive	(>1 : 100)	Positive ≤ I

[a]Indicates a pathognomonic parameter for an acute or chronic *C. burnetii* infection [1, 23].

The formerly used complement fixation reaction can no longer be recommended, as a result of its lack of sensitivity in the early phase of infection and technical variability in its performance [25]. Instead of this, highly specific and sensitive ELISAs are recommended for screening purposes, where reactive samples should be confirmed by IF tests with class-specific anti-immunoglobulin conjugates [26]. Despite this, it must be remembered that all serological tests could potentially cross react with antigenically related species such as *Legionella*, *Bartonella*, and *Chlamydiaceae*.

5.4 Pathogenesis

The two most common routes for acquiring *C. burnetii* infection are by aerosols, and less commonly by ingestion. Depending upon the infectious dose, incubation can vary between one and three weeks [27]. In many, the infection remains asymptomatic or may manifest as a nonspecific flu-like illness. In others, bacteremia can result in hematogenous spread to different organs such as liver, spleen, lungs, and bone marrow, with associated clinical signs accompanied by *C. burnetii* multiplication [28].

C. burnetii infection is usually controlled by the T-cell mediated immune response aided by the production of specific antibodies. After a receptor-mediated internalization of the SCV by phagocytic monocytes or macrophages, *C. burnetii* is able to multiply within the acid milieu of the fused phagolysosome, filling large vacuoles with LCV prior to cell lysis. Thereby, endospores can develop from LCV, which are responsible for the pathogens infectivity and environmental stability [29].

Since the T-cell mediated immune response is ineffective and does not lead to complete eradication of *C. burnetii*, some patients develop a chronic infection. Monocytes and macrophages are activated in order to control the infection, but *C. burnetii* seems to be able to overcome phagocytic activation. The exact role of the production of gamma interferon (IFN-γ), tumor necrosis factor alpha (TNF-α), transforming growth factor beta (TGF-β), interleukin 10 (IL-10), and prostaglandin E_2 (PGE_2) in the Q fever pathogenesis remains unclear.

Four factors have been hypothesized explaining the variations in the clinical presentation of Q fever: the route of infection, the inoculated dose, the infecting strain, and the immune status of the host [30].

5.5 Clinical and Pathological Findings

Almost 60% of Q fever cases are believed to be asymptomatic, the remaining 40% showing mild flu-like clinical symptoms, respiratory signs, and/or hepatitis. Only 2–5% of these symptomatic infected individuals need to be hospitalized. Thus, in most cases, Q fever manifests in an acute, self-limited febrile illness, atypical pneumonia, or hepatitis, whereas the chronic form, particularly in those with pre-existing cardiac damage, can manifest as a life-threatening endocarditis.

5.5.1 Acute Q Fever

After one to three weeks incubation, the infection either remains asymptomatic, as in the majority of cases, or first clinical signs appear as a nonspecific flu-like illness. Acute Q fever may present as prolonged fever, followed by an atypical pneumonia or hepatitis, originally proposed to correlate with the route of acquisition [31]. Symptomatic patients in addition to fever, may show signs of fatigue, chills, severe retroorbital headaches, myalgia, and sweats [32].

Pneumonia, when present, is usually mild, as typically seen for virus-induced pneumonia. Radiographic findings upon chest X-ray characteristically show multiple rounded opacities of both lungs, again mirroring viral pneumonia, or atypical bacterial pneumonia such as seen following infection with *Mycoplasma pneumoniae, Chlamydia pneumoniae,* or *Legionella pneumophila.* Q fever hepatitis presents as hepatomegaly accompanied by a mild increase of hepatic enzymes including alkaline phosphatase (AP), aspartate aminotransferase (AST), and alanine aminotransferase (ALT), but jaundice is rare.

During pregnancy Q fever has been associated with premature birth, abortion, and neonatal death [33]. Intriguingly, though seroepidemiological studies show that children are often exposed to *C. burnetii,* they are less frequently symptomatic than their adult counterparts, and when symptomatic, generally have a milder disease [34].

In some acute cases myocarditis, pericarditis, skin rash, or meningoencephalitis are reported. Myocarditis and pericarditis are quite rare (only approximately 1% of diagnosed Q fever cases), but is likely to represent an underestimate as a result of nonspecific thoracic pain in this clinical group. Whereas pericarditis usually resolves spontaneously, myocarditis can lead to cardiac failure in acute Q fever cases. Both forms may progress to endocarditis in chronically infected untreated patients [30].

5.5.2
Chronic Q Fever

The subacute endocarditis is the most common and most dangerous long-term complication of chronic Q fever, typically occurring in patients with previous valvulopathies, immunocompromised persons, or pregnant women. *Coxiella*-endocarditis can prove fatal in up to 25% of cases. However, *Coxiella*-associated chronic vascular, osteoarticular, or pulmonary infections as well as chronic granulomatous hepatitis, osteomyelitis, chronic fatigue syndrome, and hemolytic uremic syndrome (HUS) have all been described in the literature [30].

5.6
Epidemiology, Including Molecular Typing

C. burnetii is distributed worldwide with a large reservoir, including many wild and domestic mammals, birds, reptiles, fish, and even arthropods such as ticks and flies. Epidemiological studies have demonstrated that the most frequent source of human *C. burnetii* infections is represented by domestic ruminants such as cattle, sheep, or goats, which may be chronically infected without showing any clinical symptoms and shed vast numbers of the bacterium into the environment mainly during parturition. Numbers of *C. burnetii* in excess of 10^9 bacteria/g of placental tissue have been recorded, but may also be high in other birth-associated products such as amniotic fluids, or in milk. Particularly high counts have been obtained from tick feces, with reports of 10^{10} living organisms/g [35]. Despite this, ticks do not appear to be a significant risk factor for the acquisition of human infection.

Given the highly infectious nature of *C. burnetii* and the variety of potential sources (e.g., natural infection as a result of proximity with livestock, consumption of their products, exposure to infected arthropods, or indeed following potential deliberate release), it is of paramount importance to be able to investigate the source of any outbreak. To this end a variety of different genotyping systems have been established for epidemiological investigation of either naturally occurring outbreaks or those from a suspected bioterroristic act. Initial typing systems described were based on: (i) plasmid types, (ii) restriction fragment length polymorphisms (RFLPs) analyzed with SDS-PAGE and pulsed-field gel electrophoresis (PFGE), (iii) multispacer sequence typing (MST), (iv) positioning of IS1111-elements, and most recently (v) using multiple loci variable number of tandem repeats (VNTR) analysis (MLVA).

5.6.1
Plasmid Types

C. burnetii strains appear with five different plasmid types, independent from the LPS-associated phases: four different plasmids (QpH1, QpRS, QpDV, QpDG) and one type with a chromosomal plasmid-homologous sequence [36–40]. The characterization of these led to a classification into five genomic groups. Some plasmid types could be associated with various geographic regions. A formerly hypothesized correlation of these genomic groups with the virulence or the clinical appearance could not be confirmed in later studies [41].

5.6.2
RFLP

Restriction fragment length polymorphism analysis using SDS-PAGE following EcoRI- and BamHI-digestion of total chromosomal DNA led to a classification of *C. burnetii* strains in six groups (I–VI) very similar to plasmid typing [42]. Analysis of NotI- and SFiI-digested DNA restriction fragments in PFGE confirmed four of these six DNA fragment patterns, representing isolates from the genomic groups I, IV, V, and VI [43]. Beside these groups, Thiele *et al.* reported five additional DNA banding patterns after endonuclease restriction with NotI [44]. By using further *C. burnetii* isolates derived from animals and humans in Europe, United States, Africa, and Asia, 20 different restriction patterns could be distinguished after NotI restriction of *C. burnetii* DNA and PFGE [45].

5.6.3
IS1111 Typing

The IS1111 insertion sequence, coding for a transposase, is seen in up to 52 copies in *C. burnetii* genomes. Therefore this element is often used as a specific target region in highly sensitive diagnostic PCRs. In 2007, Denison *et al.* established a genotyping system based on four of these regions, which were analyzed by PCR, using an antisense primer binding inside of the IS1111-elements in combination with an upstream sense primer specific for each element [46]. The algorithm, proposed by Denison *et al.*, allowed a classification into genomic groups I–V, according to the six clusters derived by different authors described above using plasmid profiles, PCR, or restriction endonuclease-digested DNA.

5.6.4
Multispacer Sequence Typing

This sequence-based typing system, MST, first introduced in 2005, could discriminate 173 isolates into 30 genotypes, showing concordance with geographical and especially plasmid clusters previously described [47]. The use of 10 markers had the advantage of high discriminatory power and transferability as this is a sequence-based method unlike the preceding gel-based methods. This makes the results easier to compare between different laboratories.

5.6.5
MLVA Typing

Once genome sequences are known they can be interrogated for small repeat units that often vary between isolates. This has become an established typing technique from other bacteria over recent years. MLVA typing schemes were established for *C. burnetii* independently by two laboratories in 2006. Up to 17 different genomic target regions could be used for the differentiation of *Coxiella* strains. The discrimination power is higher than in MST (42 isolates yielded 37 genotypes), but the comparison of results from different laboratories remains problematic. According to different platforms used (gel-based versus capillary sequencer-based) to determine the length of the PCR products, adjustment to a standard is mandatory. In addition exact definitions of consensus sequences of the single targets are still missing, making harmonization of the assigned repeat number difficult [48–50].

5.7
Conclusion

During the course of this chapter, we have given a brief overview of the clinical, pathological, diagnostic methods, and typing approaches that represent current thinking on *C. burnetii*. What have probably become apparent are our extensive knowledge gaps. Despite the significant investment into *Coxiella* research over recent years through its inclusion on the CDC list of agents of potential bioterror, there is a lack of detailed understanding of the pathophysiological mechanisms utilized by this unique obligate intracellular pathogen. The queries of Q fever now manifest in questions: Why does this pathogen exert its detrimental effects during human pregnancy? Why do we see so little infection among children? Why do clinical cases that have been treated and clinically recovered, sometimes show persisting DNA? If indeed viable organisms do persist, in which sites do they sequester and what is the probability of reactivation?

These issues are of paramount importance if we are going to try to control or prevent infection through the use of vaccination.

In livestock, why do we see cattle as the source of human infection in some countries and small ruminants in others? Do untreated animals remain latently infected, thus a potential source of infection for others? Are the isolates of this supposedly monophyletic species really equal, or do they show different host specificities or differential clinical sequelae? Clear phylogenetic clades exist among cultivable *C. burnetii* isolates, should these be given species status?

With the introduction of high-resolution typing methods, we can now start to address some of these issues, shedding light upon this enigmatic pathogen.

References

1. Fournier, P., Marrie, T., and Raoult, D. (1998) Diagnosis of Q fever. *J. Clin. Microbiol.*, **36**, 1823–1834.

2. Davis, G. and Cox, H. (1938) A filter passing infectious agent isolated from ticks. I. Isolation from Dermacentor andersoni, reactions in animals and filtration experiments. *Public Health Rep.*, **53**, 2259–2267.
3. Seshadri, R., Paulsen, I.T., Eisen, J.A., Read, T.D., Nelson, K.E., Nelson, W.C., Ward, N.L., Tettelin, H., Davidsen, T.M., Beanan, M.J., Deboy, R.T., Daugherty, S.C., Brinkac, L.M., Madupu, R., Dodson, R.J., Khouri, H.M., Lee, K.H., Carty, H.A., Scanlan, D., Heinzen, R.A., Thompson, H.A., Samuel, J.E., Fraser, C.M., and Heidelberg, J.F. (2003) Complete genome sequence of the Q-fever pathogen Coxiella burnetii. *Proc. Natl. Acad. Sci. U.S.A.*, **100**(9), 5455–5460.
4. Williams, J.C. and Thompson, H.A. (1991) *Q Fever: The Biology of Coxiella Burnetii*, CRC Press, New York.
5. Tissot-Dupont, H., Torres, S., Nezri, M., and Raoult, D. (1999) Hyperendemic focus of Q fever related to sheep and wind. *Am.J. Epidemiol.*, **150** (1), 67–74.
6. Scott, G.H. and Williams, J.C. (1990) Susceptibility of Coxiella burnetii to chemical disinfectants. *Ann. N. Y. Acad. Sci.*, **590**, 291–296.
7. Ormsbee, R.A. (1952) The growth of Coxiella burnetii in embryonated eggs. *J. Bacteriol.*, **63**(1), 73–86.
8. Gil-Grande, R., Aguado, J.M., Pastor, C., García-Bravo, M., Gómez-Pellico, C., Soriano, F., and Noriega, A.R. (1995) Conventional viral cultures and shell vial assay for diagnosis of apparently culture-negative Coxiella burnetii endocarditis. *Eur.J. Clin. Microbiol. Infect. Dis.*, **14**(1), 64–67.
9. Raoult, D., Vestris, G., and Enea, M. (1990) Isolation of 16 strains of Coxiella burnetii from patients by using a sensitive centrifugation cell culture system and establishment of the strains in HEL cells. *J. Clin. Microbiol.*, **28**(11), 2482–2484.
10. Omsland, A., Cockrell, D.C., Howe, D., Fischer, E.R., Virtaneva, K., Sturdevant, D.E., Porcella, S.F., and Heinzen, R.A. (2009) Host cell-free growth of the Q fever bacterium Coxiella burnetii. *Proc. Natl. Acad. Sci. U.S.A.*, **106**(11), 4430–4434.
11. Stein, A. and Raoult, D. (1992) Detection of Coxiella burnetti by DNA amplification using polymerase chain reaction. *J. Clin. Microbiol.*, **30**(9), 2462–2466.
12. Stemmler, M. and Meyer, H. (2002) Rapid and specific detection of Coxiella burnetii by LightCycler-PCR, in *Rapid Cycle Real-Time PCR: Methods and Applications; Microbiology and Food Analysis* (eds U. Reischl, C. Wittwer, and F. Cockerill), Springer, Berlin, Heidelberg, New York, pp. 149–154.
13. Fenollar, F., Fournier, P.E., and Raoult, D. (2004) Molecular detection of Coxiella burnetii in the sera of patients with Q fever endocarditis or vascular infection. *J. Clin. Microbiol.*, **42**(11), 4919–4924.
14. Sekeyová, Z., Roux, V., and Raoult, D. (1999) Intraspecies diversity of Coxiella burnetii as revealed by com1 and mucZ sequence comparison. *FEMS Microbiol. Lett.*, **180**(1), 61–67.
15. Muramatsu, Y., Yanase, T., Okabayashi, T., Ueno, H., and Morita, C. (1997) Detection of Coxiella burnetii in cow's milk by PCR-enzyme-linked immunosorbent assay combined with a novel sample preparation method. *Appl. Environ. Microbiol.*, **63**(6), 2142–2146.
16. Nguyen, S.V. and Hirai, K. (1999) Differentiation of Coxiella burnetii isolates by sequence determination and PCR-restriction fragment length polymorphism analysis of isocitrate dehydrogenase gene. *FEMS Microbiol. Lett.*, **180**(2), 249–254.
17. Panning, M., Kilwinski, J., Greiner-Fischer, S., Peters, M., Kramme, S., Frangoulidis, D., Meyer, H., Henning, K., and Drosten, C. (2008) High throughput detection of Coxiella burnetii by real-time PCR with internal control system and automated DNA preparation. *BMC Microbiol.*, **8**, 77.
18. Fournier, P.E. and Raoult, D. (2003) Comparison of PCR and serology assays for early diagnosis of acute Q fever. *J. Clin. Microbiol.*, **41**(11), 5094–5098.
19. Turra, M., Chang, G., Whybrow, D., Higgins, G., and Qiao, M. (2006) Diagnosis of acute Q fever by PCR on sera

during a recent outbreak in rural south Australia. *Ann. N. Y. Acad. Sci.*, **1078**, 566–569.

20. Toman, R., Skultety, L., and Ihnatko, R. (2009) *Coxiella burnetii* glycomics and proteomics – tools for linking structure to function. *Ann. N. Y. Acad. Sci.*, **1166**, 67–78.
21. Vadovic, P., Slaba, K., Fodorova, M., Skultety, L., and Toman, R. (2005) Structural and functional characterization of the glycan antigens involved in immunobiology of Q fever. *Ann. N.Y. Acad. Sci.*, **1063**, 149–153.
22. Guigno, D., Coupland, B., Smith, E.G., Farrell, I.D., Desselberger, U., and Caul, E.O. (1992) Primary humoral antibody response to Coxiella burnetii, the causative agent of Q fever. *J. Clin. Microbiol.*, **30**(8), 1958–1967.
23. Dupont, H.T., Thirion, X., and Raoult, D. (1994) Q fever serology: cutoff determination for microimmunofluorescence. *Clin. Diagn. Lab. Immunol.*, **1**(2), 189–196.
24. Scola, B.L. (2002) Current laboratory diagnosis of Q fever. *Semin. Pediatr. Infect. Dis.*, **13**(4), 257–262.
25. Péter, O., Dupuis, G., Peacock, M.G., and Burgdorfer, W. (1987) Comparison of enzyme-linked immunosorbent assay and complement fixation and indirect fluorescent-antibody tests for detection of *Coxiella burnetii* antibody. *J. Clin. Microbiol.*, **25**(6), 1063–1067.
26. Frangoulidis, D., Schröpfer, E., Al Dahouk, S., Tomaso, H., and Meyer, H. (2006) Comparison of four commercially available assays for the detection of IgM phase II antibodies to *Coxiella burnetii* in the diagnosis of acute Q fever. *Ann. N. Y. Acad. Sci.*, **1078**, 561–562.
27. Marrie, T.J. (1990) in *Q Fever*, The Disease, Vol. 1 (ed. T.J. Marrie), CRC Press, Inc., Boca Raton, FL, pp. 125–160.
28. Dupuis, G., Petite, J., Péter, O., and Vouilloz, M. (1987) An important outbreak of human Q fever in a Swiss Alpine valley. *Int. J. Epidemiol.*, **16**(2), 282–287.
29. McCaul, T.F. (1991) in *Q Fever: The Biology of Coxiella Burnetii* (eds J.C. Williams and H.A. Thompson), CRC Press, Inc., Boca Raton, FL, pp. 223–258.
30. Maurin, M. and Raoult, D. (1999) Q fever. *Clin. Microbiol. Rev.*, **12**(4), 518–553.
31. Marrie, T.J., Durant, H., Williams, J.C., Mintz, E., and Waag, D.M. (1988) Exposure to parturient cats: a risk factor for acquisition of Q fever in Maritime Canada. *J. Infect. Dis.*, **158**(1), 101–108.
32. Marrie, T.J. (1995) Coxiella burnetii (Q fever) pneumonia. *Clin. Infect. Dis.*, **21** (Suppl. 3), S253–S264.
33. Marrie, T.J. (1988) Liver involvement in acute Q fever. *Chest*, **94**, 896–898.
34. Maltezou, H.C. and Raoult, D. (2002) Q fever in children. *Lancet Infect. Dis.*, **2**(11), 686–691.
35. Babudieri, B. (1959) Q fever: a zoonosis. *Adv. Vet. Sci.*, **5**, 81.
36. Thiele, D. and Willems, H. (1994) Is plasmid based differentiation of *Coxiella burnetii* in 'acute' and 'chronic' isolates still valid? *Eur. J. Epidemiol.*, **10**(4), 427–434.
37. Valková, D. and Kazár, J. (1995) A new plasmid (QpDV) common to Coxiella burnetii isolates associated with acute and chronic Q fever. *FEMS Microbiol. Lett.*, **125**(2–3), 275–280.
38. Willems, H., Ritter, M., Jäger, C., and Thiele, D. (1997) Plasmid-homologous sequences in the chromosome of plasmidless Coxiella burnetii Scurry Q217. *J. Bacteriol.*, **179**(10), 3293–3297.
39. Lautenschläger, S., Willems, H., Jäger, C., and Baljer, G. (2000) Sequencing and characterization of the cryptic plasmid QpRS from Coxiella burnetii. *Plasmid*, **44**(1), 85–88.
40. Jäger, C., Lautenschläger, S., Willems, H., and Baljer, G. (2002) Coxiella burnetii plasmid types QpDG and QpH1 are closely related and likely identical. *Vet. Microbiol.*, **89**(2–3), 161–166.
41. Thiele, D., Willems, H., Haas, M., and Krauss, H. (1994) Analysis of the entire nucleotide sequence of the cryptic plasmid QpH1 from Coxiella burnetti. *Eur. J. Epidemiol.*, **10**(4), 413–420.
42. Hendrix, L., Samuel, J., and Mallavia, L. (1991) Differentiation of Coxiella burnetii isolates by analysis of

restriction-endonuclease-digested DNA separated by SDS-PAGE. *J. Gen. Microbiol.*, **137**, 269–276.

43. Heinzen, R., Stiegler, G.L., Whiting, L.L., Schmitt, S.A., Mallavia, L.P., and Frazier, M.E. (1990) Use of pulsed field gel electrophoresis to differentiate Coxiella burnetii strains. *Ann. N. Y. Acad. Sci.*, **590**, 504–513.

44. Thiele, D., Willems, H., Köpf, G., and Krauss, H. (1993) Polymorphism in DNA restriction patterns of Coxiella burnetii isolates investigated by pulsed field gel electrophoresis and image analysis. *Eur. J. Epidemiol.*, **9**(4), 419–425.

45. Jäger, C., Willems, H., Thiele, D., and Baljer, G. (1998) Molecular characterization of Coxiella burnetii isolates. *Epidemiol. Infect.*, **120**(2), 157–164.

46. Denison, A., Thompson, H., and Massung, R. (2007) IS1111 insertion sequences of Coxiella burnetii: characterization and use for repetitive element PCR-based differentiation of Coxiella burnetii isolates. *BMC Microbiol.*, **7**, 91.

47. Glazunova, O., Roux, V., Freylikman, O., Sekeyova, Z., Fournous, G., Tyczka, J., Tokarevich, N., Kovacava, E., Marrie, T., and Raoult, D. (2005) Coxiella burnetii genotyping. *Emerg. Infect. Dis.*, **11**, 1211–1217.

48. Arricau-Bouvery, N., Hauck, Y., Bejaoui, A., Frangoulidis, D., Bodier, C.C., Souriau, A., Meyer, H., Neubauer, H., Rodolakis, A., and Vergnaud, G. (2006) Molecular characterization of Coxiella burnetii isolates by infrequent restriction site-PCR and MLVA typing. *BMC Microbiol.*, **6**, 38.

49. Svraka, S., Toman, R., Skultety, L., Slaba, K., and Homan, W.L. (2006) Establishment of a genotyping scheme for Coxiella burnetii. *FEMS Microbiol. Lett.*, **254**(2), 268–274.

50. Hackstadt, T., Peacock, M.G., Hitchcock, P.J., and Cole, R.L. (1985) Lipopolysaccharide variation in Coxiella burnetti: intrastrain heterogeneity in structure and antigenicity. *Infect. Immunol.*, **48**(2), 359–365.

6
Francisella tularensis: Tularemia

Anders Johansson, Herbert Tomaso, Plamen Padeshki, Anders Sjöstedt, Nigel Silman, and Paola Pilo

6.1
Introduction

Due to the ongoing threat of intentional biological release as well as the (re)-emergence of certain critical diseases, some microorganisms are of major concern for public health officials. Among them is *Francisella tularensis*, the causative agent of tularemia, because of its past development as a biological weapon and the potential high rate of morbidity and mortality following human infection. Consequently, this microorganism is listed as one of the top six biological agents that would pose the greatest impact on public health if used as a bioweapon, according to the United States Centers for Disease Control and Prevention (CDC) and the European Medicines Agency (EMEA) [1].

The discovery of this disease is attributed to McCoy; in 1911, he described a plague-like illness spread among rodents in Tulare County (Calif., USA) [2]. One year later, McCoy and Chapin isolated the causative agent from squirrels [3] and the first case description in humans was reported in 1914 by Wherry and Lamb [4]. These early studies suggested that the disease may be limited to the northern American continent but human infections were subsequently also reported from Eurasia [5–7]. The scientist Edward Francis dedicated his lifetime's work to the investigation of tularemia and its causative agent in the United States, and later the bacterium was named after him.

Differences in virulence among the strains circulating in the United States were observed by Bell and collaborators in 1955 [8]. However, Olsufjev and collaborators first linked this variation to taxonomy [9]. Nowadays, *F. tularensis* is divided into three subspecies: *F. tularensis* subsp. *mediasiatica*, *F. tularensis* subsp. *holarctica*, and *F. tularensis* subsp. *tularensis*. Only the latter two are clinically important [10]. A closely related species, named *F. novicida*, has been described. Some authors consider *F. novicida* to be a fourth subspecies of *F. tularensis*, although this change in nomenclature has not been widely accepted [11].

BSL3 and BSL4 Agents: Epidemiology, Microbiology, and Practical Guidelines, First Edition.
Edited by Mandy C. Elschner, Sally J. Cutler, Manfred Weidmann, and Patrick Butaye.

6.2
Characteristics of the Agent

F. tularensis is a facultative intracellular, nonmotile, nonsporulating, Gram-negative, aerobic bacterium. Cells are pleomorphic, rod shaped, or coccoid, and their size ranges from 0.2 to 0.7 μm [12].

Growth on synthetic media is fastidious and requires the addition of cysteine, even though atypical strains without this requirement have been isolated [13]. Furthermore, *in vitro* growth necessitates specific media like enriched chocolate agar, cysteine heart agar blood (CHAB) or Thayer–Martin agar [14, 15]. Because of its slow generation time, incubation parameters are at least 48 h at 37 °C with 5% CO_2.

6.3
Diagnosis

6.3.1
Serology

Since *F. tularensis* is highly infectious [16] and difficult to grow, serology is the most commonly used method to diagnose tularemia.

An agglutination test was established very early [17] and later on, micro-agglutination assays proved to be more rapid, easier to handle, and cheaper [18–20]. However, enzyme linked immunosorbent assay (ELISA) technology improved serology tests and was demonstrated to be more rapid and provided the possibility to determine the titer of different immunoglobulin (Ig) classes separately [21–23]. Various antigens have been tested to evaluate sensitivity and specificity, such as inactivated whole cells, lipopolysaccharides (LPSs), and purified outer membrane proteins [21–25].

Antibody detection has also drawbacks. It is effective only after an average time of two weeks [22, 24, 26]; moreover, discrimination between an ongoing phase of the disease or a past infection can be difficult to assess because of the persistence of antibodies in the blood [25].

6.3.2
Direct Isolation

Direct isolation of this fastidious bacterium may be difficult and dangerous because of the risk of laboratory acquired infection, particularly by the aerosol route. Microbiological laboratory staff should be notified of the arrival of clinical samples where there is a likelihood that the sample may contain *F. tularensis*, to allow implementation of the appropriate measures for the protection of laboratory personnel. In most countries, handling of *F. tularensis* cultures is required by law to be performed in biosafety level (BSL)3 containment. Strains of *F. tularensis* can be recovered from skin lesions, blood, respiratory specimens, lymph node aspirates, or other tissue biopsies. However, the transport of biological material should be as rapid as possible because of the risk of contamination with bacteria and fungi that grow faster than *F. tularensis* [27]. If direct culture is successful, colonies appear as small (the size of colonies ranging between 1 and 2 mm in diameter), smooth, and white-gray to blue-gray in color.

6.3.3
Phenotypical Characteristics

Phenotypic characteristics comprise [12]:

- Weakly catalase positive;
- Oxidase negative;
- Catabolism of carbohydrates is slow with the production of acid but no gas (acid is produced from maltose but not sucrose; acid is produced from glycerol by *F. tularensis* subsp. *tularensis* but not by *F. tularensis* subsp. *holarctica*);
- Production of H_2S;
- Cefinase positive;
- Urease negative.

Due to the risk of transmission to laboratory personnel and the time required to receive results, extensive phenotypical characterization is often avoided. Nowadays, methods based on molecular biology like polymerase chain reaction (PCR) are preferred.

6.3.4
Molecular Biology Tools for Identification

Because of the ease of performance and speed of molecular biology tools, a plethora of PCR protocols have been developed. They are largely based on the amplification of the *fopA* gene (encoding a 43-kDa outer membrane protein, FopA) [28] and the *tul4* gene (encoding a 17-kDa lipoprotein, LpnA) [29]. Since the aim of these methods was to avoid culture, specific techniques have been tested in order to amplify *F. tularensis* DNA directly from biological material. At first, conventional PCR procedures were developed [28, 30–32] and subsequently clinically evaluated for diagnostics in humans, followed by real-time PCR assays [33, 34] to reduce the hands-on time, to increase the sensitivity, and to reduce the risk of contamination with amplification products.

6.4
Pathogenesis

F. tularensis is a highly virulent microorganism capable of infecting a wide range of animal species [35]. Tularemia is primarily a disease of wild mammals (lagomorphs, rodents) but the ecology and the life cycle of this bacterium are still not clear. Various reservoirs have been proposed among wild animals but scientific proof is still generally missing. Rodents and lagomorphs are often designated as reservoirs of tularemia but they are also particularly susceptible and sensitive to the disease [36].

Humans become infected mainly through the bite of hematophagous arthropods or by the inhalation of infected aerosols. The predominant vectors involved in the transmission of these bacteria depend on the species present in the respective region, but particularly ticks and mosquitoes are known to be of relevance for disease in humans. Infection can also occur though the skin and mucous membranes by direct contact with infected animals (mostly hares) or by the ingestion of contaminated food or water.

After entry into the host, *F. tularensis* cells multiply locally, spread to regional lymph nodes, and further spread to organs all over the body. The most affected tissues are the skin, lymph nodes, spleen, liver, kidney, and lung [37, 38].

6.5 Clinical and Pathological Findings

6.5.1 Animals

In animals, clinical findings mainly come from *post mortem* observations [27, 39]. Some information is also provided by experimental infections [40–43]. Primates, lagomorphs, and rodents are very susceptible and sensitive to the disease. They develop a systemic disease and succumb quickly [42].

Animals can present a cachectic nutritional status at necropsy. The liver and spleen are swollen (Figure 6.1). Often, lungs are congested, edematous, and have miliary consolidated areas. Histology of the lungs reveals diffuse edema, multifocal acute necrosis (sometimes in association with rod-shaped bacteria), multifocal fibrin accumulation in alveolar spaces, and early formation of microthrombi (Figure 6.2a). Multifocal acute necrosis associated with rod-shaped bacteria is also present in the liver (Figure 6.2b) and spleen.

6.5.2 Humans

In humans, clinical manifestations are dependent upon several factors. First, the virulence of the infecting bacteria is of major importance. *F. tularensis* subsp. *holarctica* (type B tularemia) produces a milder disease than the one caused by *F. tularensis* subsp. *tularensis* (type A tularemia), but clinical signs and symptoms are mostly similar.

The route of infection determines the course of the disease. The ulcero-glandular form is the most frequent manifestation and is the consequence of the bite of

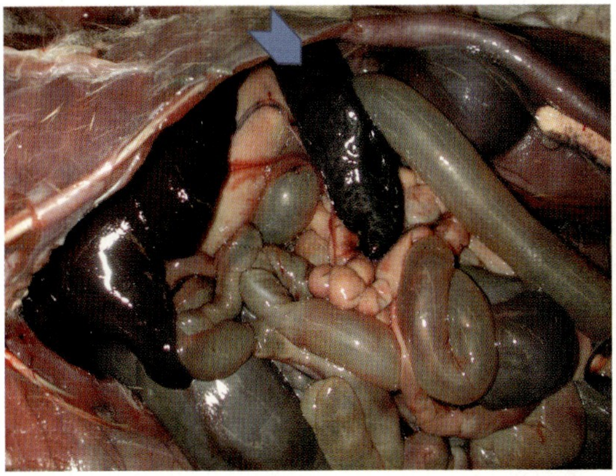

Figure 6.1 Typical swollen spleen of a hare with tularemia. The arrow indicates the spleen (Centre for Fish and Wildlife Health, Institute of Animal Pathology, Vetsuisse Faculty, Universität Bern, Switzerland). Reprinted with kind permission from Marie Pierre Ryser.

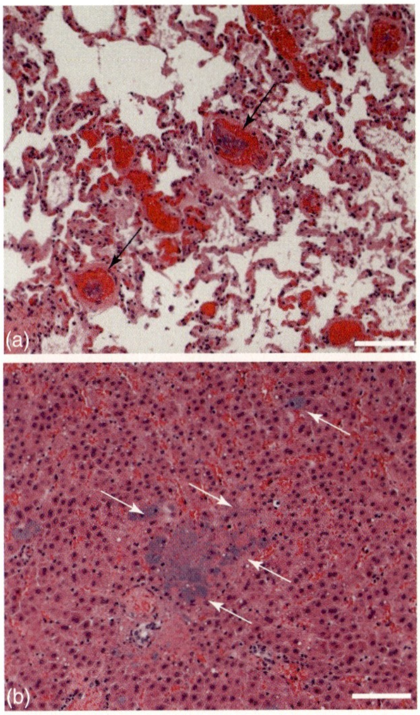

Figure 6.2 Examples of histology of the liver of a monkey infected with *F. tularensis*. Hematoxylin–eosin staining. (a) Black arrows show the formation of early microthrombi associated with clusters of bacteria. (b) White narrows designate acute necrosis in the liver section. Bar: 100 μm (Centre for Fish and Wildlife Health, Institute of Animal Pathology, Vetsuisse Faculty, Universität Bern, Switzerland). Reprinted with kind permission from Marie Pierre Ryser.

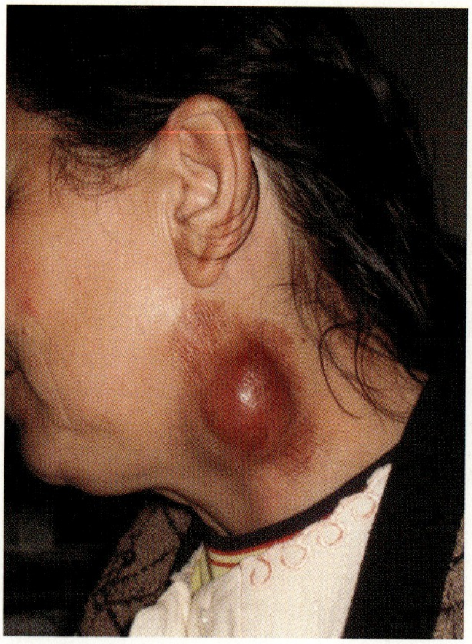

Figure 6.3 Swollen cervical lymph node in a tularemia patient with delayed diagnosis and treatment (National Center for Infectious Diseases, Sofia, Bulgaria. Reprinted with kind permission from Plamen Padeshki).

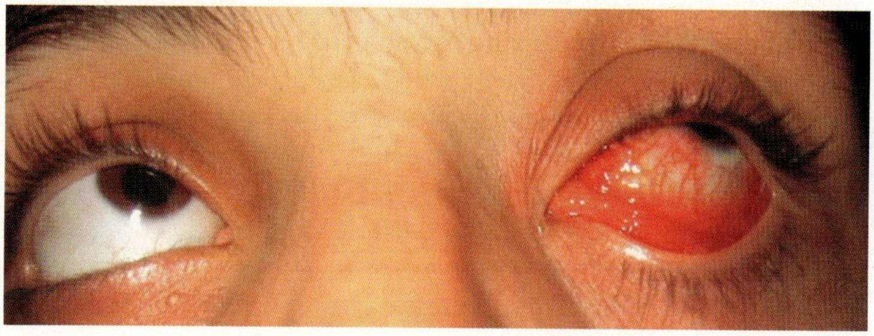

Figure 6.4 Human unilateral conjunctivitis (National Center for Infectious Diseases, Sofia, Bulgaria. Reprinted with kind permission from Plamen Padeshki).

an infected hematophagous arthropod (tick, mosquito, fly), direct contact with an infected animal, or more rarely penetration of the bacteria through the damaged skin. An ulcerated cutaneous lesion is associated with lymphadenopathy localized in the zone of drainage [14]. The glandular form involves a lymph node but without ulceration of the inoculation site (Figure 6.3). The oculo-glandular form results from an infection through the conjunctiva and the outcome is a unilateral conjunctivitis (Figure 6.4) [14, 44]. The oro-pharyngeal form results from the ingestion of contaminated food or water and provokes pharyngitis and/or gastro-enteritis (Figure 6.5). The respiratory form is caused by inhalation of the bacteria [38].

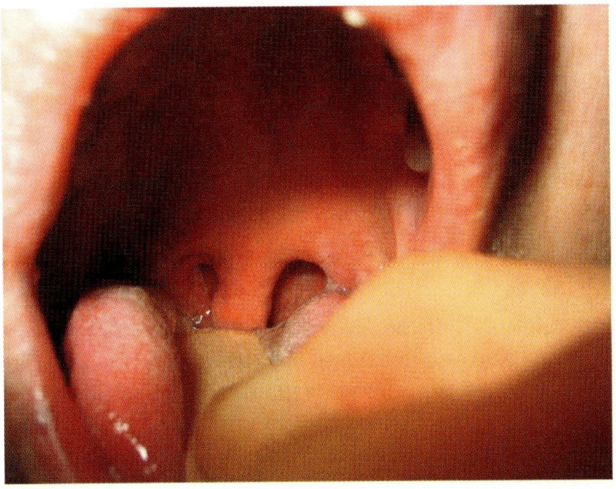

Figure 6.5 Human oro-pharyngeal tularemia (National Center for Infectious Diseases, Sofia, Bulgaria. Reprinted with kind permission from Plamen Padeshki).

The typhoid form is characterized by a systemic disease without any visible point of entry of the infectious organism. All clinical forms of tularemia may lead to septicemia. Numerous organs can be affected resulting in hepatitis, myositis, nephritis, pericarditis, and meningo-encephalitis. Delayed diagnosis can cause a fatal outcome. Mild or unapparent forms of tularemia with sero-conversion seem to be more common than previously thought even in low-prevalence areas according to a recent sero-epidemiological study in Germany [45, 46].

6.6 Epidemiology and Molecular Typing

Tularemia seems to be a disease of the northern hemisphere. Case reports of *F. novicida* infections within the southern hemisphere resulting in mild disease in a previously healthy man and a septic disease in a woman with cancer have, however, recently been reported in Australia and Thailand, respectively [47, 48]. *F. tularensis* subsp. *mediasiatica* has been reported only from the Central Asian Republics of the former Soviet Union. *F. tularensis* subsp. *tularensis* seems to be constrained to North America while *F. tularensis* subsp. *holarctica* is prevalent in many areas of the northern hemisphere including Europe [14]. So far, only one publication reports isolation of *F. tularensis* subsp. *tularensis* outside North America [49], while a second indicates the very close similarities with the North American strain Schu S4 [50]. Large geographically confined tularemia outbreaks involving several hundred humans have been reported from Finland, Kosovo, Spain, Sweden, and Turkey, while many more countries regularly report smaller outbreaks of disease in humans and animals [6].

Recently, an abundance of molecular tools have been applied to characterize strains and explain the spread of the disease. Various methods have been developed to detect the species, to differentiate among the subspecies, and for the typing of strains. The most common DNA-based techniques are *rrs* (16S rRNA) gene sequencing, DNA–DNA hybridization, restriction fragment length polymorphism (RFLP), pulsed field gel electrophoresis (PFGE), amplified fragment length polymorphism (AFLP), and real-time PCR [51], but the resolution of these approaches is restricted and allows mainly discrimination among the subspecies [52].

Indeed, *F. tularensis* shows little intra-species genetic variability, which leads to the necessity of more precise tools to subtype strains. As with several other pathogens exhibiting restricted inter-strain genetic variation, multiple locus variable number tandem repeat analysis (MLVA) targeting rapidly evolving sequences is useful for high-resolution typing of strains [53–55]. However, a drawback of MLVA typing schemes is that the rapid mutation rate of bacterial repeated sequences may decrease the robustness of the typing system, that is, the high mutation rate may add complexity in identifying major genetic lineages (homoplasy). To address this limitation, more slowly mutating DNA markers denoting canonical indels were recently added to the *F. tularensis* MLVA typing system, thereby providing a combination of unambiguous identification of major genetic lineages and high-resolution strain typing [56].

MLVA clearly reveals the phylogeny of *F. tularensis*. This method clusters the three subspecies of *F. tularensis* and *F. novicida* in four different groups [53]. Moreover, it differentiates *F. tularensis* subsp. *tularensis* in two distinct groups: *F. tularensis* subsp. *tularensis* A1 (eastern USA) and *F. tularensis* subsp. *tularensis* A2 (western USA). Furthermore, Slovakian strains of *F. tularensis* subsp. *tularensis* [49] seem to be closely related to the laboratory strain Schu S4, suggesting that human activities led to dispersal in the environment [57]. The latter finding was recently supported by genome sequencing and single nucleotide polymorphism (SNP) analyses [50, 58]. The genetic groups A1 and A2 appear to have important clinical correlates in the United States. In studies using PFGE for strain typing, significant differences were found in case-fatality rates between A1 (14%) and A2 (0%) [59]. Moreover, a further genetic division of genetic group A1 strains into A1a and A1b revealed a striking difference in disease outcome between these two groups of strains, with mortality rates of 4 and 24%, respectively [60].

F. tularensis subsp. *holarctica* is genetically more homogeneous than *F. tularensis* subsp. *tularensis* [53, 58]. This low variability led to the proposal that *F tularensis* subsp. *holarctica* is the most recent and most successful clone within the species. North American strains revealed more genetic variability than the Eurasian ones, leading Johansson and collaborators to suppose that *F. tularensis* subsp. *holarctica* evolved and spread from the American continent to Europe and Asia [53]. However, this issue needs more data to be clarified, as described recently by Vogler and colleagues [58]. Japanese strains of *F. tularensis* subsp. *holarctica* are genetically distant from all other *F tularensis* subsp. *holarctica* strains and this has been repeatedly confirmed using several typing systems, most recently by Fujita and collaborators using MLVA [61]. The situation concerning *F tularensis* subsp. *mediasiatica* is not yet clear and needs further investigation. It should be noted that analysis of this subspecies is limited by the small number of available strains.

Recent outbreak investigations using high-resolution methods are advancing the knowledge of tularemia epidemiology both in North America and Europe. In a survey of ticks on Martha's Vineyard (outside Boston, Mass., USA), Telford and colleagues were able to show that *F. tularensis* was maintained over several years in an environmental micro-focus in an area where tularemia is endemic [62]. Studies undertaken in Europe have shown that there is a specific *F. tularensis* subsp. *holarctica* clone present in Spain, France, and Switzerland and that this clone had been in western Europe for decades before the advent of the large human outbreaks that started in Spain in 1997. The main genetic characteristic of this clone is a deletion in the region of difference 23 (RD23) as well as one in variable-number tandem repeat (VNTR) marker M24 [63, 64]. Recent studies in Sweden indicate that tularemia is a disease that exhibits an epidemiologically definite localization. This has been shown by combining geographical data collected from patient interviews, with high-resolution genotyping of *F. tularensis* subsp. *holarctica* strains recovered from the same patients. The geographic distribution of specific *F. tularensis* subsp. *holarctica* sub-populations persisted over years and were highly confined during outbreaks (infections with some genotypes are restricted to areas as small as $2\,km^2$) indicating distinct point sources of infection [65].

6.7 Conclusion

Tularemia is a rarely reported disease that can be debilitating for affected people. Past experiences from animal and human outbreaks showed its dissemination potential. However, a long time is often needed to identify the cause of this disease. There is also a clear possibility that the "true" tularemia prevalence is underestimated in many part of the world because of the lack of precise knowledge about *F. tularensis*. There are many reasons for the lack of awareness about *F. tularensis*. First, since the disease is not common in most countries except for sporadic outbreaks, it is infrequently included in a differential diagnosis. There is evidently a requirement to familiarize the medical community with this pathogen. Moreover, as described above, the laboratory detection may be tricky because of the biological characteristics and specificities of this bacterium, such as its fastidious growth conditions.

An important aspect of understanding a disease is the ecological cycle of the causative agent. The lack of knowledge regarding tularemia ecology is a major limitation to understanding the prerequisites for outbreaks. The life cycle pattern in vectors and reservoirs is not yet established. Lagomorphs and rodents have been shown to be involved in transmission of the disease but other animal species may also be implicated, such as birds [66]. Information is also lacking about how *F. tularensis* survives in the environment and if it is dependent on other microorganisms for its survival. It has been shown that *F. tularensis* can replicate with protozoa and therefore it has been suggested that this may have implications for its maintenance and spread in the environment [67]. This potential ecological cycle is scientifically interesting but requires further investigation before it is possible to assess its relevance for infections in humans and animals.

Recently, an increasing number of laboratories invested time and money to study the epidemiology and molecular typing of *F. tularensis.* This led to the appearance of a large body of information that will be useful to unveil the biology and transmission patterns of this bacterium. A better understanding of natural tularemia is needed to distinguish between natural infections and those that may be caused by bioterrorists. In the event of a suspected intentional release of tularemia, novel technology will be invaluable for identification of the causative *F. tularensis* strain and novel epidemiological knowledge will be needed to unravel the mechanism behind dissemination of the disease.

References

1. The European Agency for the Evaluation of Medicinal Products (2007) EMEA/CPMP Guidance Document on Use of Medicinal Products for Treatment and Prophylasis of Biological Agents that Might be Used as Weapons of Bioterrorism, *http://www.emea.europa.eu/pdfs/human/bioterror/404801.pdf* (accessed July 21 2011).
2. McCoy, G.W. (1911) A plague-like disease of rodents. *Public Health Bull., Washington*, **43**, 53–71.
3. McCoy, G.W. and Chapin, C.W. (1912) Further observations on a plague-like disease of rodents with a preliminary note on the causative agent, *Bacterium tularense. J. Infect. Dis.*, **10**, 61–72.
4. Wherry, W.B. and Lamb, B.H. (1914) Infection of man with *Bacterium tularense. J. Infect. Dis.*, **15**, 331–340.
5. Ohara, S., Sato, T., and Homma, M. (1974) Serological studies on *Francisella tularensis, Francisella novicida, Yersinia philomiragia,* and *Brucella abortus. Int. J. Syst. Bacteriol.*, **24**, 191–196.
6. Petersen, J.M. and Schriefer, M.E. (2005) Tularemia: emergence/re-emergence. *Vet. Res.*, **36**, 455–467.
7. Thjotta, T. (1931) Tularemia in Norway. *J. Infect. Dis.*, **49**, 99–103.
8. Bell, J.F., Owen, C.R., and Larson, C.L. (1955) Virulence of *Bacterium tularense.* I. A study of the virulence of *Bacterium tularense* in mice, guinea pigs, and rabbits. *J. Infect. Dis.*, **97**, 162–166.
9. Olsufiev, N.G., Emelyanova, O.S., and Dunayeva, T.N. (1959) Comparative study of strains of *B. tularense* in the old and new world and their taxonomy. *J. Hyg. Epidemiol. Microbiol. Immunol.*, **3**, 138–149.
10. Brenner, D.J., Krieg, N.R., Staley, J.T., and Garrity, G.M. (2005) *Bergey's Manual of Systematic Bacteriology*, The Proteobacteria, Vol. **2**, Springer, New York.
11. Euzéby, J.P. (2009) List of Prokaryotic Names with Standing in Nomenclature, *http://www.bacterio.net* (accessed July 21 2011).
12. Holt, J.G., Krieg, N.R., Sneath, P.H.A., Staley, J.T., and Williams, S.T. (1994) *Bergey's Manual of Determinative Bacteriology*, Williams & Wilkins, Baltimore, MD.
13. Bernard, K., Tessier, S., Winstanley, J., Chang, D., and Borczyk, A. (1994) Early recognition of atypical *Francisella tularensis* strains lacking a cysteine requirement. *J. Clin. Microbiol.*, **32**, 551–553.
14. Ellis, J., Oyston, P.C., Green, M., and Titball, R.W. (2002) Tularemia. *Clin. Microbiol. Rev.*, **15**, 631–646.
15. Petersen, J.M., Carlson, J., Yockey, B., Pillai, S., Kuske, C., Garbalena, G.,

Pottumarthy, S., and Chalcraft, L. (2009) Direct isolation of *Francisella* spp. from environmental samples. *Lett. Appl. Microbiol.*, **48**, 663–667.

16. Burke, D.S. (1977) Immunization against tularemia: analysis of the effectiveness of live *Francisella tularensis* vaccine in prevention of laboratory-acquired tularemia. *J. Infect. Dis.*, **135**, 55–60.
17. Francis, E. (1926) Agglutination, cross-agglutination, and agglutinin absorption in tularemia. *Public Health Rep.*, **41**, 1273–1339.
18. Brown, S.L., McKinney, F.T., Klein, G.C., and Jones, W.L. (1980) Evaluation of a safranin-O-stained antigen microagglutination test for *Francisella tularensis* antibodies. *J. Clin. Microbiol.*, **11**, 146–148.
19. Massey, E.D. and Mangiafico, J.A. (1974) Microagglutination test for detecting and measuring serum agglutinins of *Francisella tularensis*. *Appl. Microbiol.*, **27**, 25–27.
20. Sato, T., Fujita, H., Ohara, Y., and Homma, M. (1990) Microagglutination test for early and specific serodiagnosis of tularemia. *J. Clin. Microbiol.*, **28**, 2372–2374.
21. Carlsson, H.E., Lindberg, A.A., Lindberg, G., Hederstedt, B., Karlsson, K.A., and Agell, B.O. (1979) Enzyme-linked immunosorbent assay for immunological diagnosis of human tularemia. *J. Clin. Microbiol.*, **10**, 615–621.
22. Koskela, P. and Salminen, A. (1985) Humoral immunity against *Francisella tularensis* after natural infection. *J. Clin. Microbiol.*, **22**, 973–979.
23. Syrjälä, H., Koskela, P., Ripatti, T., Salminen, A., and Herva, E. (1986) Agglutination and ELISA methods in the diagnosis of tularemia in different clinical forms and severities of the disease. *J. Infect. Dis.*, **153**, 142–145.
24. Bevanger, L., Maeland, J.A., and Naess, A.I. (1989) Competitive enzyme immunoassay for antibodies to a 43,000-molecular-weight *Francisella tularensis* outer membrane protein for the diagnosis of tularemia. *J. Clin. Microbiol.*, **27**, 922–926.
25. Viljanen, M.K., Nurmi, T., and Salminen, A. (1983) Enzyme-linked immunosorbent assay (ELISA) with bacterial sonicate antigen for IgM, IgA, and IgG antibodies to *Francisella tularensis*: comparison with bacterial agglutination test and ELISA with lipopolysaccharide antigen. *J. Infect. Dis.*, **148**, 715–720.
26. Waag, D.M., McKee, K.T., Sandstrom, G., Pratt, L.L., Bolt, C.R., England, M.J., Nelson, G.O., and Williams, J.C. Jr. (1995) Cell-mediated and humoral immune responses after vaccination of human volunteers with the live vaccine strain of Francisella tularensis. *Clin. Diagn. Lab. Immunol.*, **2**, 143–148.
27. Abril, C., Nimmervoll, H., Pilo, P., Brodard, I., Korczak, B., Seiler, M., Miserez, R., and Frey, J. (2007) Rapid diagnosis and quantification of *Francisella tularensis* in organs of naturally infected common squirrel monkeys (*Saimiri sciureus*). *Vet. Microbiol.*, **127**, 203–208.
28. Fulop, M., Leslie, D., and Titball, R. (1996) A rapid, highly sensitive method for the detection of *Francisella tularensis* in clinical samples using the polymerase chain reaction. *Am. J. Trop. Med. Hyg.*, **54**, 364–366.
29. Sjöstedt, A., Kuoppa, K., Johansson, T., and Sandström, G. (1992) The 17 kDa lipoprotein and encoding gene of *Francisella tularensis* LVS are conserved in strains of *Francisella tularensis*. *Microb. Pathog.*, **13**, 243–249.
30. Eliasson, H., Sjostedt, A., and Back, E. (2005) Clinical use of a diagnostic PCR for *Francisella tularensis* in patients with suspected ulceroglandular tularaemia. *Scand. J. Infect. Dis.*, **37**, 833–837.
31. Johansson, A., Berglund, L., Eriksson, U., Goransson, I., Wollin, R., Forsman, M., Tarnvik, A., and Sjostedt, A. (2000) Comparative analysis of PCR versus culture for diagnosis of ulceroglandular tularemia. *J. Clin. Microbiol.*, **38**, 22–26.
32. Sjostedt, A., Eriksson, U., Berglund, L., and Tarnvik, A. (1997) Detection of *Francisella tularensis* in ulcers of patients with tularemia by PCR. *J. Clin. Microbiol.*, **35**, 1045–1048.

33. Versage, J.L., Severin, D.D., Chu, M.C., and Petersen, J.M. (2003) Development of a multitarget real-time TaqMan PCR assay for enhanced detection of *Francisella tularensis* in complex specimens. *J. Clin. Microbiol.*, **41**, 5492–5499.

34. Wicki, R., Sauter, P., Mettler, C., Natsch, A., Enzler, T., Pusterla, N., Kuhnert, P., Egli, G., Bernasconi, M., Lienhard, R., Lutz, H., and Leutenegger, C.M. (2000) Swiss Army Survey in Switzerland to determine the prevalence of *Francisella tularensis*, members of the *Ehrlichia phagocytophila* genogroup, *Borrelia burgdorferi* sensu lato, and tick-borne encephalitis virus in ticks. *Eur. J. Clin. Microbiol. Infect. Dis.*, **19**, 427–432.

35. Grunow, R., Splettstoesser, W., McDonald, S., Otterbein, C., O'Brien, T., Morgan, C., Aldrich, J., Hofer, E., Finke, E.J., and Meyer, H. (2000) Detection of *Francisella tularensis* in biological specimens using a capture enzyme-linked immunosorbent assay, an immunochromatographic handheld assay, and a PCR. *Clin. Diagn. Lab. Immunol.*, **7**, 86–90.

36. Bell, J.F. and Stewart, S.J. (1975) Chronic shedding tularemia nephritis in rodents: possible relation to occurrence of *Francisella tularensis* in lotic waters. *J. Wildl. Dis.*, **11**, 421–430.

37. Dennis, D.T., Inglesby, T.V., Henderson, D.A., Bartlett, J.G., Ascher, M.S., Eitzen, E., Fine, A.D., Friedlander, A.M., Hauer, J., Layton, M., Lillibridge, S.R., McDade, J.E., Osterholm, M.T., O'Toole, T., Parker, G., Perl, T.M., Russell, P.K., and Tonat, K. (2001) Tularemia as a biological weapon: medical and public health management. *J. Am. Med. Assoc.*, **285**, 2763–2773.

38. Tarnvik, A. and Berglund, L. (2003) Tularaemia. *Eur. Respir. J.*, **21**, 361–373.

39. Matz-Rensing, K., Floto, A., Schrod, A., Becker, T., Finke, E.J., Seibold, E., Splettstoesser, W.D., and Kaup, F.J. (2007) Epizootic of tularemia in an outdoor housed group of cynomolgus monkeys (*Macaca fascicularis*). *Vet. Pathol.*, **44**, 327–334.

40. Baskerville, A., Hambleton, P., and Dowsett, A.B. (1978) The pathology of untreated and antibiotic-treated experimental tularaemia in monkeys. *Br. J. Exp. Pathol.*, **59**, 615–623.

41. Nelson, M., Lever, M.S., Savage, V.L., Salguero, F.J., Pearce, P.C., Stevens, D.J., and Simpson, A.J. (2009) Establishment of lethal inhalational infection with *Francisella tularensis* (tularaemia) in the common marmoset (*Callithrix jacchus*). *Int. J. Exp. Pathol.*, **90**, 109–118.

42. Schricker, R.L., Eigelsbach, H.T., Mitten, J.Q., and Hall, W.C. (1972) Pathogenesis of tularemia in monkeys aerogenically exposed to *Francisella tularensis* 425. *Infect. Immun.*, **5**, 734–744.

43. Twenhafel, N.A., Alves, D.A., and Purcell, B.K. (2009) Pathology of inhalational *Francisella tularensis* spp. *tularensis* SCHU S4 infection in African green monkeys (*Chlorocebus aethiops*). *Vet. Pathol.*, **46**, 698–706.

44. Kantardjiev, T., Padeshki, P., and Ivanov, I.N. (2007) Diagnostic approaches for oculoglandular tularemia: advantages of PCR. *Br. J. Ophthalmol.*, **91**, 1206–1208.

45. Dembek, Z.F., Kortepeter, M.G., and Pavlin, J.A. (2007) Discernment between deliberate and natural infectious disease outbreaks. *Epidemiol. Infect.*, **135**, 353–371.

46. Splettstoesser, W.D., Piechotowski, I., Buckendahl, A., Frangoulidis, D., Kaysser, P., Kratzer, W., Kimmig, P., Seibold, E., and Brockmann, S.O. (2009) Tularemia in Germany: the tip of the iceberg? *Epidemiol. Infect.*, **137**, 736–743.

47. Leelaporn, A., Yongyod, S., Limsrivanichakorn, S., Yungyuen, T., and Kiratisin, P. (2008) *Francisella novicida* bacteremia, Thailand. *Emerg. Infect. Dis.*, **14**, 1935–1937.

48. Whipp, M.J., Davis, J.M., Lum, G., de Boer, J., Zhou, Y., Bearden, S.W., Petersen, J.M., Chu, M.C., and Hogg, G. (2003) Characterization of a *novicida*-like subspecies of *Francisella tularensis* isolated in Australia. *J. Med. Microbiol.*, **52**, 839–842.

49. Gurycová, D. (1998) First isolation of *Francisella tularensis* subsp. *tularensis* in Europe. *Eur. J. Epidemiol.*, **14**, 797–802.

50. Chaudhuri, R.R., Ren, C.P., Desmond, L., Vincent, G.A., Silman, N.J., Brehm, J.K., Elmore, M.J., Hudson, M.J., Forsman, M., Isherwood, K.E., Gurycová, D., Minton, N.P., Titball, R.W., Pallen, M.J., and Vipond, R. (2007) Genome sequencing shows that European isolates of *Francisella tularensis* subspecies *tularensis* are almost identical to US laboratory strain Schu S4. *PLoS ONE*, **2**, e352.

51. Tomaso, H., Scholz, H.C., Neubauer, H., Al Dahouk, S., Seibold, E., Landt, O., Forsman, M., and Splettstoesser, W.D. (2007) Real-time PCR using hybridization probes for the rapid and specific identification of *Francisella tularensis* subspecies tularensis. *Mol. Cell. Probes*, **21**, 12–16.

52. Johansson, A., Forsman, M., and Sjöstedt, A. (2004) The development of tools for diagnosis of tularemia and typing of *Francisella tularensis. APMIS*, **112**, 898–907.

53. Johansson, A., Farlow, J., Larsson, P., Dukerich, M., Chambers, E., Byström, M., Fox, J., Chu, M., Forsman, M., Sjöstedt, A., and Keim, P. (2004) Worldwide genetic relationships among *Francisella tularensis* isolates determined by multiple-locus variable-number tandem repeat analysis. *J. Bacteriol.*, **186**, 5808–5818.

54. Klevytska, A.M., Price, L.B., Schupp, J.M., Worsham, P.L., Wong, J., and Keim, P. (2001) Identification and characterization of variable-number tandem repeats in the *Yersinia pestis* genome. *J. Clin. Microbiol.*, **39**, 3179–3185.

55. Lista, F., Faggioni, G., Valjevac, S., Ciammaruconi, A., Vaissaire, J., Le Doujet, C., Gorge, O., De Santis, R., Carattoli, A., Ciervo, A., Fasanella, A., Orsini, F., D'Amelio, R., Pourcel, C., Cassone, A., and Vergnaud, G. (2006) Genotyping of *Bacillus anthracis* strains based on automated capillary 25-loci multiple locus variable-number tandem repeats analysis. *BMC Microbiol.*, **6**, 33.

56. Larsson, P., Svensson, K., Karlsson, L., Guala, D., Granberg, M., Forsman, M., and Johansson, A. (2007) Canonical insertion-deletion markers for rapid DNA typing of *Francisella tularensis. Emerg. Infect. Dis.*, **13**, 1725–1732.

57. Farlow, J., Wagner, D.M., Dukerich, M., Stanley, M., Chu, M., Kubota, K., Petersen, J., and Keim, P. (2005) *Francisella tularensis* in the United States. *Emerg. Infect. Dis.*, **11**, 1835–1841.

58. Vogler, A.J., Birdsell, D., Price, L.B., Bowers, J.R., Beckstrom-Sternberg, S.M., Auerbach, R.K., Beckstrom-Sternberg, J.S., Johansson, A., Clare, A., Buchhagen, J.L., Petersen, J.M., Pearson, T., Vaissaire, J., Dempsey, M.P., Foxall, P., Engelthaler, D.M., Wagner, D.M., and Keim, P. (2009) Phylogeography of *Francisella tularensis*: global expansion of a highly fit clone. *J. Bacteriol.*, doi: 10.1128/JB.01786-08.

59. Staples, J.E., Kubota, K.A., Chalcraft, L.G., Mead, P.S., and Petersen, J.M. (2006) Epidemiologic and molecular analysis of human tularemia, United States, 1964–2004. *Emerg. Infect. Dis.*, **12**, 1113–1118.

60. Kugeler, K.J., Mead, P.S., Janusz, A.M., Staples, J.E., Kubota, K.A., Chalcraft, L.G., and Petersen, J.M. (2009) Molecular epidemiology of *Francisella tularensis* in the United States. *Clin. Infect. Dis.*, **48**, 863–870.

61. Fujita, O., Uda, A., Hotta, A., Okutani, A., Inoue, S., Tanabayashi, K., and Yamada, A. (2008) Genetic diversity of *Francisella tularensis* subspecies *holarctica* strains isolated in Japan. *Microbiol. Immunol.*, **52**, 270–276.

62. Goethert, H.K. and Telford, S.R. III (2009) Nonrandom distribution of vector ticks (*Dermacentor variabilis*) infected by *Francisella tularensis. PLoS Pathog.*, **5**, e1000319.

63. Dempsey, M.P., Dobson, M., Zhang, C., Zhang, M., Lion, C., Gutiérrez-Martín, C.B., Iwen, P., Fey, P., Olson, M., Niemeyer, D., Francesconi, S., Crawford, R., Stanley, M., Rhodes, J., Wagner, D.M., Vogler, A.J., Birdsell, D., Keim, P., Johansson, A., Hinrichs, S., and Benson, A.K. (2007) Genomic deletion marking an emerging subclone of *Francisella tularensis* subsp. *holarctica* in France and the Iberian Peninsula. *Appl. Environ. Microbiol.*, **73**, 7465–7470.

64. Pilo, P., Johansson, A., and Frey, J. (2009) Genotyping of Swiss *Francisella tularensis* isolates reveals a widespread central and western European cluster. *Emerg. Infect. Dis.*, **15**, 193–197.
65. Svensson, K., Bäck, E., Eliasson, H., Berglund, L., Larsson, P., Granberg, M., Karlsson, L., Forsman, M., and Johansson, A. (2009) A high-resolution landscape epidemiology of tularemia exposed by genetic analysis of the causative agent isolated from infected humans. *Emerg. Infect. Dis.*, **15**, 135–137.
66. Padeshki, P.I., Ivanov, I.N., Popov, B., and Kantardjiev, T.V. (2009) The role of birds in dissemination of *Francisella tularensis*: first direct molecular evidence for bird-to-human transmission. *Epidemiol. Infect.*, doi: 10.1017/S0950268809990513.
67. Abd, H., Johansson, T., Golovliov, I., Sandstrom, G., and Forsman, M. (2003) Survival and growth of *Francisella tularensis* in *Acanthamoeba castellanii*. *Appl. Environ. Microbiol.*, **69**, 600–606.

7
Yersinia pestis: Plague

Anne Laudisoit, Werner Ruppitsch, Anna Stoeger, and Ariane Pietzka

7.1
Introduction

Plague is an acute, often fatal, and potentially epidemic re-"emerging[1)]" disease caused by infection with the Gram-negative bacterium *Yersinia pestis* Lehmann and Neumann 1896. More specifically, plague is primarily a rodent-associated flea-borne zoonosis transmitted between rodents by infected rodent fleas. Humans are thus accidental hosts and are extremely susceptible to plague. Because of the high case-fatality rate and the epidemic potential of this disease, plague is now designated a Class I quarantinable disease and is subject to International Health Regulations, especially in countries with no or past history of plague [2]. Consequently, despite the fact that plague is a neglected infection, it remains an epidemiological threat and a disease of major public health importance, especially in a world that is becoming a global village where goods and people are moving fast. The importation of plague in an urban setting is a major concern of WHO [3, 4]. Indeed, the 1994 outbreak in Bombay, India, led to national chaos and prompted the international community to enforce an embargo that cost millions of dollars [5]. In the modern imagination and in many languages, the word "plague" sounds like a dreadful ancient disease, a nuisance of any kind or an evil spell to be avoided. Plague remains the oldest infection in human history and was responsible for many deaths in the three great pandemics in Europe and Asia [6, 7]. The Black Death, the second plague pandemic, was the greatest medical disaster of the Middle Ages and one of the most deadly pandemics in human history. While in the past plague crawled in huge devastating waves on the world, it has now settled in selected areas and is currently distributed in localized pockets called *"endemic or enzootic foci*[2)]" even though this term is subject to discussion. Indeed, the ecology

1) Emerging diseases: an emerging disease is one that has appeared in a population for the first time, or that may have existed previously but is rapidly increasing in incidence or geographic range (WHO).
2) Endemic or enzootic plague focus: a limited geographic area where the plague bacillus *Y. pestis* circulates in relatively resistant sylvatic rodent populations, from which it can spread to urban or commensal rodent populations and/or humans.

BSL3 and BSL4 Agents: Epidemiology, Microbiology, and Practical Guidelines, First Edition.
Edited by Mandy C. Elschner, Sally J. Cutler, Manfred Weidmann, and Patrick Butaye.

of plague and the genetic plasticity of *Y. pestis* in those endemic foci are much more complex than previously thought (see [8] for a review).

Globally, the current distribution of human plague corresponds to the distribution of those natural foci where plague only persists in wild habitat in one of two states: enzootic or epizootic. The *enzootic state* is defined as a stable rodent–flea infection cycle in a relatively resistant host population, without excessive rodent mortality. During an epizootic, however, plague bacilli have been introduced into moderately or highly susceptible mammals. In general, the natural cycle of infection is characterized by relatively stable periods of enzootic activity where *Y. pestis* circulates at low levels within the "maintenance" host community, followed by explosive epizootics involving one or more species of "amplifying" host that often experience high mortality [9].

With the development of technologies and molecular tools, the etiologic agent of plague has been sequenced and the function of many virulence genes described [10–12]. The biochemical mechanisms of the plague bacillus and interactions within mammalian hosts [13–15] and flea vectors [16] are now well documented. On the whole, since the discovery of the plague bacillus by Alexander Yersin in 1894, the establishment of the link between rat, flea, and men by Paul Louis Simond in 1898 in India, and the discovery of an efficient mode of transmission by the flea vector [17], some progress has been made in plague ecology and plague dynamics worldwide but a fundamental question remains unanswered. Indeed, from Vietnam to Argentina crossing the Kazakh steppes to the desert areas of Algeria and along the slopes of the mountains of East Africa, researchers are still looking for the mechanisms and factors that allow *Y. pestis* to maintain itself in those endemic foci. In fact we still do not know what those conditions are. Whether *Y. pestis* remains in competent (or resistant) hosts, in fleas, lice, ticks, associated with plants, or whether it survives in soil during interepizootics and interepidemics is still a major issue since plague ecology involves so many biotic and abiotic factors. Indeed, plague devastates most host species, making *Y. pestis* an exception to the general rule that a successful pathogen does not kill its host. In the classical plague cycle, *Y. pestis* infection begins with a flea bite, after which the bacteria soon colonize the lymphatic system and spread into the blood. In epizootic hosts, this high bacteremia, which occurs late in the course of the disease but results in high mortality, is necessary to ensure transmission back into fleas, and that is why presumably hypervirulence of *Y. pestis* might compensate for the poor vector efficiency of most fleas [18]. For the most part, the shuttling between mammals and insects occurs in rodents and their fleas, and this sylvatic plague cycle is endemic in many regions worldwide without reaching the human population [19]. Recently, the distribution of plague foci has also been associated with calcium- and iron-enriched soils in agreement with the known role of those chemical elements and others in the regulation of *Y. pestis* virulence factors [20, 21]. Similarly, iron and manganese are metallic ions that are critical in determining full *Yersinia pestis* virulence in mammalian host tissues since the bacillus requires exogenous iron for growth [22]. Highly pathogenic *Yersinia* secrete a siderophore called *yersiniabactin* which chelates iron molecules bound to eukaryotic proteins and transports them back into the bacterium [23]. *Y. pestis* can only grow if the serum or plasma Fe^{++} concentration is sufficient to give an excess of the cation over the unbound Fe^{++}-binding capacity of transferin [24].

If *Y. pestis* is able to survive outside a mammalian host, some physical and chemical properties of soil might still play a role in the maintenance of *Y. pestis* and thus could account for its "selective" distribution. But, plague foci are not temporally stable and the factors that allow for plague maintenance and resurgence in a particular, geographically limited area deserve further studies. Some foci may become quiescent, others may appear, and even at a local scale, within a known focus, some villages may be affected while others never get plague. In conclusion, the set of factors and their interactions that allow for the persistence of plague and for its re-emergence in those so-called endemic plague foci are still major questions in plague epidemiology. Moreover, many of the natural plague foci are geographically not connected, resulting in considerable ecological differences that *Y. pestis* has to face to survive and be transmitted in these different environments [25]. An obvious common feature between those endemic foci is the plague bacillus itself, *Y. pestis*, but the host niches in different natural plague foci have a unique natural selection pressure and direct the parallel adaptation of *Y. pestis* to the corresponding hosts, vectors, and environment [8, 26]. The adaptation of *Y. pestis* to various niches drove this lineage to diversify into different biovars[3] with not less than 16 ribotypes[4] that were associated with historical plague pandemics [28–30]. This structural variation presumably accounts for the bacterium's ability to resist innate immunity in both fleas and mammals. While there is only one *Y. pestis* lineage with – yet – genetic variability or clones among strains, the diversity, biology, ecology, and dynamics of the potential reservoirs and epizootic hosts and their fleas as well as the physical and chemical features of the environment are apparently seldom similar among the known foci. However, despite major differences in latitude and altitude, open steppes and high plateaus crossed by river beds appear to be "plaguey" areas in all plague endemic known zones. Similarly, there is growing support showing that "el Niño" and "la Niña" oscillations [31] and seasonal rainfall abundance patterns [32, 33] further increase the risk of plague outbreaks wherever plague is currently endemic [1, 34]. Epizootics and human epidemics also are associated with local climate and seasonal peaks in flea abundance [35–38]. This has two direct consequences. First, new plague foci might also arise within certain geographic areas with similar landscape features subject to similar hydric pressure. Second, the apparent biocenotic and "above ground" differences between plague foci might actually suggest a common, yet

3) Biovar: a group (infrasubspecific) of bacterial strains distinguishable from other strains of the same species on the basis of physiological characters. This basic classification was recently reviewed and indicates that modern typing of world strains and some of their characteristics could constitute a strong phylogenetic signal (see [27]). Formerly called *biotype*. Origin: bio- +variant.

4) Ribotype: the RNA complement of a cell – by analogy with phenotype or genotype. A clear correlation between the history of the three plague pandemics and the ribotypes of the strains has also been established (see [28]).

unraveled, mechanism of plague maintenance in the soil. Indeed, despite a nearly worldwide distribution, one can hardly imagine that the mechanism for plague maintenance is distinct in the various known foci. This question can also be asked in another way: why are there non-plague areas as big as Western Europe and why did plague disappear from Europe where it had spread so far for more than two centuries? Even though the spread of medieval plague responded to a different dynamic, progressing in waves, while the current pandemic is clearly contained in endemic pockets, the absence of plague in Europe remains a mystery (see further). Surely, answering this question could help us to understand which characteristics make a favorable niche for *Y. pestis*. Nevertheless, despite many hypotheses, none seem satisfactory to explain the modern distribution of plague.

Over 200 different species of mammals and at least 80 different species of fleas have been implicated in maintaining *Y. pestis* in enzootic foci throughout the world. Unfortunately, in most foci, the specific biology, ecology, and behavior of each potential reservoir and vector as well as the level of resistance of mammalian hosts and the vector competency of flea species are not known. For example, new flea species have recently been discovered within the endemic plague focus of Lushoto in northern Tanzania, adding new candidates to the pool of potential plague vectors [39, 40]. Eskey and Haas [41] postulated that knowing which species of fleas are efficient plague vectors could allow the eventual forecast of the likelihood of infection becoming established in a particular locality from studies of the rodent population and their flea infestation. Another aspect of plague surveillance and epidemiology is to understand if socio-cultural factors can favor the re-emergence of plague or the spread of infection at a local scale within the human population. Indeed, modern bubonic plague is a rural disease hitting harder the small villages in the countryside where ancient beliefs and traditions are still alive, whereas Black Death was a devastating epidemic that indiscriminately attacked large urban centers and the countryside alike. In those rural, remote and sometimes politically unstable areas like in Uganda, Tanzania, Madagascar, or the Democratic Republic of (DR) Congo, people live in simple mud houses and are daily exposed to both wild and domestic rodents, their fleas, and the pathogens they carry and transmit.

7.2 Characteristics of the Agent

7.2.1 The Plague Bacterium: *Yersinia pestis*

Yersinia pestis [1], formerly *Pasteurella pestis*, is an enterobacteriaceae with a unique vectorial mode of transmission – although infection can sometimes occur by ingestion of the bacillus – compared to the two pathogenic *Yersinia* species commonly infecting humans, *Y. enterocolitica* and *Y. pseudotuberculosis* that both use the fecal–oral route and cause gastrointestinal syndromes of moderate intensity. As a matter of fact, *Y. pestis* is thought to be a recently emerged clone of *Y. pseudotuberculosis*, and the two species remain closely related [10]. Transmission factors identified to date suggest that the rapid evolutionary transition of *Y. pestis* to flea-borne transmission within the last 1500–20 000 years involved the acquisition of two *Y. pestis*-specific plasmids by horizontal gene transfer, with the recruitment of endogenous chromosomal genes for new functions [42]. Even more, *Y. pseudotuberculosis*, unlike *Y. pestis*, is orally toxic to fleas, which indicates that the loss of one or more insect gut toxins was a critical step in the recent evolution of flea-borne transmission in the genus *Yersinia* [43].

7.2.1.1 *Yersinia pestis* Microbiology

Yersinia pestis is a pleomorphic Gram-negative coccobacillus (0.5–0.8 μm in diameter and 1–3 μm long) appearing as single cells or short chains in direct smears. The bacillus has a typical bipolar ("closed safety pin") staining with Wright's, or Wayson stains and may not always be visible on Gram stain. Yersin's bacillus is a facultative anaerobe, nonmotile, nonsporulating, nonlactose fermenter that is slow-growing in culture. At an optimal growth temperature of 25–28 °C, *Y. pestis* grows as gray-white, translucent colonies, usually too small to be seen as individual colonies at 24 h on selective culture medium. After incubation for 48 h, colonies are about 1–2 mm in diameter, gray-white to slightly yellow, and opaque. After 48–72 h of incubation, colonies have a raised, irregular "fried egg" appearance (Figure 7.1), which becomes more prominent as the culture ages. Colonies also can be described as having a "hammered copper," shiny surface [44]. The optimum pH for growth ranges between 7.2 and 7.6; however, extremes of pH 5.0–9.6 are tolerated [45]. The bacteria can also survive, in proper culture medium, at 4 °C, a temperature at which, paradoxically, a high mutation rate is observed, while at 37 °C it loses its ability to ferment glycerol. This latter characteristic has several consequences on its mode of colonization and multiplication in both its potential flea vectors and hosts.

7.2.1.2 *Yersinia pestis* Virulence Markers and Pathogenesis

The complete genome of several strains of *Y. pestis* has been sequenced and consists, for the fully virulent CO92 strain, of a 4.65-Mb chromosome, and three plasmids of 96.2, 70.3, and 9.6 kb; the size of the plasmids may vary slightly according to the strain [46]. The genome is unusually rich in insertion sequences and contains around 150 pseudogenes, many of which are remnants of a redundant entero-pathogenic lifestyle [47]. Virulence factors for *Y. pestis* are encoded on the chromosome and on three plasmids, namely the pesticin or *Pst* plasmid (±9.5 kb), the low calcium response or *Lcr* plasmid (±70 kb) and the *pFra* plasmid (±110 kb). The *Lcr* plasmid, also called *pYV*, is common to the three human pathogenic *Yersinia*, displaying 95% genetic homology between *Y. pestis*, *Y. pseudotuberculosis*, and *Y. enterocolitica* [48]. Plasmids *Pst*

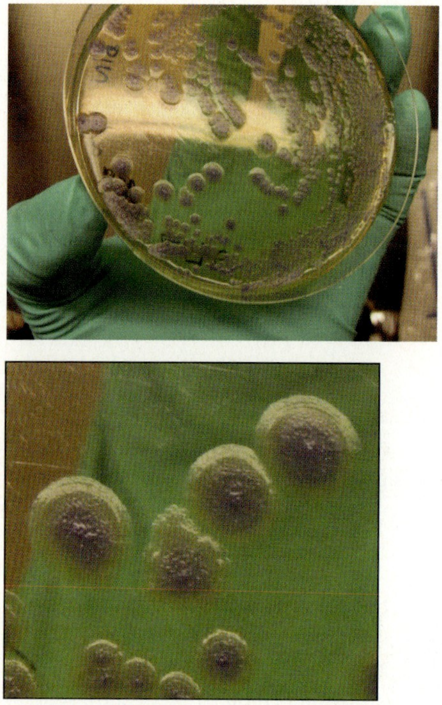

Figure 7.1 Colonies of *Y. pestis* on BIN agar.

and *pFra* are unique to *Y. pestis* and are thought to promote deep tissue penetration by *Y. pestis* and to contribute to the acute infection associated with this species [49]. Many factors thus play direct and indirect roles in *Y. pestis* pathogenesis.

7.2.1.3 **Chromosomal Virulence Genes**

Three chromosomally encoded virulence genes (*yst, psa, invA*) have been recognized along with a region called the high pathogenicity island (HPI) located within a much larger 102 kb chromosomal fragment called the *pgm locus* [50, 51]. The *yst* gene encodes a heat-stable enterotoxin Yst, *psaA* encodes the main structural component of an antigen fimbriae and *invA* codes for an invasin that is disrupted by an inserted sequence [23, 52]. Recently, a set of chromosomal genes (*yadBC* operon) that encodes proteins likely to have an adhesion/invasion function in host epithelioid cells has been described as a new virulence property of plague [53]. The *Yersinia* HPI carries a cluster of genes involved in the biosynthesis, transport, and regulation of the siderophore yersiniabactin. The major function of this island is to acquire iron molecules essential for *in vivo* bacterial growth and dissemination [23]. The physiological function of genes located on the HPI appears to be the blockage of the flea proventriculus (PV; foregut) through the production of a *Y. pestis hms*-dependent thermoregulated biofilm[5)] [57, 58].

7.2.1.3.1 Plasmidic Genes

The Pesticin or pPla Plasmid (also Designated pYP, pPCP1, or pPst; ±9.5 kb) The plasmid pPCP1 encodes a bacteriocin termed *pesticin*, a pesticin immunity protein and a plasminogen activator activity. The fibrinolytic (plasminogen activator) and coagulase activities of *Y. pestis* coded by the *pla* gene on the *pPla* plasmid play a significant role in the pathogenesis of plague [59]. Coagulase is responsible for the formation of micro thrombi and the plasminogen activator promotes the dissemination of the organism from the site of the initial fleabite [60].

The Low Calcium Response or LCR Plasmid (or pCD1; pYV ±70 kb) The plasmid encodes a complex virulence property called the LCR that enables the organism to grow in a low Ca^{++} (intracellular) environment. The pCD1 plasmid also encodes a type III secretion system (TTSS) that coordinates the production of several other virulence factors, such as *Yops* (targeting and effector *Yersinia* outer protein), LcrV antigen, and W protein that are actually associated with rapid proliferation and septicemia. Expression of the LCR has a profound immunosuppressive effect that results from the interference with innate defenses at the site of infection and the host organism's inability to mobilize an effective cell-mediated immune response [61].

The pFra Plasmid (also Designated pMT1 or pYT; ±110 kb) Two important virulence factors are encoded on pFra plasmid : the phospholipase D (previously accepted as a mouse toxin) and the F1 capsular antigen [62] that plays a major role in survival of plague bacteria in fleas (anti-phagocytic). Indeed, the survival of the bacterium in the flea and its ability to be transmitted, require the phospholipase D [63] that protects the bacteria from lyses in the flea mid-gut and also depends on the *Y. pestis* hemin storage gene products (*hms* gene on *pgm* locus), which are required for colonization and blockage of the PV of the flea [58]. The *pla*, *yopM*, and *caf1* are the targets of multiplex PCR for retrospective plague diagnostic in ethanol fixed tissues.

5) A biofilm is a population of microbes that adheres tenaciously to a surface by means of a polysaccharide-rich extracellular matrix that the microbes themselves produce [54]. The dynamics of biofilm formation facilitates the transmission of pathogens by providing a stable protective environment and acting as a nidus for the dissemination of large numbers of microorganisms; both as detached biofilm clumps and by the fluid-driven dispersal of biofilm clusters along surfaces [55, 56]. Emerging evidence indicates that biofilm formation conveys a selective advantage to certain pathogens by increasing their ability to persist under diverse environmental conditions. Biofilms are the predominant form of microbial growth in nature and occur in myriad environments, but the use of a biofilm to colonize an insect vector and promote transmission could be unique among bacteria, as it has not been described for any pathogen except *Y. pestis* [19].

Table 7.1 *Yersinia pestis* biovar characteristics and today's distribution.

Biovar	Pandemic	Present distribution	Nitrate reduction	Glycerol fermentation
Antiqua	Thought to be the cause of the first or Justinian plague pandemic	Africa, south-eastern Russia, central Asia	Yes	Yes
Medievalis	Thought to be the cause of the second or Black death pandemic	Caspian Sea	No	Yes
Orientalis	Cause of the third or modern pandemic	Asia, western hemisphere	Yes	No

7.2.1.4 *Yersinia pestis* Variants

Y. pestis is divided into three biovars classified according to their implication in each of the three major plague pandemics of the Christian era and that can be distinguished by their enzymatic properties (nitrate reduction, glycerol fermentation) as listed in Table 7.1 and various molecular markers (see [64]). Strains of *Y. pestis* isolated in DR Congo from man, rodents, or fleas were all var. *Antiqua* (except for some isolates from the Isiro region which belong to the *Orientalis* biovar; Bearden, personal communication), and the same biovar also occurs in Uganda [65], Kenya, Tanzania [66], and Zambia [67]. A nonvirulent strain, *Y. microtus*, has been proposed as a fourth biovar [30]. Studies in the former Soviet Union suggest that greater strain diversity exists in Asia and eastern Europe than is recognized in the west [8]. Atypical plague isolates can be found in the ancient reservoirs in the territory of the former USSR as well as typical epidemic forms of *Y. pestis*. This heterogeneous atypical group received the common name *pestoides* [68, 69].

However, the subdivision of the *Y. pestis* lineage into these three biovar each one being associated with one pandemic was recently questioned. Indeed, combining two molecular approaches (multiple spacer typing, sequencing of the glpD gene) on *Y. pestis* DNA extracted from the dental pulp of persons thought to have died of plague ("plague teeth") researchers found that the genotype *Orientalis* – which now occurs worldwide – was involved in all three pandemics [64, 70–73]. However, this conclusion is criticized and triggered a controversy due to the lack of evidence of *Y. pestis* DNA in the dental pulps of suspected plague victims in other European grave sites using the same methods and several other sets of primers [74].

7.2.2
Molecular Typing

Yersinia spp. show very low sequence variations between each other. However, diverse molecular methods have been used for the genotyping of *Y. pestis* in the past few years. These include PCR-based methods such as single nucleotide polymorphism (SNP) analysis, insertion sequence (IS) screening [10, 75], and multiple locus variable number of tandem repeat analysis (MLVA). Motin *et al.* [69] describes a genotyping method for detecting divergences of the location of insertion sequences in the genome, such as IS*100*. Previous work [76] characterizes clustered regularly interspaced short palindromic repeats (CRISPRs) as suitable utilities for typing of even degraded DNA. Complete genome sequence analysis of *Y. pestis* strains revealed several single nucleotide polymorphisms as well as rearrangements, insertions, and deletions, providing insight into the relationship within the species [29]. The comparison of complete genome sequences of various strains highlighted multiple inverted and rearranged segments that suggest a greater than expected diversity and environment-driven evolution [77]. Other genotyping methods such as pulsed field gel electrophoresis (PFGE), ribotyping (rRNA gene restriction patterns), and restriction fragment length polymorphism (RFLP) are based on the detection of DNA fragments with different lengths gained by restriction enzyme digestion. The obtained DNA fragments can be separated using electrophoretic methods [28]. The ribotyping of *Y. pestis* strains show that at least 16 ribotypes exist and that a clear correlation between the history of the three plague pandemics and the ribotypes of the strains can be found. Strains of biovar *Orientalis* are of ribotypes A to G, those of biovar *Antiqua* are of ribotypes F to O, and those of biovar *Medievalis* are of ribotypes O and P [28].The locations of strains of ribotype O correspond to the remaining plague foci of the first and second pandemics (central Asia, central Africa), while the distribution of ribotype B (worldwide) reflects the geographical spread of *Y. pestis* during the third pandemic. The greatest ribotype heterogeneity is observed in Africa [28] with, for example, the Q, R, and T ribotypes that emerged in Madagascar at the beginning of the 1980s in highly active rural plague zones [78].

Recently, a unique chromosomal sequence of *Y. pestis* has been described [79]. A 25-bp insertion site, located in the *yp48* gene of *Y. pseudotuberculosis*, can be applied for a real-time PCR assay, which allows the specific detection of *Y. pestis* and its differentiation from its closely related neighbor *Y. pseudotuberculosis* [80]. Variable number tandem repeats (VNTRs) analysis promises a satisfying strain discrimination capability by targeting genomic regions with high sequence variation [80]. Extended to a considerable number of loci, MLVA analysis constitutes to be a useful tool to compare results from different laboratories all over the world [27, 81].

7.3 Pathogenesis

7.3.1 Clinical and Pathological Signs

After an incubation period of three to seven days, patients typically experience "flu-like" symptoms characterized by a sudden onset of fever, chills, headaches,

Figure 7.2 Biologist puncturing an inguinal bubo in a suspected plague-infected child, in Zaa, Ituri, DR Congo.

body aches, weakness, vomiting, and nausea. Following this, the infection can take on several forms, with the three principal clinical presentations of bubonic, septicemic, and pneumonic plague. Other uncommon plague infections listed by WHO are meningeal plague, cellulocutaneous plague, pharyngeal plague, abortive plague, *pestis minor*, and asymptomatic plague [82].

The bubonic form is the most common form of plague resulting from the bite of an infective flea and is likely to be fatal (50–60% mortality) if left untreated. The plague bacillus enters the skin from the site of the bite and travels through the lymphatic system to the nearest lymph node. The lymph node then becomes swollen and inflamed because the plague bacterium replicates in high numbers and is called a "*bubo*" which is very painful and can become suppurated as an open sore in the advanced stage of infection (Figure 7.2). The infection usually involves the lymph nodes of the groin, although the axillae and neck may also be affected.

The septicemic form of plague occurs when infection spreads directly through the bloodstream without evidence of a bubo. Septicemic plague may result from flea bites resulting in the presence of *Y. pestis* in the blood and from direct contact with infective materials through cracks in the skin (e.g., veterinarians, hunters). The infection spreads into the blood and septic shock ensues. Patients may have bleeding from the skin and mucous membranes and hemorrhages into organs due to disseminated intravascular coagulation (the skin turns deep shades of purple, hence the name Black Death). They may also develop tender nodules on the skin with a white center. Complications of this form of plague include septic shock, consumptive coagulopathy, meningitis, and coma [84]. Symptoms can appear on the day and patients can die within 24 h if not promptly treated.

The pneumonic form of plague is the most virulent and least common form of plague, although recently important pneumonic plague outbreaks have been reported from Uganda (in 2004) [85] and DR Congo (in 2005) [86]. Typically, the pneumonic form is due to a secondary spread from the advanced infection of an initial bubonic form. Primary pneumonic plague results from the inhalation of aerosolized infective droplets and can be transmitted from human to human without the involvement of fleas or animals. The mortality and contagibility of pneumonic plague is very high. The infection may present just as any bronchopneumonic illness with chest pain, cough, breathlessness, and hemoptysis. Complications include disseminated intravascular coagulation, multiorgan failure, and acute respiratory distress syndrome. A chest X-ray will show consolidation, and pneumonic plague can progress rapidly to septicemia. The incubation period can range from 2 h to 4 days. Untreated pneumonic plague has a very high case-fatality ratio (100% if untreated within 18–24 h after onset of respiratory symptoms).

7.3.2 Diagnosis

From material punctured at an infected site, such as a bubo, sputum, blood, or a lymph node, or crushed fleas in saline, a diagnosis of plague can normally be established either directly, by either observation of adequately stained plague bacteria on a microscopic slide ("safety-pin" appearance), growth on solid agar-based selective media such as sheep blood agar, MacConkey, cefsulodin–irgasan–novobiocin (CIN) or irgasan, cholate salts, crystal violet, and nystatin (BIN; [87]) or selective liquid culture media such as brain heart infusion (BHI) broth and Luria Bertani (LB) broth. A new rapid diagnostic test (RDT) based on the antigenic properties of the plague bacillus (immuno-chromatographic detection of the F1 antigen specific to *Y. pestis*; [88]) is now available and besides being very easy to use, sensitive, and specific (100% in trials performed in Madagascar; [88]), it offers an immediate result without electric appliances or heavy equipment, which makes it appropriate for field work. These tests can be applied to serum, the bubo pus, and urine with a specificity of 98.4% and a sensitivity of 90.1% for the serum and 100% for bubo pus.

Indirect diagnostic tools refer to classical and recent serologic and immunologic tests. Those are the enzyme-like immunosorbent assay (ELISA) for the detection of antibody IgG anti-F1 in rat and human sera [89], passive hemagglutination test (PHA) consisting in complex formation of specific antibodies with purified antigens fixed on a support (human, sheep, or turkey red blood cells, and inert particles between 0.1 and 3.0 μm), a rapid diagnostic test based on a protein A–gold conjugate and the detection of anti-F1 IgG [90], and a immunohistochemical (IHC) assay using a monoclonal anti-F1 *Y. pestis* antibody for formalin-fixed tissues. The latter allows for retrospective plague case study and localizes bacteria while retaining tissue morphologic features [91]. Several multiplex-PCR, classical, and real-time PCR, targeting specific segments of *Y. pestis* DNA and associated plasmids, have been developed to detect *Y. pestis* in patient blood [92], rodent spleen biopsies [93], and in fleas [94, 95]. In recent years, the RDT based on F1 antigen detection proved its efficacy and detected 41.6 and 31.0% more positive clinical specimens than did more expensive bacteriological methods and ELISA, respectively [88]. However, PCR can be useful for the detection and characterization of *Y. pestis* in old and recent samples even when the bacteria are no longer viable and when culture diagnosis has been hampered by the growth of contaminants [96].

Direct culture from body fluids remains the "gold standard" which – together with at least a fourfold increase in antibody titer to *Y. pestis* – allows for unequivocal confirmation of a human case of plague. However, the new RDTs might provide a new and cheaper reference for field workers. Additional to the positive growth of the bacteria on solid media, a typical bacteriophage lysis of the colonies is needed to confirm their identity as *Y. pestis*. Thereafter a microscopic observation after staining and the antibiotic resistance screening can be achieved. Serology, including anti-F1 antibody detection (ELISA) or agglutination confirms plague diagnosis [2, 97]. However diagnosis remains a challenge because of the lack of health care and the non-availability of trained personnel in affected regions. Moreover culture of *Y. pestis* demands an adequate biosafety equipment as it is defined as a biohazard class 3 pathogen.

7.3.2.1 Plague Prevention and Treatment

Although treatable, plague still causes fear, panic, and even mass hysteria, as demonstrated by international mobilization during the 1994 plague outbreak in India (which had been free of plague since 1966), leading to the imposition of travel and trade restrictions by a number of countries [98]. Indeed, plague's notoriety comes largely from its role as the cause of the three massive pandemics, and it remains the standard by which the effects of AIDS, severe acute respiratory syndrome (SARS), or other new diseases are measured [45].

7.3.2.1.1 Isolation of Confirmed and Suspected Plague Patients Because of its epidemic potential, and in the view of the increase in the number of human cases and outbreaks, all patients suspected of having bubonic plague should be placed in isolation until two days after starting antibiotic treatment to prevent the potential spread of the disease should the patient develop secondary plague pneumonia. Patients suspected of or diagnosed with pneumonic plague should be placed in respiratory droplet isolation for two to four days while receiving antibiotic therapy until clinical improvement is noted. Those who have come in close contact with infected individuals should receive antibiotic prophylaxis. *Y. pestis* is very sensitive to heat and sunlight and does not survive very long outside the host, there is thus minimal risk of environmental transmission 1 h after the release of aerosolized *Y. pestis* [99].

For more than 45 years, the recommended antibiotic against *Y. pestis* was streptomycin, and this still remains the drug of choice. However, due to the risk of endotoxic shock during long-term use (>10 days), current treatment and prophylaxis have now been adopted. Alternative antibiotics include gentamycin, chloramphenicol, doxycylcine, tetracycline, and oxytetracycline in doses and modes of inoculation varying according to the antibiotic elicited (for dosage, see [2]). A multiple antibiotic resistance strain of *Y. pestis* was found in Madagascar in 1995; the multiresistance was mediated by a transferable plasmid that was probably acquired either by contact with other enterobacteria, or in the bloodstream of an infected host (man or rodent), or in the gut of the flea vector [100].

7.3.2.1.2 Plague Vaccine Various plague vaccine formulations have been devised, employed, and discarded over the last century. In the first acknowledged attempt, Haffkine used a killed whole-cell preparation. However, inconsistent dosing suggestions and severe reactions in the vaccinated led to a cessation of its use [101]. There are two types of plague vaccine currently used in various parts of the world. The live vaccine is derived from an attenuated strain, usually related to EV76, while the killed vaccine uses a formalin-fixed virulent strain of *Y. pestis*. The former has been in use since 1908. A first killed whole-cell plague vaccine was also developed in the United States in 1941 [102]. The "Army Vaccine," now commercially known as United States Plague (USP) vaccine (Cutter Laboratory, Inc./Greer Laboratory), is a formaldehyde-killed preparation of the highly virulent strain 195/P *Y. pestis*. The vaccine was administered to military personnel during both World War II and the Vietnam War [103]. The current formaldehyde-killed *Y. pestis* vaccine (Sampar; produced in Indonesia) provides protection against flea-transmitted plague, but it requires multiple doses and does not provide protection against pneumonic plague [104]. However, all the above vaccines are highly reactogenic and have considerable side effects.

Most of the vaccines under development are composed of a combination of two antigens: the F1 antigen which is *Y. pestis*-specific, produced in large amounts and not essential to strain virulence; and the type III secretion system component LcrV, which is common to the three pathogenic *Yersinia* species. The latter component is necessary for virulence. Various forms and galenic formulations of these vaccines have been tested successfully in mice, but the results in primates vary depending on the species [2].

The research for a new generation of plague vaccines has recently been rewarded with the possibility of using attenuated *Y. pseudotuberculosis* as a live vaccine against plague because it shares high genetic identity with *Y. pestis* while being much less virulent, genetically more stable, inexpensive, safe, and easy to produce for oral immunization against bubonic plague [105].

7.3.3
Plague as a Biological Weapon

In 2001, after September 11 and the anthrax cases associated with the intentional distribution of *Bacillus anthracis* spores through the postal system, it was clear that a new era had begun for high-priority agents. From then on, bacteria and viruses with epidemic potential, namely anthrax (*Bacillus anthracis*), botulism (*Clostridium botulinum* toxin), smallpox (*Variola major*), tularemia (*Francisella tularensis*), viral hemorrhagic fevers, and plague (*Yersinia pestis*) were brought back into the world spotlight and the fear about terrorist attacks with biological weapons progressively grew. Those agents are all classified as "category A" diseases because they threaten national security for three major reasons, namely: (i) they can be easily disseminated or transmitted from person to person, (ii) they result in high mortality rates, and (iii) they have the potential for major public health impact and might cause public panic and social disruption because they require special action for public health preparedness. However, *Y. pestis* has no spore stage and is difficult to process, handle, and disperse [106]. A bioterrorist attack would most probably involve aerosol dissemination of *Y. pestis*, and the ensuing outbreak would consist almost exclusively of pneumonic plague cases. The size of such an epidemic would depend on the amount of agent used, the meteorological conditions, the methods of aerosolization and dissemination, and the number of people initially infected [107]. In 1970, WHO calculated that, in the worst case scenario, if 50 kg of *Y. pestis* were released as an aerosol over a city of five million, pneumonic plague could occur in as many as 150 000 persons, 36 000 of whom would be expected to die. The plague bacilli would remain viable as an aerosol for 1 h for a distance of up to 10 km. Significant numbers of city inhabitants might attempt to flee, further spreading the disease [99].

7.4 Epidemiology

Plague is still a major concern of public health authorities and should not be discounted as a threat of the past [2]. During the last few years, plague cases have been reported, especially from Africa, Asia, and the Americas. African countries such as DR Congo, Madagascar, Mozambique, Uganda, Tanzania, and Algeria have been heavily affected during the last decade. Cases from several Asian regions like Kazakhstan, Turkmenistan, Uzbekistan, Mongolia, and China have also been reported. In the Americas, plague foci have been reported in Bolivia, Brazil, Ecuador, Peru, and also in the United States. European countries have been affected from the "Black Death" between the fourteenth and eighteenth centuries, but actually there have been no reports for over half a century. The occurrence of recent outbreaks in areas previously free of plague and in others where plague was absent for more than 20 years showed us that plague can (re-)emerge after a long period of silence (in India) [108]. For example, in 2003 and in 2008 plague re-emerged in Algeria, in villages south of Oran, in a country for which no plague cases had been reported since 1950. During the outbreak of 2003, epidemiological and molecular investigations showed that the strains were unrelated to any other *Y. pestis* strain studied [2]. In 2007, a small bubonic plague outbreak was reported from the Mbulu district in Tanzania, and again in 2008, in an ancient known plague focus that had no history of plague since 1977 [109]. A few years earlier, in Madagascar, the isolation of a multiresistant strain and a streptomycin-resistant strain raised concern for future therapeutic and prophylactic treatments since no harmless vaccine is currently available [2, 100, 110]. The observation that an *Orientalis* variant has restored the ability to ferment glycerol and that such a genetic lesion might be repaired as part of the natural evolutionary process further suggests the existence of genetic exchange between different *Yersinia* strains in nature [69]. These facts not only underline the capacity of *Y. pestis* to persist in enzootic cycles for decades and the genetic plasticity of the bacillus, but raise questions about the mechanisms and factors that favor plague mid- to long-term maintenance and resurgence.

7.4.1
Plague Distribution Today

In 1927, Ricardo Jorge proposed the explanation for the endemic occurrence of sporadic cases and outbreaks of plague, arguing that wild or sylvatic rodents – as opposed to domestic rodents – were the reservoirs of endemic plague [111, 112]. This type of naturally occurring plague was subsequently termed "*sylvatic plague*" or "*natural plague*" and was extensively studied in the 1950s in the United States by Burroughs [113] and Holdenried [114, 115], in South America by Macchiavello [116–119] and in the former USSR by Feynuk [120]. In the following years, plague endemic foci were described from southern Europe in fringe areas of the Caspian depression and the eastern slopes of the Caucasus [82], in Africa, the Americas, and Asia. Since then, plague has never been eradicated but also has never established endemic foci or re-emerged in some parts of the world (most of Europe and Australia). In fact, today, the distribution of plague coincides with the natural distribution of *Y. pestis* in rodent populations (Figure 7.3) and covers 6–7% of the earth in a variety of habitats such as grassland, native forests, altitude rainforests, deserts, and steppes [121, 122]. Natural foci of plague are situated in all continents except Australia, within a broad belt in tropical, subtropical, and warmer temperate climates, between the parallels 55°N and 40°S. But plague foci are not fixed and can change in response to shifts in factors such as climate, landscape, and the population dynamics of rodents, fleas, and human migrations [45, 123, 124].

During the period 1954–1997, WHO received reports of plague from 38 countries with a total of 80 613 cases and 6587 deaths. The maximum number of cases (6004) was reported in 1967 and the minimum (200) in 1981; the latter figure accounts

Figure 7.3 World distribution of endemic plague in rodent populations (with kind permission from WHO, 2000, WHO/CDS/CSR/EDC/99.2 Page 15 – Map 1).

mostly for the cases in a plague focus that had emerged the year before: the Lushoto district in Tanzania. Seven countries (Brazil, DR Congo, Madagascar, Myanmar, Peru, United States, Vietnam) have been affected by plague virtually every year during the second half of the twentieth century.

During this period, there were three periods of increased plague activity:

- During the mid-1960s;
- Between 1973 and 1978;
- From the mid-1980s to the present.

In 2003, nine countries reported 2118 cases, including 182 deaths. These figures represent a decrease when compared with the annual average figures (2895 cases, 206 deaths) for the previous 10 years (1992–2001), when 28 956 cases with 2064 deaths were reported from 22 countries. During that decade, 80.3% of cases and 84.5% of deaths were reported from Africa [125]. Those figures are probably underestimated because, on the one hand, some countries like Kazakhstan and Mongolia have only been notifying their cases since 1989 and, on the other hand, countries fail to report their plague cases because of their fear for a commercial and tourist embargo, and hence a risk of an important economic loss for the country [126]. Finally, no (accurate) data on plague cases might be available in endemic regions suffering wars and civil unrest like in DRCongo for the period 1997–2003.

7.4.2 Plague in Its Historical Perspective

Known from Egypt in the year 1320 before the Christian era [127], plague later returned in three major pandemics [128]. In 541 AC, plague was first observed around a seaport at the mouth of the Eastern branch of the Nile delta in Egypt [129] and it then spread to reach Europe as far north as Denmark and west to Ireland [130]. This first pandemic (Figure 7.4a), or so-called Justinian plague, lasted until about 750 AC and primarily consisted of the bubonic and septicemic forms of the disease [131].

The second pandemic, the notorious Black Death, started in 1346 at the mouth of the river Volga and spread via the river Don to the shores of the Black Sea, the endpoint of the northern branch of the Trans-Asiatic silk route (Figure 7.4b). A key incident in the introduction of the disease into Europe was the irruption of the plague into the Genoese settlement of Kaffa in the Crimean peninsula. In 1347, the city was besieged by the Tartars in whose camp an epidemic of plague broke out. The Tartars catapulted bodies of their own comrades who died of the disease over the walls of the city, making *Y. pestis* the first bioterrorist agent [101, 132]. The Genoese ships withdrew with cases of plague on board. Their crew went ashore at various places in Constantinople, Cyprus, Sicily, southern France, and Italy, after which a major epidemic broke out in December 1347. In June 1348, plague reached Paris. Black Death moved as a tide northward through Europe and killed around 40% of the population [133]. Between 1663 and 1670, plague returned to Amsterdam, and spread to England, Belgium, and northern France, making its

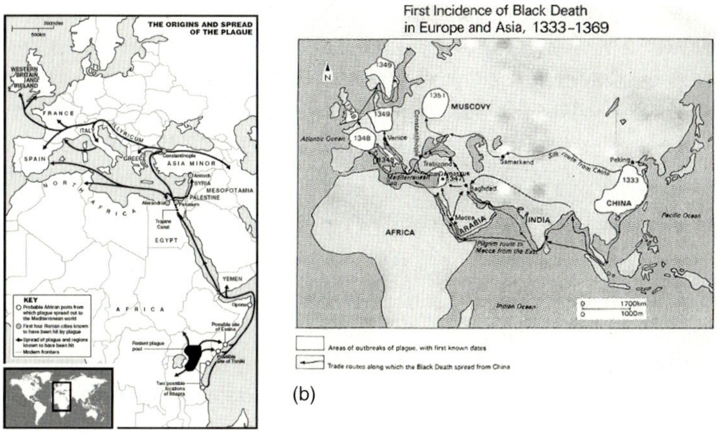

(a) (b)

Figure 7.4 (a) The origin and spread of Justinian plague. (b) Routes of the Black Death (From the Decameron Web, *http://www.brown.edu/Departments/Italian_Studies/dweb/plague/*).

way to Switzerland along the Rhine and to the northern German ports on the North and Baltic Seas. By 1670, plague had virtually disappeared from western Europe. Subsequently, smaller outbreaks occurred along trade routes and high-traffic areas such as Marseilles (1720), Messina (1743), and Malta (1813) via ships coming from plague endemic countries [134], but plague did not become endemic in western Europe. In 1860, a new plague epidemic arose in Yunnan, China, which later spread, first to Canton (Guangzhou), before subsequently traveling downstream and reaching Hong Kong in 1894, Calcutta in 1895, and Bombay in 1896, and the pandemic of the twentieth century had begun (Figure 7.5).

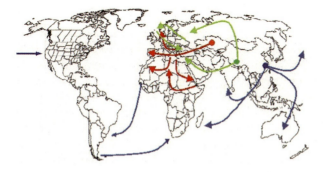

Figure 7.5 Geographic origin and routes of spreading of the three historical plague pandemics labeled in red (Justinian plague), green (Black Death), and blue (modern; after [65]).

It was during the Hong Kong epidemic that the plague microorganism was discovered in 1894, presumably independently by two scientists, Alexander Yersin and Shibasaburo Kitasato [135]. Steamships carried infected rats and fleas from the infested warehouses of the Chinese ports to many of the warmer parts of the world, including Australia that had been spared until then, wherever suitable rodent hosts could be found. In the Second World War, the Japanese Imperial Army killed thousands of prisoners in Japanese-occupied Manchuria in experiments to develop plague as a biological weapon. Infected *Pulex irritans*, the human flea, was bred and released in a few Chinese towns, resulting in small epidemics of bubonic plague. One should note that this latter historical fact is of prime importance since the vector status of *P. irritans* is still currently debated [136]. In total, this last worldwide plague episode ultimately killed more than 12 million people in India and China alone and, between the years 1894 and 1903, plague had entered 77 ports on five continents. Since then, smaller outbreaks have occurred around the globe, and, according to the World Health Organisation, the pandemic was considered active until 1959, when worldwide cases dropped to less then 200 per year.

7.4.3
An Updated Plague Cycle?

Plague is a zoonotic infection that is spread to humans from natural rodent reservoirs, commonly via the bite of an infected flea. In the past, black rats, *Rattus rattus*, were incriminated as being mostly responsible for the spread of the infection, while *Xenospylla cheopis*, the oriental rat flea, was considered the most efficient plague vector.

During the third pandemic, the classical urban plague scenario where *R. rattus* and *X. cheopis* were brought to land by ships led to all the major discoveries in plague epidemiology. However, after the third pandemic, it became clear that this species was a cosmopolitan plague carrier in most cases, that it was able to import the infection to other continents and transfer it into local wild rodent populations with their fleas, and that infection was subsequently maintained by the new endemic host population. In Madagascar, *R. rattus* and *X. cheopis* appear to be mainly responsible for the persistence of plague in the highlands where no endemic rodent reservoir has been found to date [137, 138]. Except for the particular situation in Madagascar where *R. rattus* is the major actor of both the domestic and wild plague cycles, *Y. pestis* mostly circulates in a large variety of endemic rodents and has been found in populations of 203 species of small mammals in 73 genera distributed over all the continents except Australia and Antarctica [8]. Evidence suggests that virtually all mammals can become infected with plague, with a wide variation in susceptibility among them. Typically, plague is thought to exist indefinitely in so-called enzootic (maintenance) cycles that cause little obvious host mortality and involve transmission between partially resistant rodents (enzootic or maintenance hosts) and their fleas. Occasionally, the disease spreads from enzootic hosts to more highly susceptible animals, termed *epizootic* or *amplifying hosts*, often causing rapidly spreading die-offs (epizootics). Humans and other highly susceptible mammals also experience high exposure risks during epizootics [45]. Approximately 80 species of fleas have been found to be susceptible to *Y. pestis* infection and 31 species are known to transmit plague [18, 123]. Fleas acquire *Y. pestis* from the blood of a bacteremic animal, and infection is restricted to the alimentary tract of the flea. Interestingly, the plague bacterium multiplies in the PV of some fleas, obstructing it totally or partially, forming a bacterial plug that slowly starves the flea to death. Such a flea is said to be "blocked" and "blockable" fleas are considered better vectors compared to those which do not have the ability to be blocked. Some have proposed that epizootic activity decreases during hot weather ($\geq$27.5 °C) because high temperatures adversely affect the blockage of fleas by *Y. pestis* [139]. *X. cheopis* is considered the most efficient flea vector and is the standard against which all other flea species are measured because of its high blocking rate. This perfect mode of transmission, discovered by Bacot and Martin in 1914, rapidly became the reference for vector efficiency but was also a drawback because it was taken for granted that blocked fleas were the only good vectors. However, recent studies have shown that an early-phase transmission mechanism not involving gut blockage could be more efficient even in *X. cheopis* and could also apply to wild and other domestic flea species [140, 141]. This new mode of plague transmission by fleas is likely to provide new insights into plague epidemiology since it clearly widens the pool of potential good plague vectors.

7.4.3.1 **Flea-Borne Plague Transmission**

7.4.3.1.1 Flea Biology Several ectoparasites have been found infected with *Y. pestis* and at least 80 flea species are involved in maintaining plague cycles, while 31 are proven vectors [123]. Fleas, also called *Siphonaptera*, consist of 16 different families and 2575 species classified recently in 25 subfamilies, 26 tribes, and 83 flea genera have been described worldwide [142, 143]. Fleas are temporary but obligate blood-sucking ectoparasites, ranging from opportunistic to highly specific, and alternating between periods when they occur on the host body and periods when they occur in its burrow or nest. Flea distribution extends to all continents, including Antarctica, and fleas inhabit a range of habitats and hosts from equatorial deserts, through tropical rainforests, to the arctic tundra. The majority of fleas feed essentially on mammals (94%) with about 74% of described species recorded from rodents, and the remaining on birds; and they are most diverse on small and medium sized species [144]. Adults are brown, wingless, laterally flattened insects about 1–6 mm long, with an efficient jumping ability. The head is usually small and shield- or helmet-shaped, compound eyes are absent, and the mouth parts are specialized for piercing and sucking [145]. The life cycle proceeds as holometabolous development, including eggs, three detritus-feeding larvae, and a pupa in a cocoon within which the formed adult flea, the imago, can rest for a long time, for example, until a bird or rodent nest is settled again. Emergence signals differ among flea species and can be vibrations, warmth, carbon dioxide, or shadowing, and the natural main host is recognized by specific odors probably linked with constituents of the blood [144]. After two moults within two or three weeks, mature larvae spin cocoons from silk produced by the salivary glands. Dust adheres to the freshly spun silk in which the larva moults to the pupa within two or three days, the latter usually completing its development within one or two weeks. The future imago remains in the cocoon and can wait long periods of time (6–12 months) until a cue indicates the presence of a host.

In most cases, preimaginal development is entirely off-host, the larvae are usually not parasitic, and feed on organic debris and digested blood of adults in the burrow or nest of the host [146].

Since parts of the development and life cycle occur on-host and off-host, flea reproduction is partly influenced by host species identity and is strongly dependent on the off-host environment, in particular the temperature, humidity, and soil texture of the host dwelling place [147]. Those microclimatic factors greatly affect both egg laying and the development of the larvae. Temperatures between 18 and 27 °C and a relative humidity of 70% are ideal for most fleas, whereas temperatures below 7 °C are deleterious to all developmental stages except the adult. The fact that fleas and flea larvae are temperature-dependent and sensitive to low humidity (>70% necessary) is important in *Y. pestis* distribution since adaptations and tolerance thresholds to withstand harsh or extreme variations in the off-host environmental conditions will vary according to their host habitat. The caterpillar-like larvae live in the nest of the host – in rural African huts, preferably in the dust floor of the bedroom – feeding on organic material and on the digested blood and drops of undigested blood deposited by the adults during feeding. This means that living plague bacillus might also be ingested by larva in rejected feces which remain infectious for up to five weeks [148]. However, even if flea larvae ingest plague bacillus contained in flea feces, there seems to be no transovarial, nor transstadial plague transmission in fleas [149]. For each flea species, host specificity has been estimated by a combination of: (i) the number of mammalian species on which the flea species is found and (ii) the specificity index which measures the average taxonomic distinctness of all host species used by a parasite species [151]. However, host specificity varies among populations of the same flea species [150, 151] and this might also be due to some physiological and biochemical differences in properties of the hosts such as odor cues, immune defense, or constituents of blood [152, 153] and circulating sex hormones [154–156] On the whole, this nonstable specificity has several consequences in the frame of plague epidemiology. Indeed, the possibility of a host switching during plague epizootics can participate to the spread of infection to sensitive rodent species and occasionally to man. However, if flea development requires a definite concentration of circulating hormones or other cues, and if *Y. pestis* requires a high level of iron and calcium to encounter the optimal conditions in both sucked blood and new blood (the new host in which it is introduced), plague transmission might be less feasible during hibernation, estivation, and outside the breeding period of most rodents. In conclusion, the degree of association between a particular flea species and a particular host species varies between flea populations of the same species, and it ranges from highly host-specific to host-opportunistic fleas, with the majority of species being specific for a particular host genus or subfamily [144].

7.4.3.1.2 Mechanisms of Plague Transmission by Fleas Fleas are hematophagous intermittent ectoparasites, taking 2–5 min blood meals – on average – three to four times a day. The blood is pumped through the epipharynx and transported through the cibarial and pharyngeal pump directly into the PV or foregut. The PV, which separates the stomach and esophagus, is a sphincter-like organ with needle-like teeth directed back toward the stomach (Figure 7.6a); it aids in the rupture of blood cells and normally prevents regurgitation of a blood meal. From then the triturated

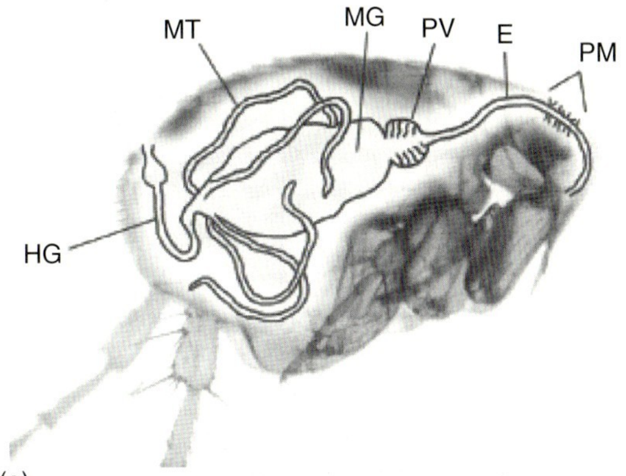

(a)

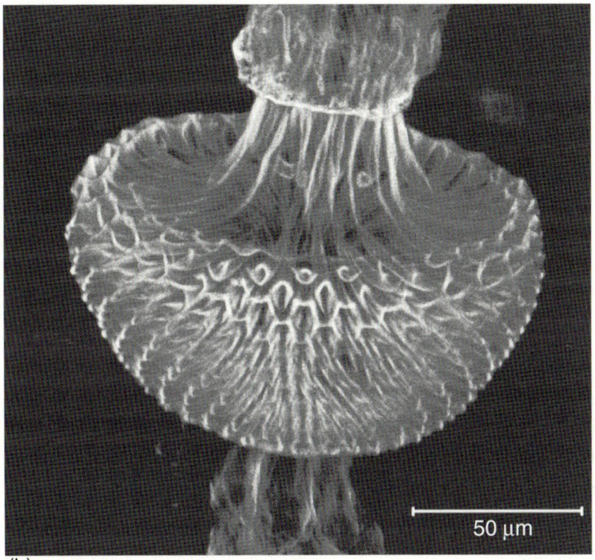

(b)

Figure 7.6 (a) Digestive tract anatomy of fleas. E = Esophagus, PV = proventriculus, MG = midgut, HG = hindgut. The muscles that pump blood into the midgut (PM) and the malpighian tubules (MTs) are also indicated [42] (with kind permission from Hinnebusch, 2005 and Horizon Scientific Press/Caister Academic Press, UK). (b) Scanned proventriculus of the flea *Nosopsyllus incisus* viewed in a laserpoint scanning microscope image browser [157].

blood is propelled to the midgut in which it is stored and digested. The number, length, shape, and size of the proventricular spines (or acanthi; Figure 7.6b) and the shape and size of the PV are anatomical features that might influence the blocking rate of a particular flea species [157].

Until very recently, two major mechanisms of plague transmission by fleas were recognized (not taking into account the oral route, following the ingestion of a plague infected flea):

- The **proventricular blockage**, where *Y. pestis* forms a biofilm (bacterial plug) that occludes the PV, was the process historically considered necessary for efficient flea-borne transmission of *Y. pestis* [17]. In this process, when a flea bites an infected, bacteremic host, it pumps up blood containing the plague bacillus. To produce a transmissible infection, *Y. pestis* grows in the flea digestive tract as a biofilm that adheres to and fills the PV. This infectious matrix of *Y. pestis* blocks the flow of blood and – in this way – overcomes the rhythmic, pulsating action of the proventricular valve and the inward flow of blood during feeding that would otherwise counteract transmission by washing the bacteria backwards into the midgut. Earlier observations showed that flea blockage was a temperature-dependant phenomenon and this has now been linked with the expression of *hms* gene products [158]. Indeed, biofilm formation is a temperature-dependant phenomenon that occurs at 26 °C, but biofilms do not form, and actually dissolve, at 37 °C and above [19, 159]. The degradation of the biofilm at the mammalian host body temperature might thus aid in the dispersal of *Y. pestis* from the site of the flea bite [160]. Experimentally, between days 3 and 9 after the infective blood meal, the bacterial masses may completely block the PV, extend into the esophagus, and prevent ingested blood from reaching the stomach. Proventricular blockage is estimated to occur in 25–50% of *X. cheopis* [57], in 0.19–1.75% of *P. irritans*, and 23% of the chicken flea *Echidnophaga gallinacea* [136, 161]. In order to feed, the flea will repeat and even attempt more frenetically to bite the available host. The sucked blood will reach the bacterial plug clogging the foregut and transmission occurs when bacteria released from the biofilm are refluxed into the bite site when blocked fleas attempt to feed [18]. The bacterium is then passed on when the fleas, in turn, bite another rodent or a human. Blockages rarely form in most North American wild rodent fleas [161], including prairie dog fleas, and "blocked" fleas rapidly die as a result of dehydration and starvation [162]. Knowing that *X. cheopis* will ingest 0.03–0.05 ml of blood, a bacteremia of 10^4 CFU/ml would ensure ingestion of at least 300 *Y. pestis* organisms [123]. Nonetheless, to cause infection levels sufficient to eventually become capable of transmitting *Y. pestis* to uninfected mice, the host blood must contain 10^7–10^9 bacteria/ml – the bacteremia level of a septicemic rodent host just before death [163]. The rapidly fatal sepsis in the mammalian host and the hypervirulence of the plague bacterium is a consequence of the high threshold bacteremia level that must be attained to complete the transmission cycle. Epidemiological modeling predicts that, to compensate for a relatively short period of infectivity of the mammalian host for the arthropod vector, plague epizootics require a high flea burden per host, even when the susceptible host population density is high [18]. Such low rates of flea-borne transmission, high flea index, and the delay between infection and blockage cannot explain the rapid spread of plague among mammalian hosts during plague epizootics or epidemics, and point to alternative routes.

- A first alternative route is the **mechanical mass transmission via soiled mouth parts** [161] that was later described in *P. irritans*, the human flea [164, 165]. Because the mechanical transmission requires at least four concomitant events in a short time, this route is considered the least plausible and the least efficient. Because survival of *Y. pestis* on the mouth parts is possible between 24 and 48 h after the last infecting bite (at most 36 h with *X. cheopis*), they can transmit the infection only three days after infection [166]. Secondly, the bacteremia in humans, sufficient to allow transmissible infection in and by fleas, occurs in the last 24 h before death, which leaves only a short period of time for enough vectors, at least 60 (for the human flea), to feed on the body and find a new host [136]. Even if such *P. irritans* densities can be observed, the dynamics of familial plague epidemics tended to show that neither blockage nor mechanical transmission were likely to explain the rapid spread of the infection, especially in the Middle Ages [136]. Those observations led scientists to the discovery of a third and tempting new mode of transmission, which mechanism is not yet fully understood, namely, the **early-phase transmission without blockage**, and more and more evidence tends to show that it can occur in many fleas [140, 141, 167]. Indeed, in contrast to the classical blocked flea model, *Oropsylla montana*, a flea infesting California ground squirrel (*Spermophilus beechyii*) and rock squirrels (*S. variegatus*), is immediately infectious, transmits efficiently for at least four days post-infection (early phase), and may remain infectious for a long time because the fleas do not suffer block-induced mortality [140]. Moreover, the early-phase transmission of *Yersinia pestis* by unblocked *X. cheopis* is as efficient as transmission by blocked fleas and also results in an infectious period considerably longer than previously thought [168]. This discovery not only widens the range of potential efficient plague vectors but also gives more credit to the human flea *P. irritans* as a possible and efficient vector of the rapidly spreading Black Death in regions of Europe where *X. cheopis* was rare or absent [141].

7.4.4
Classical Plague Cycle

A complete plague cycle was recently proposed (Figure 7.7; [124]). It involves, as already mentioned, an urban or domestic cycle and a wild cycle, in other words, a quiescent pool of *Y. pestis* in the wild. This figure illustrates the fact that plague may be regarded as falling into the following four categories – epidemic, endemic, epizootic, and enzootic, and it can also be classified as human plague and domestic- or wild-small mammal plague [169].

Each of these categories or classification is dependent upon several intermingled ecological and environmental factors. Mammalian species that are sufficient to

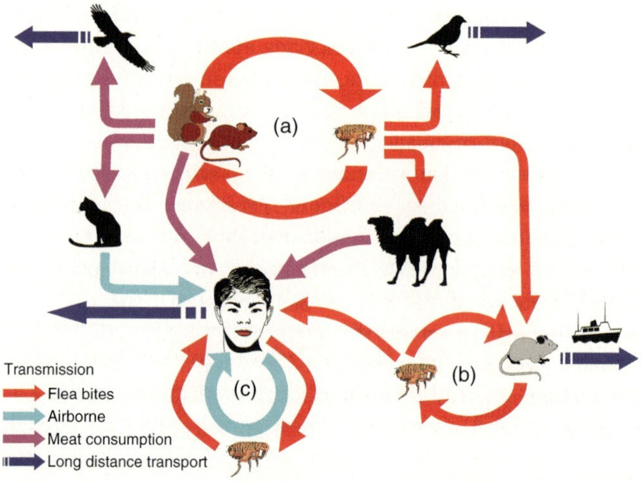

Figure 7.7 Possible transmission pathways for the plague agent, *Y. pestis*. These pathways include wild rodent–flea cycles (a), commensal rodent–flea cycles (b), and pneumonic transmission in humans (c). The color of the arrows indicates the mechanism (flea bites, air particles, meat consumption) through which the bacteria are transferred from one host to another. Dark blue arrows indicate ways in which plague can move to other areas [124].

maintain plague in natural systems were defined as enzootic or primary hosts [170]. The ability of these animals to sustain plague is based on their susceptibility and sensitivity to plague, numbers, distribution, and patterns of behavior.

Epizootic or secondary hosts may facilitate the spread of plague but are not able to maintain plague for extended periods without primary hosts [171]. The maintenance of plague foci thus depends on a whole suite of rodent hosts and their associated fleas (Figure 7.7a). Under favorable conditions, the plague bacillus might survive in the environment, essentially in the soil of rodent burrows [143, [172], in the litter of dead animals [173, 174], or even in complex symbiosis with soil microorganisms, including the cysts of protozoa [175].

When an infected flea happens to feed on a commensal rodent, the cycle continues in the latter (Figure 7.7b). As the commensal rodents die, their fleas are forced to move to alternative hosts, for example, humans. Indeed, blocked fleas might be less specific in their host choice and this might contribute to a distant host switching which would not have been likely otherwise. If humans develop pneumonic plague, the infection may be transmitted from person to person through respiratory droplets spread by coughing (Figure 7.7c). Humans might also be contaminated, but less frequently, by contact with infectious body fluids or by ingesting raw or uncooked infected meat. Occasional plague cases have been reported from hunters during the skinning of cotton-tail rabbit in the United States [176], or *Marmotta bobac* in Manchuria [177] and Mongolia (Bolormaa, personal communication) [178], veterinarians handling cats in the United States [179], and people handling and consuming guinea pigs in Peru and Ecuador [180, 181], or goat and camel meat in Lybia, Jordan, and Kazakhstan [182, 183].

In Zimbabwe, epidemiological analysis after a human plague outbreak indicated that the presence of sick cats was a risk factor for contracting the disease [184]. Indeed, cats become acutely ill and experience high mortality in experimental studies [185]. Canada lynx (*Felis canadensis*) and pumas (*F. concolor*) in the United States appear susceptible to the pneumonic form of the disease [143] but show various degrees of resistance [186]. Dogs may also directly be involved by transferring infected fleas into homes [187, 188]. Carnivores are exposed to plague by eating infected rodents or by being bitten by rodent fleas [189, 190]. Most carnivore species are resistant to infection, show little clinical evidence of disease [191], typically seroconvert when exposed to *Y. pestis*, and rarely die [192]. In Lushoto district (Tanzania), 5.5% of the dogs were seropositive for plague and 93.8% of their flea fauna consisted of the cat flea *Ctenocephalides felis strongylus* [193]. This suggests that dogs – and other carnivores – could serve as serological sentinel animals for the detection of plague activity, especially where they roam freely and may be in frequent contact with wild rodent populations [15, 194, 195]. Due to their resistance to plague, showing persistent signs of infection, or at least exposure, and since they share spatial proximity to other susceptible species, carnivores should also be regarded as possible enzootic reservoirs of plague [196, 197]. While infection can occur by direct contact or ingestion, these routes do not normally play a role in the maintenance of *Y. pestis* in animal reservoirs. Mammal predators such as swift fox (*Vulpes velox*) in the United States [192, 197], birds of prey [198, 199], and other birds (such as the isabelline wheatears, *Oenanthe isabellina*) that use rodent burrows for nesting may move over larger areas than the rodents themselves, spreading the infection over longer distances.

7.5
Conclusion

Plague is a modern infection in the Southern hemisphere and is a serious public health threat, especially in rural environments where rodents live commensally in flea infested houses. Today, due to the recent development of rapid diagnostic tests and molecular tools, clinicians working in the field are able to faster detect, report, and treat plague. Indeed, for an effective treatment of patients infected with the plague bacillus, it is essential to start an antibiotic therapy within 24 h post infection. Thus, fast detection methods and molecular techniques are indispensable. Molecular typing methods constitute a powerful tool for the diagnosis of plague as for epidemiological investigations, since they are fast, sensitive, and afford low costs concerning consumables. Furthermore the resulting data can be compared easily between different laboratories all over the world. Today, a rapid response to a sudden plague outbreak can also be favored by the advances of model research (ecological niche modeling, combining geographic tools with population, teledetection, satellite data) which allow epidemiologists to establish risk maps for plague re-emergence. Those maps could in turn be useful to field medical staff when facing an unidentified disease outbreak in an area with a past history of plague, even if the disease has not reemerged in the spot for decades.

After looking at the general plague cycle, considering the human involvement, the genetic diversity, the basic biochemical characteristics of *Y. pestis*, the modes of plague transmission by fleas, the plague bacillus worldwide and continental distribution, and the factors that influence plague re-emergence, we know that one cannot speak about plague ecology but about the ecologies of plague. In general, while the persistence of plague outbreaks has been associated with multiple factors (biocenotic, vector morphological, environmental, microclimatic, landscape related, topographical, economic, socio-cultural factors), none of those factors – either considered alone or integrated – still succeed to explain how and why plague persists and re-emerges in the long term. And despite the ecology of plague proper to each focus, one is tempted to suggest that a similar mechanism is involved in the long-term maintenance of the plague bacterium in the ecosystem. If the enzootic host theory, the potential host, the flea dynamics *sensu lato*, and the climatic, sociocultural, environmental, and other abiotic factors cannot explain the long-term focality and unpredictability of plague re-emergence, then the soil ecosystem appears to be the ultimate refuge where to look for the missing piece of the plague puzzle. From the preceding discussion which presents gaps in our understanding in the natural plague cycle and which only gathers some findings from more than 200 years of plague research, one can see that either we are still missing an element of the whole cycle or the interpretation of the epidemiological facts still lacks a unifying thought.

References

1. Ben Ari, T., Neerinckx S., Gage, K.L., Kreppel, K., Laudisoit, A., *et al.* (2011) Plague and climate: scales matter. *PLoS Pathog*, **7** (9), e1002160, doi: 10.1371/journal.ppat.1002160
2. WHO (2008) Interregional Meeting on Prevention and Control of Plague. Antananarivo, Madagascar, April 1–11, 2006, 65 p.
3. Bertherat, E., Lamine, K.M., Formenty, P., Thullier, P., Mondonge, V., Mitifu, A., and Rahalison, L. (2005) Sur le front des émergences. *Med. Trop.*, **65**, 511–514.
4. Laudisoit, A., Kisasa, R., and Kidimbu, A. (2006) Les ectoparasites des micromammifères de la ville de Kinshasa: un facteur de risques pour la santé publique? in Recherche Scientifique et Développement des Pays Africains Edité par ASETIB, FUSAGx, Gembloux, 2005. 176 p.
5. Gubler, D.J. (1998) Resurgent vector-borne diseases as a global health problem. *Emerg. Infect. Dis.*, **4** (3), 442–450.
6. Dennis, D.T., Gage, K.L., Gratz, N., Poland, J.D., and Tikhonov, I. (1999) Plague Manual: Epidemiology, Distribution, Surveillance and Control, WHO/CDS/CSR/EDC/99.2, World Health Organization, Geneva.
7. WHO (2000) PLAGUE. Report on Global Surveillance of Epidemic-Prone Infectious Diseases, WHO/CDS/CSR/ISR/2000.1.
8. Anisimov, A.P., Lindler, L.E., and Pier, G.B. (2004) Intraspecific diversity of *Yersinia pestis*. *Clin. Microbiol. Rev.*, **17** (2), 434–464. [Erratum in: *Clin. Microbiol. Rev.*, **17** (3), 695.].

9. Holmes, B.E., Foresman, K.R., and Matchett, M.R. (2006) No evidence of persistent *Yersinia pestis* infection at prairie dog colonies in North Central Montana. *J. Wildl. Dis.*, **42** (1), 164–169.
10. Achtman, M., Morelli, G., Zhu, P., Wirth, T., Diehl, I., Kusecek, B., Vogler, A.J., Wagner, D.M., Allender, C.J., Easterday, W.R., Chenal-Francisque, V., Worsham, P., Thomson, N.R., Parkhill, J., Lindler, L.E., Carniel, E., and Keim, P. (2004) Microevolution and history of the plague bacillus, *Yersinia pestis*. *Proc. Natl. Acad. Sci. U.S.A.*, **101** (51), 17837–17842.
11. Derbise, A., Chenal-Francisque, V., Pouillot, F., Fayolle, C., Prévost, M.C., Médigue, C., Hinnebusch, B.J., and Carniel, E. (2007) A horizontally acquired filamentous phage contributes to the pathogenicity of the plague bacillus. *Mol. Microbiol.*, **63** (4), 1145–1157.
12. Pouillot, F., Fayolle, C., and Carniel, E. (2008) Characterization of chromosomal regions conserved in *Yersinia pseudotuberculosis* and lost by *Yersinia pestis*. *Infect. Immun.*, **76** (10), 4592–4599.
13. Chauvaux, S., Rosso, M.L., Frangeul, L., Lacroix, C., Labarre, L., Schiavo, A., Marceau, M., Dillies, M.A., Foulon, J., Coppée, J.Y., Médigue, C., Simonet, M., and Carniel, E. (2007) Transcriptome analysis of *Yersinia pestis* in human plasma: an approach for discovering bacterial genes involved in septicaemic plague. *Microbiology*, **153** (9), 3112–3124.
14. Guinet, F., Avé, P., Jones, L., Huerre, M., and Carniel, E. (2008) Defective innate cell response and lymph node infiltration specify *Yersinia pestis* infection. *PLoS ONE*, **3** (2), e1688.
15. Li, B. and Yang, R. (2008) Interaction between *Yersinia pestis* and the host immune system. *Infect. Immun.*, **76** (5), 1804–1811.
16. Hinnebusch, B.J., Rosso, M.L., Schwan, T.G., and Carniel, E. (2002) High-frequency conjugative transfer of antibiotic resistance genes to *Yersinia pestis* in the flea midgut. *Mol. Microbiol.*, **46** (2), 349–354.
17. Bacot, A.W. and Martin, C.J. (1914) Observations on the mechanism of the transmission of plague by fleas. *J. Hyg. (Lond).*, **3** (Suppl), 423–439.
18. Lorange, E.A., Race, B.L., Sebbane, F., and Hinnebusch, J.B. (2005) Poor vector competence of fleas and the evolution of hypervirulence in *Yersinia pestis J. Infect. Dis.*, **191**, 1907–1912.
19. Darby, C. (2008) Uniquely insidious: *Yersinia pestis* biofilms. *Trends Microbiol.*, **16** (4), 158–164.
20. Tan, J., Liu, Y., Shen, E., Zhu, W., Li, R., and Yang, L. (2002) Towards "the atlas of plague and its environment in the People's Republic of China": idea, principle and methodology of design and research results. *Huan Jing Ke Xue*, **23**, 1–8.
21. Drancourt, M., Houhamdi, L., and Raoult, D. (2006) *Yersinia pestis* as a telluric, human ectoparasite-borne organism. *Lancet Infect. Dis.*, **6**, 234–241.
22. Bearden, S.W. and Perry, R.D. (1999) The Yfe system of *Yersinia pestis* transports iron and manganese and is required for full virulence of plague. *Mol. Microbiol.*, **32**, 403–414.
23. Carniel, E. (1999) The *Yersinia* high-pathogenicity island. *Int. Microbiol.*, **2** (3), 161–167.
24. Weinberg, E.D. (1966) Roles of metallic ions in host-parasite interactions. *Bacteriol. Rev.*, **30** (1), 136–151.
25. Girard, J.M., Wagner, D.M., Vogler, A.J., Keys, C., Allender, C.J., Drickamer, L.C., and Keim, P. (2004) Differential plague-transmission dynamics determine *Yersinia pestis* population genetic structure on local, regional, and global scales. *Proc. Natl. Acad. Sci. U.S.A.*, **101** (22), 8408–8413.
26. Oyston, P.C., and Isherwood, K.E.. (2005). The many and varied niches occupied by *Yersinia pestis* as an arthropod-vectored zoonotic pathogen. *Antonie Van Leeuwenhoek*, **87** (3), 171–177.
27. Li, Y., Cui, Y., Hauck, Y., Platonov, M.E., Dai, E., Song, Y., Guo, Z., Pourcel, C., Dentovskaya, S.V., Anisimov, A.P., Yang, R., and

Vergnaud, G. (2009) Genotyping and phylogenetic analysis of Yersinia pestis by MLVA: insights into the worldwide expansion of central Asia plague foci. *PLoS ONE*, **4** (6), e6000, doi: 10.1371/journal.pone.0006000

28. Guiyoule, A., Grimont, F., Iteman, I., Grimont, P.A., Lefévre, M., and Carniel, E. (1994) Plague pandemics investigated by ribotyping of *Yersinia pestis* strains. *J. Clin. Microbiol.*, **32** (3), 634–641.
29. Chain, P.S.G., Hu, P., Malfatti, S.A., *et al.* (2006). Complete genome sequence of *Yersinia pestis* strains Antiqua and Nepal516: evidence of gene reduction in an emerging pathogen. *J. Bacteriol.*, **188** (12), 44–53. doi: 10.1128/JB.00124-06
30. Zhou, D., Han, Y., Song, Y., Tong, Z., Wang, J., Guo, Z., Pei, D., Pang, X., Zhai, J., Li, M., Cui, B., Qi, Z., Jin, L., Dai, R., Du, Z., Bao, J., Zhang, X., Yu, J., Wang, J., Huang, P., and Yang, R. (2004) DNA microarray analysis of genome dynamics in *Yersinia pestis*: insights into bacterial genome microevolution and niche adaptation. *J. Bacteriol.*, **186** (15), 5138–5146.
31. Dávalos, V.A., Torres, M.A., Mauricci, C.O., Laguna-Torres, V.A., and Chinarro, M.P. (2001) Surto de peste bubônica na localidade de Jacocha, Huancabamba, Perú. *Rev. Soc. Bras. Med. Trop.*, **34** (1), 87–90.
32. Parmenter, R.R., Yadav, E.P., Parmenter, C.A., Ettestad, P., and Gage, K. (1999) Incidence of plague associated with increased winter-spring precipitation in New Mexico. *Am. J. Trop. Med. Hyg.*, **61** (5), 814–821.
33. Enscore, R.E., Biggerstaff, B.J., Brown, T.L., Fulgham, R.E., Reynolds, P.J., Engelthaler, D.M., Levy, C.E., Parmenter, R.R., Montenieri, J.A., Cheek, J.E., Grinnell, R.K., Ettestad, P.J., and Gage, K.L. (2002) Modeling relationships between climate and the frequency of human plague cases in the southwestern United States, 1960–1997. *Am. J. Trop. Med. Hyg.*, **66** (2), 186–196.
34. Ben Ari, T., Gershunov, A., Gage, K.L., Snäll, T., Ettestad, P., Kausrud, K.L., and Stenseth, N.C. (2008) Human plague in the USA: the importance of regional and local climate. *Biol. Lett.*, **4** (6), 737–740.
35. Davis, D.H. (1953) Plague in South Africa: a study of the epizootic cycle in gerbils (*Tatera brantsi*) in the northern Orange Free State. *J. Hyg. (Lond)*, **51** (4), 427–449.
36. Pollitzer, R. (1954) La peste, Série de Monographies. Organisation mondiale de la Santé, Genève, 737 p.
37. Lang, J.D. (1996) Factors affecting the seasonal abundance of ground squirrel and wood rat fleas (Siphonaptera) in San Diego County, California. *J. Med. Entomol.*, **33** (5), 790–804.
38. Collinge, S.K., Johnson, W.C., Ray, C., Matchett, R., Grensten, J., Cully, J.F. *et al.* (2005) Landscape structure and plague occurrence in black-tailed prairie dogs on grasslands of the western USA. *Landsc. Ecol.*, **20**, 941–955.
39. Laudisoit, A. and Beaucournu, J.C. (2007a) Two new *Ctenophthalmus* (Insecta: Siphonaptera: Ctenophthalmidae) from Tanzania. *Parasite*, **14** (2), 101–105.
40. Laudisoit, A. and Beaucournu, J.C. (2007b) *Ctenophthalmus (Ethioctenophthalmus) kemmelberg* n. sp. (Insecta: Siphonaptera: Ctenophthalmidae), a new flea from Tanzania and description of unknown small structures in the mecopteroids. *Parasite*, **14** (3), 213–217.
41. Eskey, C.R. and Haas, V.H. (1940) Plague in the western part of the United States. *Public Health Bull.*, **254**, 1–83.
42. Hinnebusch, B.J. (2005) The evolution of flea-borne transmission in *Yersinia pestis*. *Curr. Issues Mol. Biol.*, **7** (2), 197–212.
43. Erickson, D.L., Waterfield, N.R., Vadyvaloo, V., Long, D., Fischer, E.R., French-Constant, R., and Hinnebusch, B.J. (2007) Acute oral toxicity of *Yersinia pseudotuberculosis* to fleas: implications for the evolution of vector-borne transmission of plague. *Cell Microbiol.*, **9** (11), 2658–2666.

44. American Society for Microbiology (2010) *Sentinel level clinical microbiology laboratory guidelines for suspected agents of bioterrorism and emerging infectious diseases: Yersinia pestis*. American Society for Microbiology, New York.
45. Gage, K.L. and Kosoy, M.Y. (2005) Natural history of plague: perspectives from more than a century of research. *Annu. Rev. Entomol.*, **50**, 505–528.
46. Golubov, A., Neubauer, H., Nölting, C., Heesemann, J., and Rakin, A. (2004) Structural organization of the pFra virulence-associated plasmid of rhamnose-positive *Yersinia pestis*. *Infect. Immun.*, **72** (10), 5613–5621.
47. Parkhill, J., Wren, B.W., Thomson, N.R., Titball, R.W., Holden, M.T., Prentice, M.B., Sebaihia, M., James, K.D., Churcher, C., Mungall, K.L., Baker, S., Basham, D., Bentley, S.D., Brooks, K., Cerdeño-Tárraga, A.M., Chillingworth, T., Cronin, A., Davies, R.M., Davis, P., Dougan, G., Feltwell, T., Hamlin, N., Holroyd, S., Jagels, K., Karlyshev, A.V., Leather, S., Moule, S., Oyston, P.C., Quail, M., Rutherford, K., Simmonds, M., Skelton, J., Stevens, K., Whitehead, S., and Barrell, B.G. (2001) Genome sequence of *Yersinia pestis*, the causative agent of plague. *Nature*, **413** (6855), 523–527.
48. Ferber, D.M. and Brubaker, R.R. (1981) Plasmids in *Yersinia pestis*. *Infect. Immun.*, **31**, 839–841.
49. Brubaker, R.R. (1991) Factors promoting acute and chronic diseases caused by Yersiniae. *Clin. Microbiol. Rev.*, **4**, 309–324.
50. Carniel, E., Guilvout, I., and Prentice, M. (1996) Characterization of a large chromosomal "high-pathogenicity island" in biotype 1B *Yersinia enterocolitica*. *J. Bacteriol.*, **178** (23), 6743–6751.
51. Fetherston, J.D., Schuetze, P., and Perry, R.D. (1992) Loss of the pigmentation phenotype in *Yersinia pestis* is due to the spontaneous deletion of 102 kb of chromosomal DNA which is flanked by a repetitive element. *Mol. Microbiol.*, **6**, 2693–2704.
52. Simonet, M., Riot, B., Fortineau, N., and Berche, P. (1996) Invasin production by *Yersinia pestis* is abolished by insertion of an IS200-like element within the inv gene. *Infect. Immun.*, **64** (1), 375–379.
53. Forman, S., Wulff, C.R., Myers-Morales, T., Cowan, C., Perry, R.D., and Straley, S.C. (2008) yadBC of *Yersinia pestis*, a new virulence determinant for bubonic plague. *Infect. Immun.*, **76** (2), 578–587.
54. Costerton, J.W. (1995) Overview of microbial biofilms. *J. Ind. Microbiol.*, **15** (3), 137–140.
55. Schembri, M.A., Givskov, M., and Klemm, P. (2002) An attractive surface: gram-negative bacterial biofilms. *Sci. STKE*, **13** (132), RE6.
56. Hall-Stoodley, L. and Stoodley, P. (2005) Biofilm formation and dispersal and the transmission of human pathogens. *Trends Microbiol.*, **13** (1), 7–10. [Erratum in: *Trends Microbiol.*, **13** (7), 300–301.].
57. Hinnebusch, B.J., Perry, R.D., and Schwan, T.G. (1996) Role of the *Yersinia pestis* hemin storage (hms) locus in the transmission of plague by fleas. *Science*, **273** (5273), 367–370.
58. Jarrett C.O., Deak, E., Isherwood, K.E., Oyston, P.C., Fischer, E.R., Whitney A.R., Kobayashi, S.D., DeLeo, F.R., and Hinnebusch, B.J. (2004) Transmission of *Yersinia pestis* from an Infectious biofilm in the flea vector. *J. Infect. Dis.*, **190** (4), 783–792.
59. Sodeinde, O.A. and Goguen, J.D. (1988) Genetic analysis of the 9.5-kilobase *virulence plasmid of Yersinia pestis*. *Infect. Immun.*, **56** (10), 2743–2748.
60. Brubaker, R.R., Beesley, E.D., and Surgalla, M.J. (1965) Pasteurella pestis: role of Pesticin I and Iron in experimental plague. *Science*, **149** (3682), 422–424.
61. Fields, K.A. and Straley, S.C. (1999) LcrV of Yersinia pestis enters infected eukaryotic cells by a virulence plasmid-independent mechanism. *Infect. Immun.*, **67** (9), 4801–4813.
62. Protsenko, O.A., Anisimov, P.I., Mozharov, O.T., Konnov, N.P., and Popov, I.A. (1983) Detection and characterization of the plasmids of the plague microbe which determine

the synthesis of pesticin I, fraction I antigen and "mouse" toxin exotoxin. *Genetika*, **19** (7), 1081–1090. (Russian).

63. Welkos, S.L., Davis, K.M., Pitt, L.M., Worsham, P.L., and Freidlander, A.M. (1995) Studies on the contribution of the F1 capsule-associated plasmid pFra to the virulence of Yersinia pestis. *Contrib. Microbiol. Immunol.*, **13**, 299–305.
64. Drancourt, M., Roux, V., Dang, L.V., Tran-Hung, L., Castex, D., Chenal-Francisque, V., Ogata, H., Fournier, P.E., Crubézy, E., and Raoult, D. (2004) Genotyping, Orientalis-like *Yersinia pestis*, and plague pandemics. *Emerg. Infect. Dis.*, **10** (9), 1585–1592.
65. Devignat, R. (1953) Geographical distribution of three species of *Pasteurella pestis*. *Schweiz. Z. Pathol. Bakteriol1.*, **16** (3), 509–514.
66. Heisch, R.B., Grainger, W.E., and D'Souza, J. (1953) Results of a plague investigation in Kenya. *Trans. R. Soc. Trop. Med. Hyg.*, **47** (6), 503–521.
67. Davis, D.H., Heisch, R.B., McNeill, D., and Meyer, K.F. (1968) Serological survey of plague in rodents and other small mammals in Kenya. *Trans. R. Soc. Trop. Med. Hyg.*, **62** (6), 838–861.
68. Martinevsky, I.L. (1969) *Biology and Genetic Features of Plague and Plague Related Microbes*, Meditsina Press, Moscow.
69. Motin, V.L., Georgescu, A.M., Elliott, J.M., Hu, P., Worsham, P.L., Ott, L.L., Slezak, T.R., Sokhansanj, B.A., Regala, W.M., Brubaker, R.R., and Garcia, E. (2002) Genetic variability of *Yersinia pestis* isolates as predicted by PCR-based IS100 genotyping and analysis of structural genes encoding glycerol-3-phosphate dehydrogenase (glpD). *J. Bacteriol.*, **184** (4), 1019–1027.
70. Drancourt, M., Aboudharam, G., Signoli, M., Dutour, O., and Raoult, D. (1998) Detection of 400-year-old *Yersinia pestis* DNA in human dental pulp: an approach to the diagnosis of ancient septicemia. *Proc. Natl. Acad. Sci. U.S.A.*, **95**, 12637–12640.
71. Raoult, D., Aboudharam, G., Crubezy, E., Larrouy, G., Ludes, B., and Drancourt, M. (2000) Molecular identification by "suicide PCR" of *Yersinia pestis* as the agent of medieval black death. *Proc. Natl. Acad. Sci. U.S.A.*, **97**, 12800–12803.
72. Wiechmann, I. and Grupe, G. (2005) Detection of *Yersinia pestis* DNA in two early medieval skeletal finds from Aschheim (Upper Bavaria, 6th century A.D.). *Am. J. Phys. Anthropol.*, **126**, 48–55.
73. Drancourt, M., Signoli, M., Dang, L., Bizot, B., Roux, V., Tzortzis, S., and Raoult, D. (2007) *Yersinia pestis* Orientalis in remains of ancient plague patients. *Emerg. Infect. Dis.*, **13** (2), 332–333.
74. Gilbert, M.T., Cuccui, J., White, W., Lynnerup, N., Titball, R.W., Cooper, A., and Prentice, M.B. (2004) Absence of *Yersinia pestis*-specific DNA in human teeth from five European excavations of putative plague victims. *Microbiology*, **150**, 341–354.
75. Tong, Z., Zhou, D., Song, Y., Zhang, L., Pei, D., Han, Y., Pang, X., Li, M., Cui, B., Wang, J., Guo, Z., Qi, Z., Jin, L., Zhai, J., Du, Z., Wang, J., Wang, X., Yu, J., Wang, J., Huang, P., Yang, H., and Yang, R. (2005) Pseudogene accumulation might promote the adaptive microevolution of Yersinia pestis. *J. Med. Microbiol.*, **54**, 259–268.
76. Pourcel, C., Salvignol, G., and Vergnaud, G. (2005) CRISPR elements in *Yersinia pestis* acquire new repeats by preferential uptake of bacteriophage DNA, and provide additional tools for evolutionary studies. *Microbiology*, **151**, 653–663.
77. Auerbach, R.K., Tuanyok, A., Probert, W.S., Kenefic, L., Vogler, A.J., Bruce, D.C., Munk, C., Brettin, T.S., Eppinger, M., Ravel, J., Wagner, D.M., and Keim, P. (2007) *Yersinia pestis* evolution on a small timescale: comparison of whole genome sequences from North America. *PLoS ONE*, **2** (1), e770.
78. Guiyoule, A., Rasoamanana, B., Buchrieser, C., Michel, P., Chanteau, S., and Carniel, E. (1997) Recent emergence of new variants of *Yersinia pestis* in Madagascar. *J. Clin. Microbiol.*, **35**, 2826–2833.

79. Chase, C.J., Ulrich, M.P., Wasieloski, L.P. Jr., Kondig, J.P., Garrison, J., Lindler, L.E., and Kulesh, D.A. (2005) Real-time PCR assays targeting a unique chromosomal sequence of Yersinia pestis. *Clin. Chem.*, **51**, 1778–1785.
80. Adair, D.M., Wosham, P.L., Hill, K.K., Klevytska, A.M., Jackson, P.J., Friedlander, A.M., and Keim, P. (2000) Diversity in a variable-number tandem repeat from Yersinia pestis. *J. Clin. Microbiol.*, **38**, 1516–1519.
81. Pourcel, C., André-Mazeaud, F., Neubauer, H., Ramisse, F., and Vergnaud, G. (2004) Tandem repeats analysis of *Yersinia pestis. BMC Microbiol.*, **4**, 22.
82. WHO (1999) Plague Manual. Epidemiology, Distribution, Surveillance and Control, WHO/CDS/CSR/EDC/99.2. 172 p.
83. Laudisoit, A., Neerinckx, S., Makundi, R.H., Leirs, H., and Krasnov, B.R. (2009) Are local plague incidence and ecological characteristics of vectors and reservoirs related? A case study in north-east Tanzania. *Curr. Zool.*, **55**, 199–221.
84. CDC (1996) Prevention of plague: recommendations of the Advisory Committee on Immunization. Practices (ACIP). *MMWR Morb. Mortal. Wkly. Rep.*, **45** (RR-14), 1–15.
85. Begier, E.M., Asiki, G., Anywaine, Z., Yockey, B., Schriefer, M.E., Aleti, P., Ogden-Odoi, A., Staples, J.E., Sexton, C., Bearden, S.W., and Kool, J.L. (2006) Pneumonic plague cluster, Uganda, 2004. *Emerg. Infect. Dis.*, **12** (3), 460–467.
86. WHO (2005) Democratic Republic of The Congo. *Water Environ. Res.*, **80** (8), 65.
87. Ber, R., Mamroud, E., Aftalion, M., Tidhar, A., Gur, D., Flashner, Y., and Cohen, S. (2003) Development of an improved selective agar medium for isolation of Yersinia pestis. *Appl. Environ. Microbiol.*, **69** (10), 5787–5792.
88. Chanteau, S., Rahalison, L., Ralafiarisoa, L., Foulon, J., and Ratsitorahina, M. (2003) Development and testing of a rapid diagnostic test for bubonic and pneumonic plague. *Lancet*, **361** (9353), 211–216.
89. Rasoamanana, B., Leroy, F., Boisier, P., Rasolomaharo, M., Buchy, P., Carniel, E., and Chanteau, S. (1997) Field evaluation of an immunoglobulin G anti-F1 enzyme-linked immunosorbent assay for serodiagnosis of human plague in Madagascar. *Clin. Diagn. Lab. Immunol.*, **4** (5), 587–591.
90. Thullier, P., Guglielmo, V., Rajerison, M., and Chanteau, S. (2003) Short report: Serodiagnosis of plague in humans and rats using a rapid test. *Am. J. Trop. Med. Hyg.*, **69** (4), 450–451.
91. Guarner, J., Shieh, W.J., Greer, P.W., Gabastou, J.M., Chu, M., Hayes, E., Nolte, K.B., and Zaki, S.R. (2002) Immunohistochemical detection of Yersinia pestis in formalin-fixed, paraffin-embedded tissue. *Am. J. Clin. Pathol.*, **117** (2), 205–209.
92. Norkina, O.V., Kulichenko, A.N., Gintsuburg, A.L., Tuchkov, I.V., Popov, Y.U.A., Aksenov, M.U., and Drosdov, I.G. (1994) Development of a diagnostic test for *Yersinia pestis* by the polymerase chain reaction. *J. Appl. Bacteriol.*, **76**, 240–245.
93. Leal, N.C., Abath, F.G., Alves, L.C., and de Almeida, A.M. (1996) A simple PCR-based procedure for plague diagnosis. *Rev. Inst. Med. Trop. (Sao Paulo)*, **38** (5), 371–373.
94. Hinnebusch, J. and Schwan, T.G. (1993) New method for plague surveillance using polymerase chain reaction to detect Yersinia pestis in fleas. *J. Clin. Microbiol.*, **31** (6), 1511–1514.
95. Tomaso, H., Reisinger, E.C., Al Dahouk, S., Frangoulidis, D., Rakin, A., Landt, O., and Neubauer, H. (2003) Rapid detection of Yersinia pestis with multiplex real-time PCR assays using fluorescent hybridisation probes. *FEMS Immunol. Med. Microbiol.*, **38** (2), 117–126.
96. Melo, A.C., Almeida, A.M., and Leal, N.C. (2003) Retrospective study of a plague outbreak by multiplex-PCR. *Lett. Appl. Microbiol.*, **37** (5), 361–364.
97. Rajerison, M., Dartevelle, S., Ralafiarisoa, L.A., Bitam, I., Tuyet, D.T.N., Andrianaivoarimanana, V.,

Nato, F., and Rahalison, L. (2009) Development and evaluation of two simple, rapid immunochromatographic tests for the detection of *Yersinia pestis* antibodies in humans and reservoirs. *PLoS Negl. Trop. Dis.*, **3** (4), e421, doi: 10.1371/journal.pntd.0000421

98. Madan, T.N. (1995) The plague in India, 1994. *Soc. Sci. Med.*, **40** (9), 1167–1168.

99. Inglesby T.V., Dennis D.T., Henderson D.A., Bartlett J.G., Ascher M.S., Eitzen E., Fine A.D., Friedlander A.M., Hauer J., Koerner J.F., Layton M., McDade J., Osterholm M.T., O'Toole T., Parker G., Perl T.M., Russell P.K., Schoch-Spana M., and Tonat K. (2000) Plague as a biological weapon: medical and public health management. *J. Am. Med. Assoc.*, **283**, 2281–2290.

100. Galimand, M., Guiyoule, A., Gerbaud, G., Rasoamanana, B., Chanteau, S., Carniel, E., and Courvalin, P. (1997) Multidrug resistance in *Yersinia pestis* mediated by a transferable plasmid. *New Engl. J. Med.*, **337**, 677–680.

101. Haffkine, W. (1897) Remarks on the plague prophylactic fluid. *Br. Med. J.*, **1**, 461.

102. Cornelius, C., Quenee, L., Anderson, D., and Schneewind, O. (2007) Protective immunity against plague. *Adv. Exp. Med. Biol.*, **603**, 415–424.

103. Williamson, E.D., Flick-Smith, H.C., Lebutt, C., Rowland, C.A., Jones, S.M., Waters, E.L., Gwyther, R.J., Miller, J., Packer, P.J., and Irving, M. (2005) Human immune response to a plague vaccine comprising recombinant F1 and V antigens. *Infect. Immun.*, **73** (6), 3598–3608.

104. Titball, R.W. and Williamson, E.D. (2001) Vaccination against bubonic and pneumonic plague. *Vaccine*, **19** (30), 4175–4184. (Review).

105. Blisnick, T., Ave, P., Huerre, M., Carniel, E., and Demeure, C.E. (2008) Oral vaccination against bubonic plague using a live avirulent Yersinia pseudotuberculosis strain. *Infect. Immun.*, **76** (8), 3808–3816. [Epub 2008 May 27].

106. Cunha, B.A. (2002) Anthrax, tularemia, plague, ebola or smallpox as agents of bioterrorism: recognition in the emergency room. *Clin. Microbiol. Infect.*, **8** (8), 489–503. (Review).

107. Gani, R. and Leach, S. (2004) Epidemiologic determinants for modeling pneumonic plague outbreaks. *Emerg. Infect. Dis.*, **10** (4), 608–614.

108. WHO (1990) Consultation on Plague, New Delhi, September 11–15, 1989, Geneva, World Health Organization, p. 7.

109. Makundi, R.H., Massawe, A.W., Mulungu, L.S., Katakweba, A., Mbise, T.J., and Mgode, G. (2008) Potential mammalian reservoirs in a bubonic plague outbreak focus in Mbulu District, northern Tanzania, in 2007. *Mammalia*, **72** (3), 253–257.

110. Guiyoule, A., Gerbaud, G., Buchrieser, C., Galimand, M., Rahalison, L., Chanteau, S., Courvalin, P., and Carniel, E. (2001) Transferable plasmid-mediated resistance to streptomycin in a clinical isolate of Yersinia pestis. *Emerg. Infect. Dis.*, **7** (1), 43–48.

111. Jorge, R. (1927a) Les faunes régionales des rongeurs et des puces dans leurs rapports avec la peste. *Bull. Off. Int. Hyg. Publ.*, **19**, 1094–1109.

112. Jorge, R. (1927) Les faunes régionales des rongeurs et des puces dans leurs rapports avec la peste. *Bull. Off. Int. Hyg. Publ.*, **19**, 1257–1288.

113. Burroughs, A.L. (1953) Sylvatic plague studies. Survival of rodent fleas in the laboratory. *Parasitology*, **43** (1–2), 35–48.

114. Holdenried, R. (1952) Sylvatic plague studies. VII. Plague transmission potentials of the fleas *Diamanus montanus* and *Polygenis gwyni* compared with *Xenopsylla cheopis*. *J. Infect. Dis.*, **90** (2), 131–140.

115. Holdenried, R. (1952) Sylvatic plague studies. VIII. Notes on the alimentary and reproductive tracts of fleas, made during experimental studies of plague. *J. Parasitol.*, **38** (4, sec.1), 289–292.

116. Macchiavello, A. (1957) Studies on sylvatic plague in South America. II. Sylvatic plague in the frontier region of Peru and Ecuador. 2. Focus of sylvatic plague of the District of Lancones, Department of Piura, Peru. *Bol. Oficina. Sanit. Panam.*, **43** (3), 225–250.

117. Macchiavello, A. (1958) Studies on the sylvatic plague in South America. III. Sylvatic plague in the mountain range of Huancabama, Peru. *Bol. Oficina. Sanit. Panam.*, **44** (6), 484–512.

118. Macchiavello, A. (1959a) Studies on sylvatic plague in South America. V. Sylvatic plague in Bolivia; general data on the geography and history of the plague. *Bol. Oficina. Sanit. Panam.*, **46** (6), 509–524.

119. Macchiavello, A. (1959b) Studies on the sylvatic plague in South America. IV. Experimental transmission of the plague by *Polygenis litargus*. *Bol. Oficina. Sanit. Panam.*, **45** (2), 122–131.

120. Feynuk, B.K. (1960) Experience in the eradication of enzootic plague in the North-West Part of the Caspian Region of the USSR. *Bull. WHO*, **23**, 263–273.

121. Clover, J.R., Hofstra, T.D., Kuluris, B.G., Schroeder, M.T., Nelson, B.C., Barnes, A.M., and Botzler, R.G. (1989) Serologic evidence of *Yersinia pestis* infection in small mammals and bears from a temperate rainforest of north coastal California. *J. Wildl. Dis.*, **25** (1), 52–60.

122. Neerinckx, S.B., Peterson, A.T., Gulinck, H., Deckers, J., and Leirs, H. (2008) Geographic distribution and ecological niche of plague in sub-Saharan Africa. *Int.J. Health Geogr.*, **13** 7–54.

123. Perry, R.D. and Fetherston, J.D. (1997) *Yersinia pestis*: etiologic agent of plague. *Clin. Microbiol. Rev.*, **10** (1), 35–66.

124. Stenseth, N.C., Atshabar, B.B., Begon, M., Belmain, S.R., Bertherat, E., Carniel, E., Gage, K.L., Leirs, H., and Rahalison, L. (2008) Plague: past, present, and future. *PloS Med.*, **5** (1), e3.

125. WHO (2004) Human plague in 2002 and 2003. *Wkly. Epidemiol. Rec.*, **33** (79), 301–306.

126. Carniel, E. (1995) Situation mondiale des infections à *Y.* pestis. *Med. Mal. Infect.*, **25** (7), 675–679.

127. Hirst, L.F. (1953) *The Conquest of Plague. A Study of the Evolution of Epidemiology*, Oxford University Press, New York, 478 p.

128. Wu, L.T., Chun, J.W., Pollitzer, R., and Wu, C.Y. (1936) *Plague: A Manual for Medical and Public Health Workers, Weihengshu*, National Quarantine Service, Shangai, 547 p.

129. Schat, M. (2005) Justinian's Foreign Policy and the Plague: did Justinian Create the First Pandemic? *http://entomology.montana.edu/historybug/YersiniaEssays/Schat.htm* (accessed July 21 2009).

130. Keys, D. (1999) *Catastrophe: An Investigation Into the Origins of the Modern World*, Ballentine Books, New York, 452 p.

131. Orent, W. (2004) *Plague: the Mysterious Past and Mystifying Future of the World's Most Dangerous Disease*, Free Press, New York.

132. Mollaret, H.H. (2002) L'arme Biologique. Bactéries, Virus Et Terrorisme, Plon, 214 p.

133. Duncan, C.J. and Scott, S. (2005) What caused the black death? *Postgrad. Med.J.*, **81** (955), 315–320.

134. Bacci, M.L., De Nardi Ipsen, C., and Ipsen, C. (2000) *The Population of Europe: A History*, Wiley-Blackwell, 240 p.

135. Bibel, D.J. and Chen, T.H. (1976) Diagnosis of plague: an analysis of the Yersin-Kitasato controversy. *Bacteriol. Rev.*, **40** (3), 633–451.

136. Audouin-Rouzeau, F. (2003) *Les Chemins de la Peste - Le Rat, la Puce et l'homme*, Presses Universitaires de Rennes, 371 p.

137. Boisier, P., Rasolomaharo, M., Ranaivoson, G., Rasoamanana, B., Rakoto, L., Andrianirina, Z., Andriamahefazafy, B., and Chanteau, S. (1997) Urban epidemic of bubonic plague in Majunga, Madagascar: epidemiological aspects Tropical. *Med. Int. Health*, **2** (5), 422–427.

138. Chanteau, S., Ratsitorahina, M., Rahalison, L., Rasoamanana, B., Chan, F., Boisier, P., Rabeson, D., and Roux, J. (2000) Current epidemiology of human plague in Madagascar. Microbes

and infection. *Inst. Pasteur.*, **2** (1), 25–31.

139. Cavanaugh, D.C. (1971) Specific effect of temperature upon transmission of the plague bacillus by the Oriental rat flea, *Xenopsylla cheopis. Am. J. Trop. Med. Hyg.*, **20**, 264–273.

140. Eisen, R.J., Bearden, S.W., Wilder, A.P., Montenieri, J.A., Antolin, M.F., and Gage, K.L. (2006) Early-phase transmission of *Yersinia pestis* by unblocked fleas as a mechanism explaining rapidly spreading plague epizootics. *Proc. Natl. Acad. Sci. U.S.A.*, **103** (42), 15380–15385.

141. Eisen, R.J., Petersen, J.M., Higgins, M.S., Wong, D., Levy, C.E., Mead, P.S., Schriefer, M.E., Griffith, K.S., Gage, K.L., and Beard, C.B. (2008) Persistence of *Yersinia pestis* in soil under natural conditions. *Emerg. Infect. Dis.*, **14**, 941–943.

142. Lewis, R.E. (1993) in *Medical Insects and Arachnids* (eds P.L. Richard and R.W. Crosskey), Chapman and Hall, p. 744.

143. Whiting, M.F., Whiting, A.S., Hastriterb, M.W., and Dittmar, K. (2008) A molecular phylogeny of fleas (Insecta: Siphonaptera): origins and host associations. *Cladistics*, **24**, 1–31.

144. Marshall, A.G. (1981) *The Ecology of Ectoparasitic Insects*, Academic Press Inc., London, 459 p.

145. Dunnet, G.M. and Mardon, D.K. (1991) *A Textbook for Students and Research Workers*, vol. II, Melbourne University Press, Melbourne, pp. 705–716.

146. Krasnov, B.R., Mouillot, D., Shenbrot, G.I., Khokhlova, I.S., and Poulin, R. (2004) Geographical variation in host specificity of fleas (Siphonaptera) parasitic on small mammals: the influence of phylogeny and local environmental conditions. *Ecography*, **27** (6), 787–797.

147. Krasnov, B.R., Khokhlova, I.S., Fielden, L.J., and Burdelova, N.V. (2001) Development rates of two *Xenopsylla* flea species in relation to air temperature and humidity. *Med. Vet. Entomol.*, **15**, 249–258.

148. Horrox R. (1994) *The Black Death*, Manchester University Press, 384 p.

149. Golov, D.A. and Ioff, I.G. (1925) On the question of the role of the fleas of spermophiles in the epidemiology of plague. *Vestn. Mikrobiol. Epidemiol. Parazitol.*, **4**, 19–48.

150. Poulin, R. and Mouillot, D. (2003) Parasite specialization from a phylogenetic perspective: a new index of host specificity. *Parasitology*, **126** (Pt 5), 473–480.

151. Poulin, R., Krasnov, B.R., Shenbrot, G.I., Mouillot, D., and Khokhlova, I.S. (2006) Evolution of host specificity in fleas: is it directional and irreversible? *Int. J. Parasitol.*, **36** (2), 185–191.

152. Krasnov, B.R., Khokhlova, I.S., Fielden, L.F., and Burdelova, N.V. (2002) The effect of substrate on survival and development of two species of desert fleas (Siphonaptera: Pulicidae). *Parasite*, **9** (2), 135–142.

153. Krasnov, B.R., Sarfati, M., Arakelyan, M.S., Khokhlova, I.S., Burdelova, N.V., and Degen, A.A. (2003) Host specificity and foraging efficiency in blood-sucking parasite: feeding patterns of the flea *Parapulex chephrenis* on two species of desert rodents. *Parasitol. Res.*, **90** (5), 393–399.

154. Rothschild, M. and Ford, B. (1973) Factors influencing the breeding of the rabbit flea (*Spilopsyllus cuniculi*): a spring-time accelerator and a kairomone in nestling rabbit urine with notes on *Cediopsylla simplex*, another "hormone bound" species. *J. Zool.*, **170** (1), 87–137.

155. Rothschild, M. (1975) Recent advances in our knowledge of the order Siphonaptera. *Ann. Rev. Entomol.*, **23**, 241–259.

156. Prasad, R.S. (1976) Studies on host-flea relationship. IV. Progesterone and cortisone do not influence the reproductive potentials of rat fleas *Xenopsylla cheopis* (Rothschild) and *X. astia* (Rothschild). *Parasitol. Res.*, **50** (1), 81–86.

157. DeRaedt, N. (2008) Morfologie van de proventriculus van verschillende vlooiensoorten in Tanzania: zijn er aanwijzingen voor vectoriële capaciteit bij de overdracht van builenpest? Eindverhandeling voorgelegd tot het bekomen

van de graad van Licentiaat in de Biologie, zwaartepunt organismen – en populatiebiologie. PhD Thesis, 55 p.

158. Hinnebusch, B.J. and Erickson, D.L. (2008) *Yersinia pestis* biofilm in the flea vector and its role in the transmission of plague. *Curr. Top. Microbiol. Immunol.*, **322**, 229–248.

159. Perry, R.D., Bobrov, A.G., Kirillina, O., Jones, H.A., Pedersen, L., Abney, J., and Fetherston, J.D. (2004) Temperature regulation of the hemin storage (Hms+) phenotype of *Yersinia pestis* is posttranscriptional. *J. Bacteriol.*, **186** (6), 1638–1647.

160. Kirillina, O., Fetherston, J.D., Bobrov, A.G., Abney, J., and Perry, R.D. (2004) HmsP, a putative phosphodiesterase, and HmsT, a putative diguanylate cyclase, control Hms-dependent biofilm formation in *Yersinia pestis. Mol. Microbiol.*, **54** (1), 75–88.

161. Burroughs, A.L. (1947) Sylvatic plague studies. The vector efficiency of nine species of fleas compared with *Xenopsylla cheopis. J. Hyg. (Lond).*, **45**, 371–396.

162. Vashchenok, V.S. and Tarakanov, N.F. (1977) Effect of the digestive process on the survival of the plague agent in *Xenopsylla gerbilli* Minax fleas. *Parazitologiia*, **11**, 474–479.

163. Engelthaler, D.M. and Gage, K.L. (2000) Quantities of *Yersinia pestis* in fleas (Siphonaptera: Pulicidae, Ceratophyllidae, and Hystrichopsyllidae) collected from areas of known or suspected plague activity. *J. Med. Entomol.*, **37** (3), 422–426.

164. Baltazard, M. (1959) New data in the interhuman transmission of plague. *Bull. Acad. Natl. Med.*, **143**, 517–522.

165. Baltazard, M. (1969) La recherche épidemiologique et son évolution. L'exemple d'un travail d'équipe sur la peste. *Bull. l'Inst. Pasteur*, **67** (2), 235–262.

166. Indian Plague Commission (1907) Reports on plague investigations in India; further observations on the transmission of plague by fleas, with special reference to the fate of the plague bacillus in the body of the rat flea (*P. cheopis*). *Indian Plague. Comm.*, 7, 395–420.

167. Wilder, A.P., Eisen, R.J., Bearden, S.W., Montenieri, J.A., Tripp, D.W., Brinkerhoff, R.J., Gage, K.L., and Antolin, M.F. (2008) *Transmission efficiency of two flea species (Oropsylla tuberculata cynomuris* and *Oropsylla hirsuta*) involved in plague epizootics among prairie dogs. *Ecohealth*, **5** (2), 205–212.

168. Eisen, R.J., Wilder, A.P., Bearden, S.W., Montenieri, J.A., and Gage, K.L. (2007) Early-phase transmission of *Yersinia pestis* by unblocked *Xenopsylla cheopis* (Siphonaptera: Pulicidae) Is as Efficient as Transmission by Blocked Fleas. *J. Med. Entomol.*, **44** (4), 678–682.

169. Garnham, P.C. (1949) Distribution of wild-rodent plague. *Bull. WHO*, **2** (2), 271–278.

170. Cully, J.F. and Williams, A.S. Jr. (2001) Interspecific comparisons of sylvatic plague in prairie dogs. *J. Mammal.*, **82**, 894–905.

171. Biggins, D.E. and Kosoy, M.Y. (2001) Influences of introduced plague on North American mammals: implications from ecology of plague in Asia. *J. Mammal.*, **82** (4), 906–916.

172. Balthazard, M., Karimi, Y., Eftekhari, M., Chamsa, M. and Mollaret, H.H. (1963) The interepizootic preservation of plague in an inveterate focus. Working hypotheses. *Bull. Soc. Pathol. Exot. Filiales*, **56**, 1230–1245.

173. Mollaret, H.H. (1963) Conservation expérimentale de la peste dans le sol. *Bull. Soc. Pathol. Exotique.*, **6**, 1169–1183.

174. Karimi, Y. (1963) Conservation naturelle de la peste dans le sol. *Bull. Soc. Pathol. Exot.*, **56**, 1183–1186.

175. Lomaradskii, I.V., Medinskii, G.M., Mishan'kin, B.N., and Suchkov, I.G. (1995) The problem of natural foci of plague: the search for ways for its resolution. *Med. Parazitol. (Mosk.)*, **13** (4), 3–9.

176. Von Reyn, C.F., Barnes, A.M., Weber, N.S., and Hodgin, U.G. (1976). Bubonic plague from exposure to a rabbit: a documented case in the

United States. *Am. J. Epidemiol.*, **104**, 81–87.

177. Gamsa, M. (2006) The epidemic of pneumonic plague in manchuria 1910–1911. *Past. Present.*, **190** (1), 147–183.

178. Demberel, J. (1997) in *Holarctic Marmots as a Factor of Biodiversity*, Abstracts, 3rd Conference on Marmots (Cheboksary, Russia, 25-30 August 1997 (eds V.Y. Rumiantsev, A.A. Nikolskii, and O.V. Brandler), ABF, Moscow, pp. 132–133.

179. Orloski, K.A. and Lathrop, S.L. (2003) Plague: a veterinary perspective. *J. Am. Vet. Med. Assoc.*, **222** (4), 444–448.

180. Gabastou, J.-M., Proano, J., Vimos, A., Jaramillo, G., Hayes, E., Gage, K., Chu, M., Guarner, J., Zaki, S., Bowers, J., Gage, K.L., and Montenieri, J.A. (1994) The role of predators in the ecology, epidemiology, and surveillance of plague in the United States. *Vertebr. Pest Conf.*, **16**, 200–206.

181. Ruiz, A. (2001) Plague in the Americas. *Emerg. Infect. Dis.*, **7**, 539–540.

182. Christie, A.B., Chen, T.H., and Elberg, S.S. (1980) Plague in camels and goats: their role in human epidemics. *J. Infect. Dis.*, **141**, 724–726.

183. Arbaji, A., Kharabsheh, S., Al-Azab, S., Al-Kayed, M., Amr, Z.S., Abu Baker, M., and Chu, M.C. (2005) A 12-case outbreak of pharyngeal plague following the consumption of camel meat, in north-eastern Jordan. *Ann. Trop. Med. Parasitol.*, **99** (8), 789–793.

184. Manungo, P., Peterson, D.E., Todd, C.H., Mthamo, N., and Pazvakavambwa, B. (1998) Risk factors for contracting plague in Nkayi district, Zimbabwe. *Cent. Afr. J. Med.*, **44** (7), 173–176.

185. Watson, R.P., Blanchard, M.G., Mense, M.G., and Gasper, P.W. (2001) Histopathology of experimental plague in cats. *Vet. Pathol.*, **38**, 165–172.

186. Wild, M.A., Shenk, T.M., and Spraker, T.R. (2006) Plague as a Mortality Factor in Canada Lynx (*Lynx canadensis*) reintroduced to Colorado. *J. Wildl. Dis.*, **42** (3), 646–650.

187. Wang, H., Jiao, B.T., Wang, G.J., Yang, Y.H., Mu, Y., Tian, T., and Lou, Y.L. (2005) Study on an epidemic of human lung in Nangqian county, Qinghai province. *Zhonghua Liu Xing Bing Xue Za Zhi (Zhonghua liuxingbingxue Zazhi)*, **26** (9), 684–686.

188. Gould, L.H., Pape, J., Ettestad, P., Griffith, K.S., and Mead, P.S. (2008) Dog-associated risk factors for human plague. *Zoonoses Public Health*, **55** (8–10), 448–454.

189. Barnes, A.M. (1982) in *Animal Disease in Relation to Conservation* (eds M.A. Edwards and U. McDonnel), Academic Press, New York, pp. 237–270.

190. Gage, K.L., Montenieri, J.A., and Thomas, R.E. (1994). The role of predators in the ecology, epidemiology, and surveillance of plague in the United States. *Proc. Vertebr. Pest Conf.*, **16**, 200–206.

191. Smith, C.R. (1994) Wild carnivores as plague indicators in California – a cooperative interagency disease surveillance program. *Proc. Vertebr. Pest. Conf.*, **16**, 192–199.

192. Gage, K.L., Ostfeld, R.S., and Olson, J.G. (1995) Nonviral vector-borne zoonoses associated with mammals in the United States. *J. Mammal.*, **76**, 695–715.

193. Kilonzo, B.S., Gisakanyi, N.D., and Sabuni, C.A. (1993) Involvement of dogs in plague epidemiology in Tanzania: serological observations in domestic animals in Lushoto District. *Scand. J. Infect. Dis.*, **25**, 503–506.

194. Taylor, R., Gordon, D.H., and Isaacson, M. (1981) The status of plague in Zimbabwe. *Ann. Trop. Med. Parasitol.*, **75**, 165–173.

195. Smith, C.R., Nelson, B.C., and Barnes, A.M. (1984) The use of wild carnivore serology in determining patterns of plague activity in rodents in California. *Proc. Vertebr. Pest Conf.*, **11**, 71–76.

196. Kilonzo, B., Mhina, J., Sabuni, C., and Mgode, G. (2005) The role of rodents and small carnivores in plague endemicity in Tanzania. *Belg. J. Zool.*, **135** (Suppl.), 119–125.

197. Salkeld, D.J. and Stapp, P. (2006) Seroprevalence rates and transmission of plague (*Yersinia pestis*) in mammalian carnivores. *Vector. Borne. Zoonotic. Dis.*, **6** (3), 231–239.

198. Antolin, M., Gober, P., Luce, B., Biggins, D., Van Pelt, W., Seery, D., Lockhart, M., and Ball, M. (2002) The influence of sylvatic plague on North American wildlife at the landscape level, with special emphasis on black-footed ferret and prairie dog conservation. *Trans. North Am. Wildl. Nat. Resour. Conf.*, **67**, 105–127.

199. Witmer, G.W. (2004) Rodent ecology and plague in North America. Proceedings of the 19th International Congress of Zoology, Beijing, August 23–27, 2004, China Zoological Society, Beijing, pp. 154–156.

8
Rickettsia Species: Rickettsioses

Alice N. Maina, Stephanie Speck, Eva Spitalska, Rudolf Toman, Gerhard Dobler, and Sally J. Cutler

8.1
Introduction

Many regard the *Rickettsia* with intrigue and apprehension. The intrigue stems from the devastating role that rickettsial disease has had on mankind, typified by the demise of the Napoleonic army, reduced from 10 000 to a mere 3000 [1]; and the rapidly changing complexity of this genus is a source for apprehension. Recently published works describe two to four different rickettsial groups, while new species names are being reported in every issue of some journals [2]. Many different arthropod vectors are potential sources of Rickattsiae. Some species are clinically relevant while others have not been associated with any untoward effects and some appear only to influence their arthropod host [3–5]. Within the space of this chapter, we aim to demystify the Rickettsiae, providing an overview that will cover the diversity of this genus through to their relevance as potential human or animal pathogens.

Rickettsioses are thought to be one of the oldest infectious diseases, as exemplified by epidemic typhus suspected to have been responsible for the Athens plague described by Thucydides during the fifth century BC [6]. Howard Taylor Ricketts later discovered the "bacillus of the Rocky Mountain spotted fever" (RMSF) in 1906. He and others characterized the basic features of the disease, including the role of tick vectors. He subsequently gave his life to his work, succumbing to a laboratory acquired infection with "typhus fever" that claimed his life in 1910. The causative agent of RMSF was named *R. rickettsii* in honor of Ricketts [7].

During this same period, in 1909, Stanislaus von Prowazek discovered the organism that became known as *R. prowazekii*, the causative agent of epidemic typhus [7]. This was soon followed by the discovery by Charles Nicolle that the epidemic form of typhus fever was louse-borne [8].

Rickettsioses have been reported in all continents, although some Rickettsiae are confined to certain geographical regions. Rickettsial illnesses seem to follow a spatial and temporal trend dictated by their arthropod vectors [9, 10]. *R. conorii* the cause of Mediterranean spotted fever (MSF) also known as *botonneuse fever*,

BSL3 and BSL4 Agents: Epidemiology, Microbiology, and Practical Guidelines, First Edition.
Edited by Mandy C. Elschner, Sally J. Cutler, Manfred Weidmann, and Patrick Butaye.

Marseilles fever, or *escaro-nodular fever*, [11] and whose vector is *Rhipicephalus sanguineus* (the brown ear tick) is prevalent in the Mediterranean region, Southern Europe [12], and Africa [9, 13]. While RMSF and *R. parkeri* are prevalent in the Americas [14, 15], African tick bite fever (ATBF), as suggested by its name, is prevalent in Africa [16] but is also endemic in the Caribbean islands [17]. Geographical confinement of rickettsial illness is also witnessed with *R. sibirica* in the USSR and *R. australis* in Australia [10].

Recently recognized tick-borne species include *R. helvetica* [18], *R. sibirica mongolotimonae* [19, 20], *R. slovaca*, and *R. raoultii* [21, 22]. Others such as *R. hoogstraalii* [23], Candidatus *R. davousti* [24], and Candidatus *R. barbariae* [21] remain of uncertain clinical significance. Recent years have seen a plethora of publications describing presence of the flea-borne *R. felis* which appears to have a remarkably wide global distribution [25–28].

Rickettsiae are parasites with a life cycle involving invertebrates, hematophagous arthropods, and vertebrate hosts (including humans), classifying them as zoonotic pathogens. Human rickettsial infections are known to cause many diseases, including epidemic typhus (*R. prowazekii*), murine typhus (*R. typhi*), RMSF (*R. rickettsii*), murine-like typhus (*R. felis*), rickettsial pox (*R. akari*), Boutonneuse fever (*R. conorii*), and North Asian tick typhus (*R. siberica*). These virulent species are of potential interest as emerging infectious diseases [29] and most recently as bioterrorism agents [30].

Rickettsial organisms have comparatively small genomes (1.1–1.3 Mb) that have arisen through reductive evolution as they developed dependence on the host cell for necessary functions [31]. As a result, their genomes are littered with pseudogenes. The Rickettsiae have a close evolutionary relationship with the progenitor of the mitochondria [32].

8.2 Characteristics of the Agent

The Rickettsiae are small (approx. 0.7–1.0 μm in length, 0.3–0.5 μm in width), highly pleomorphic cocco-bacillary Gram-negative rods belonging to the class *Alphaproteobacteria*. Within *Alphaproteobacteria*, the order *Rickettsiales* comprises three families: *Holosporaceae*, *Anaplasmataceae*, and *Rickettsiaceae* [33]. They are obligate intracellular pathogens, probably arising from their highly reductive genomes of only approximately 1.2 Mb [2]. In common with many intracellular pathogens, they utilize a "zipper" mechanism to mediate their cellular internalization. Following their binding to host cells and associated cytoskeletal rearrangements, the bacterium is engulfed by a cellular vacuole (see further details in Section 8.5). Unlike the *Coxiella* with which they are frequently compared, *Rickettsia* spp. rapidly escape this vacuole entering the cytoplasm of the host cell. It is believed that rickettsial phospholipases and hemolysin C play an as yet undetermined role in escape from the vacuole. This escape is remarkably quick, with *R. conorii* exiting the vacuole in an estimated 12 min. It is likely that this rapid escape is instrumental in the survival of these microbes, prior to the fusion of lysosomes. Once within the cytoplasm, in common with other cytosolic bacteria, they demonstrate actin-based motility, probably facilitating their spread. The notable exceptions are the typhus group (TG): *R. prowazakii*, *R. typhi*, and *R. peacockii* of the spotted fever group (SFG); these do not possess functional rickA protein (*ompA*) essential for interaction with the multi-subunits of the eukaryotic Arp 2/3 complex, among other genes, regulating the actin cytoskeleton responsible for their actin-based motility [32].

To understand the interrelationships of the Rickettsiae, many different gene targets have been utilized. Initial studies using 16s rDNA showed low discriminatory power for this target to be useful, instead better resolution was obtained using citrate synthase (*gltA*) [34] outer membrane proteins *ompA* and *ompB* [35] and the *sca* [36] (cell surface genes) sequence. Although largely concurrent, minor differences were obtained. The validity of multi-locus sequence typing (MLST) has been proven for many genera of bacteria, similarly a MLST scheme has been applied to study population structure among the Rickettsiae [37]. Some would argue that the few sequences analyzed by MLST could still skew the data, thus the best comparative method should be comparison of whole genomes. While this is a valid point, the restrictions imposed by using the few isolates for which whole genomic sequence is available, must also be considered. Currently 11 fully sequenced genomes are available [38]. Comparison of these confirmed the similarity between *R. felis* and *R. akari* and their dissimilarity with other members of the SFG. Indeed, this divergence has resulted in the use of a new term, "transitional group" (TRG), to include these divergent species [39].

The arthropod associations of various Rickettsiae are diverse and varied. The specificity of these relationships has been called into question with the finding of the normally louse-borne *R. prowazekii* in arthropods primarily feeding upon flying squirrels [40]. These pathogens are usually highly adapted to transmission by their arthropod vectors, many showing highly efficient transovarial transmission. Others appear to exist as endosymbionts within their vector with no documented deleterious effects upon vertebrate hosts [41, 42]. Indeed, some have evolved to become the essential mycetomic partner necessary for the survival of their arthropod host [4]. It has been postulated that mitochondria essential for the survival of mammalian cells have their origins within the Rickettsiaceae [43]. Some *Rickettsia* spp. however have a noticeable detrimental effect upon the survival of their arthropod vectors, typified by the "red louse" seen following infection of these vertebrates with *R. prowazekii* [44]. Other rickettsial species display differential effects upon cell lines derived from different tick species that could account for the competence of ticks for their transmission [41]. Interestingly, different rickettsial species coexisting in the same environment have shown a competitive exclusion effect, with *R. peacockii* showing the ability to interfere with the multiplication of the pathogenic *R. rickettsii* in *Dermacentor andersoni* ticks [31].

8.3 Phylogenetic Classification of Rickettsiae

Rickettsioses are caused by bacteria in the order *Rickettsiales*, family *Rickettsiaceae*, genus *Rickettsia*. Historically, the order *Rickettsiales* was divided into three families, namely: *Rickettsiaceae*, *Bartonellaceae*, and *Anaplasmataceae*. The family *Rickettsiaceae* was made up of three tribes: *Rickettsieae*, *Ehrlichieae*, and *Wolbachieae*. The genera *Coxiella*, *Rickettsia*, and *Rochalimaea* belonged to the tribe *Rickettsieae*. Following recent taxonomic studies based on 16S rRNA the two genera *Coxiella* and *Rochalimaea* were removed from the tribe *Rickettsiaceae*, leaving the stand-alone genus *Rickettsia* [45].

Traditionally, members of the genus *Rickettsia* were divided into SFG, TG, and the scrub typhus group (STG) on the basis of phenotyphic characteristics [6, 34]. The monophyletic STG containing *R. tsutsugamushi* was subsequently transferred into the genus, *Orientia* [46]. More recently phylogenetic analyses including some of the newer members described within this genus and those utilizing comparative genomic methods based on alignment of orthologous groups have produced a more robust phylogenetic reconstruction revealing four biotypes/lineages. These include: (i) the TG comprising *R. typhi* and *R. prowazekii*, (ii) the SFG comprises over 20 species with *R. conorii*, *R. africae*, and *R. rickettsii* being among them, (iii) the TRG which consist of *R. felis* and *R. akari*, and (iv) the ancestral group (AG) which has *R. bellii* and *R. canadensis* [2, 39]. Inclusion of *Rickettsia* spp. with close association with their arthropod vectors, but little clinically relevant impact has made us rethink our understanding of this group of organisms. *Rickettsia* spp. have been found in several insect species that could impact on our current taxonomic thinking [47].

8.3.1 Typhus Fever Group

The TG contains probably the most notorious members of the Rickettsiae. This group cause typhus in humans and comprises of *R prowazekii* and *R. typhi*. Historically, *R. prowazekii* is believed to have accounted for over 30 million cases with three million deaths during epidemic outbreaks of louse-borne typhus prevalent following the end of World War I. Over the years the associated illness has maskaraded under several names including "*Fleckfieber*" (German), "*typhus exanthematique*" (French), and "*tabardillo*" (Spanish). More recently, *R. prowazekii* has shown a resurgence affecting over 45 000 refugees in one year alone in Burundi [48]. This serves as a stark reminder against complacency for control of rickettsial disease, especially during times of socio-political upheaval.

R. prowazekii, unlike the other Rickettsiae, is transmitted from person to person by the human clothing louse *Pediculus humanus humanus*. The louse (see Figure 8.1) is highly host-specific; however over period of years can adapt to feed from rabbits, facilitating the establishment of models to study microbial vector interactions. These models have been instrumental in establishing that transmission occurs

(a)

(b)

Figure 8.1 The human clothing louse, *Pediculus humanus*, vector of *R. prowazekii*. (a) Adult lice and larval louse. (b) Larval louse and eggs cemented to clothing.

not only through the crushing of lice into skin abrasions, but also through aerosolized louse feces or the inoculation of louse fecal material [44]. Once these Rickettsiae gain entry to the louse, they promote destruction of the louse gut epithelium, leading to the appearance of a "red louse" and concomitant death of the louse vector. In contrast to other arthropod–*Rickettsia* interactions, there is no transovarial transmission, thus propagation of this infection is entirely dependent upon acquisition of the Rickettsiae from an infected host and its subsequent transmission. Lice have a strong temperature preference, thus the onset of fever facilitates their rapid dispersal to new hosts, promoting epidemic person to person transmission. Additionally, *R. prowazekii* is able to establish a latent infection within their host, despite apparent clinical recovery and strong host immunity. This latent infection can reactivate causing a recrudescent infection, known as *Brill–Zinsser*, many years after the initial exposure. Thus these individuals can serve as a source for re-emergence of infection during times conducive for lice [49].

Unlike the other rickettsial microorganisms, humans are the primary reservoir of *R. prowazekii*. The bacterium is transmitted to humans by the body louse, *Pediculus humanus humanus*, a strict human parasite, living and multiplying in clothing [50]. When an infected louse bites a human, it defecates and the bacteria are found in the feces. Irritation caused by the bite causes the person to scratch the bite and thereby to inoculate the bacteria into abraded skin. Transovarian transmission in the louse does not occur since lice die several weeks after being infected. Another explanation for the fact that *R. prowazekii* is not transovarially spread in its vector might be its intracellular motility [50]. Whereas *R. typhi* keeps a part of its intracellular motility, *R. prowazekii* does not. Thus, the latter is the only *Rickettsia* species unable to be transmitted transovarially to its progeny within its arthropod host. More recently possible slyvatic cycles for *R. prowazekii* have been postulated in both ticks in Mexico [51] while, in the United States, flying squirrels have been documented as reservoirs for *R. prowazekii* [52].

Rickettsia typhi causes murine typhus and is transmitted by rat fleas (*Xenopsylla cheopis*). The vertebrate reservoirs are *Rattus norvegicus* and *R. rattus* both of which frequently dwell close to humans. Murine typhus has a global distribution, however most cases arise from those locations with more humid climates [9, 53]. It is believed that the mild clinical course of this infection results in a vast underestimation of case burden [9]. Estimates suggest that up to 10% of untreated cases require hospitalization, with an associated mortality of up to 2% [54]. Transmission to humans occurs through contaminated skin, conjuctiva, or via the respiratory route by aerosols of contaminated flea feces [6], or through the flea bite itself. Intriguingly, *R. typhi* can be transmitted by other vertebrate vectors experimentally; however, it remains to be established whether these alternative vectors could serve a significant role in the ecology of this organism [55].

The TG Rickettsiae contain, among many other proteins and genes, an OmpB, a Sca protein family which is a surface array protein similar to that present in other Gram-negative bacteria, a gene D, and a gene that encodes a 17-kDa predicted lipoprotein. These features are usually used for molecular typing of the TG agents. Collectively this group lack RickA, an essential prerequisite for the actin-based motility seen among the SFG of Rickettsiae.

8.3.2
Spotted Fever Group

The SFG is associated with spotted fever in humans. This group comprises of over 20 different species [6]. Characteristically this group are transmitted by various tick species that act as reservoirs for rickettsial infection being maintained through transovarial and transstadial transmission [56].

Rickettsia conorii causes MSF, also known as *botonneuse fever*, *Marseilles fever*, or *escaro-nodular fever* [11]. It is transmitted by ticks in the genus *Rhipicephalus* (mainly the brown dog tick). *R. conorii* appears to be heterogeneous, containing several serotypes and genotypes leading to the proposal that this species be divided into subspecies [57, 58].

Meanwhile, in the United States, *R. rickettsii* causes RMSF and is transmitted by *Ixodes* ticks. Reservoirs include rabbits, rodents, dogs, and birds [29].

Rickettsia africae causes ATBF and is transmitted by particularly aggressive ticks of the genus *Amblyomma* (*A. variegatum, A. hebraeum*). Other SFG *Rickettsia* spp. reported in Africa include *R. aeschlimanii, R. sibirica* subsp. *mongolotimonae* [6, 20].

8.3.3
Transitional Group Rickettsiae

Members of TRG Rickettsiae were previously grouped together in the SFG, but use various arthropod vectors. *R. felis* has been isolated in cat fleas (*Ctenocephalides felis*) [25] while the vector for *R. akari*, the cause of rickettsialpox, are mites (*Liponyssoides sanguineus*) [59].

R. felis has challenged the previously existing phylogenetic understanding of the Rickettsiae as it displays genotypic and phenotyphic characteristics of both the spotted fever and typhus fever groups. These manifested in serological cross-reactivity, transovarial maintenance in vectors, actin-based motility, hemolytic activity, and association with insects [39]. *R. felis* has been the source of much research effort over recent years, disclosing a global prevalence for this species. As it is effectively transmitted transovarially, the requirement for vertebrate maintenance hosts remains uncertain, however flea–opossum cycles have been described [60]. The clinical significance is similarly unclear. Despite clinically documented cases from certain regions, many others fail to see cases even in the presence of infected fleas and human exposure [26]. Furthermore, its vector specificity is less stringent than many Rickettsiae, with recent reports of *R. felis* in ticks [61].

Intriguingly, *R. felis* was also found to harbor plasmids (pRF) and conjugative pili [62], subsequently corroborated among other members of the genus *Rickettsia*. It was hypothesized that members of the AG of Rickettsiae once contained functional plasmids that might have been lost through evolutionary decay or that plasmidless *Rickettsia* may have incorporated once plasmid-encoded genes within their chromosomes [39]. The polyphyletic nature of the pRF and chromosomal genes in *R. felis* suggests that the pRF genes may not have been vertically passed over time, but rather were inherited horizontally from other sources [39]. Acquisition of plasmids as a result of lateral gene transfer is further supported by possession of a type IV secretion system (T4SS), present in all Rickettsiae [63].

8.3.4
Ancestral Group Rickettsiae

This group comprises *R. bellii* and *R. canadensis*. This group is speculated to have diverged prior to the separation of the TG and SFG. It is basal in location to other pathogenic Rickettsiae in the phylogenetic tree [39]. *R. bellii* is found in both soft and hard ticks and exhibits the largest arthropod host range among known members of the genus *Rickettsia* [64]. Its large size (1 522 076 bp) is comparable to that of *R. felis* (1 587 240 bp) [39] and the early branching in phylogenic position may suggest that it might have retained ancestral features lost in other Rickettsiae

[64]. A conjugative sex pili-like appendage has been documented to occur in *R. bellii* that could allow conjugative DNA transfer.

The differences in the gene assemblages across the rickettsial genome are mainly due to differential gene losses from the ancestors. The *Rickettsia* genus is an excellent paradigm for understanding the process of reductive evolution. The number of genes varies within the genus *Rickettsia*, with the TG having the fewest genes (both *R. prowazekii* and *R. typhi* having a predicted 877 genes). *R. felis* has the highest number of genes, followed by *R. bellii* [39]. This variation in genome size is likely to reflect differential gene loss (and limited acquisition) from different rickettsial lineages [65]. Evolutionary reductive gene loss when compared with ancestral counterparts is well documented among other organisms, with large gene deletions hypothesized to have occurred soon after these microbes acquired their endosymbiotic lifestyle. Many of these gene losses are among those required for basic cellular processes and biosynthesis of metabolic intermediates, no longer necessary as a result of their intracellular lifestyle.

8.4 Diagnosis

Rickettsial diagnosis is based on both clinical symptoms and epidemiologic history, with laboratory confirmation provided by the development of specific convalescent phase antibodies, a positive PCR result, immunohistological detection, and rare isolation of a rickettsial pathogen. A preliminary diagnosis of murine typhus is based on compatible clinical signs, such as fever, headache, and rash in a patient within an endemic zone, particularly who remembers a flea contact. This presumptive diagnosis can be confirmed by serology [23]. *R. typhi* can be detected in rash eschar biopsy, blood, and in fleas by cultivation in a specialized laboratory with containment level 3 security, on cell cultures [43], by immunohistochemistry with poly- or monoclonal antibodies, and by PCR. Similarly, the SFG rickettsiosis can also be diagnosed by characteristic rash and scar, accompanied by the non-specific signs of rickettsioses and the aforementioned laboratory confirmatory methods. Molecular amplification with PCR targets different genes (citrate synthase *gltA*, *ompB*, gene D; see Section 8.4.2) [40, 46, 66, 67].

8.4.1 Clinical Diagnosis

Clinical signs of rickettsioses may vary depending on the rickettsial species involved and therefore are not pathognomonic. Furthermore, diagnosis might additionally be complicated by the emergence of a new species of unknown pathogenicity and clinical outcome that cross-reacts in serological diagnostic assays. Characteristically the tick-borne SFG rickettsioses present with fever and headache, often accompanied by a characteristic inoculation eschar at the site of tick bite. Characteristic rickettsial rash is only present in 60–80% of patients. As seen with SFG rickettsioses, murine typhus (caused by *R. typhi*) is associated with fever, headache, and rash; however the rash can be nonspecific and only apparent in half of the patients. In contrast, the onset of disease due to *R. prowazekii* infection (epidemic typhus) is severe. Patients present with high fever, headache, severe myalgias, and rash that can be purpuric. Fatality can be high, approaching 10–30% of patients depending on host factors such as underlying diseases [6].

8.4.2 Laboratory Diagnostics

Diagnosis of rickettsioses relies on the detection of the pathogen by isolation or molecular methods, or indirectly through the hosts' production of specific antibodies to infection. Conventional bacteriological diagnostic approaches cannot be applied to Rickettsiae because of their fastidious nature and strictly intracellular growth [68].

Laboratory support is invaluable for confirmation of rickettsial diagnosis but harmonized diagnostics are still lacking [69]. For most clinical laboratories, assays for antibody detection are the only tests performed. On a cautionary note, it must be remembered that early treatment may abrogate antibody production. These assays include the "gold standard" indirect immunofluorescence assay, indirect immunoperoxidase assay, latex agglutination, enzyme immunoassay, reverse line blotting, Western immunoblotting, and rapid flow assays. Although still used in developing countries, the classical Weil–Felix test is no longer recommended because of its poor sensitivity and specificity [10, 69]. Detection of antibodies becomes useful during the second week of illness and blood should be collected early during the course of disease and where possible later samples additionally collected. Paired samples are important to confirm increased IgM- and/or IgG-antibody titers in acute and convalescent sera. Interpretation of serological results is complicated by cross-reactivity encountered between the SFG Rickettsiae and those belonging to the TG Rickettsiae and even between pathogens of other bacterial genera. Epidemic and endemic typhus cannot yet be differentiated by serology, and similarly, SFG Rickettsiae are significantly cross-reactive, requiring that either Western blot and/or cross-adsorptions of sera are done [10]. So far, most of the aforementioned assays are not commercially available and restricted to a few reference laboratories worldwide [70].

The culture of Rickettsiae is demanding and laborious, but remains the most definitive diagnostic method. Furthermore, it is a prerequisite for the characterization of rickettsial species and delineation of rickettsial diseases. With the exception of the isolation of *O. tsutsugamushi*, cell cultures (Vero, L929, HEL, XTC-2, MRC5) have been used to isolate Rickettsiae from human samples and arthropods [10, 69, 71]. Best isolation results have been achieved using skin biopsy specimens (preferably taken at the margin of an eschar), heparin- or citrate-anticoagulated plasma, buffy coat, and even arthropods [69, 72]. Overnight storage of samples at 4 °C prior to culture may severely reduce the isolation of Rickettsiae from human samples [72]. The centrifugation shell vial technique originally has been adjusted

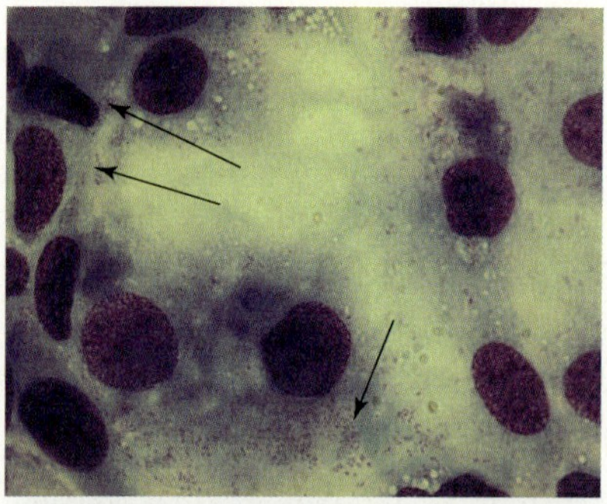

Figure 8.2 *R. africae* (arrows) grown in L929 cells isolated at the Bundeswehr Institute of Microbiology. Gomori stain, magnification ×1000.

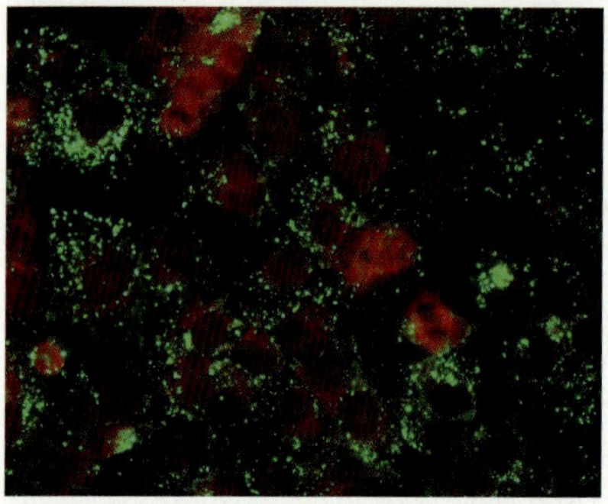

Figure 8.3 *R. africae* (arrows) grown in Vero cells isolated at the Bundeswehr Institute of Microbiology. Immunofluorescence staining, magnification ×400.

to culture *R. conorii* [73] but proved to be an effective method also for the isolation of other Rickettsiae [10, 74]. Detection of rickettsial agents within the cell cultures can be achieved microscopically using various cell-staining methods, for example, Romanowsky–Giemsa, Gimenez, Gomori (Figure 8.2); immunofluorescence (Figure 8.3), or molecular–biological methods. Isolation requires at least 7–14 days in antibiotic-free cell cultures and a culture is considered negative if immunofluorescence remains negative after 20 days of incubation [72].

The development of molecular methods for the detection of Rickettsiae in clinical and biological specimens has proven to be a rapid, sensitive, and specific diagnostic procedure. Indeed, this has resulted in the description of several new rickettsial species within the last decade, highlighting the need for reliable tools to infer robust phylogenetic analysis and reliable classification within the group Rickettsiae. Various DNA-based techniques have been developed for species identification and differentiation. Several PCR primer sets are commonly used, for example, targeting the 16S rRNA gene, a 17-kDa protein, citrate synthase (*gltA*), *ompA* and *ompB*, gene D (*sca4*), surface cell antigen 1 (*sca1*), and the 60-kDa heat shock protein (*groEL*) [67, 75]. Selected primer sets are given in Table 8.1.

Although frequently used in species identification, differentiation of *Rickettsia* based on 16S rRNA gene sequences lacks discriminatory power at species level [79]. Comparison of *gltA* sequences is complementary to 16S rDNA sequence analysis but also lacks discriminatory power [34]. The *sca* family genes have been most useful for species identification. Of these, *ompA* encodes an omp specific to the

Table 8.1 Oligonucleotide primers used for PCR amplification of rickettsial species.

Genomic target	Primer	References
16S rRNA	Ric: TCTAGAACGAACGCTATCGGTAT	[76]
	Ric U8: TGCGTTAGCTCACCACCTTCAGG	
gltA	CS1d: ATGACTAATGGCAATAATAA	[34]
	RpCS1258: ATTGCAAAAAGTACAGTGAACA	
ompA	190-3588: AACAGTGAATGTAGGAGCAG	[35]
	190-4406: ACTATACCCTCATCGTCATT	
	190-4338: TTCAGGAAACGACCGTACG	
	190-5238: ACTATTAAAGGCTAGGCTATT	
	190-5125: GCGGTTACTTTAGCCAAAGG	
	190-6092: AGATCATCGCTAACGAAGCC	
	190-5917: TCAGGGAATAAAGGTCCTG	
	190-6808: CACGAACTTTCACACTACC	
"Suicide PCR"	AF1F: CACTCGGTGTTGCTGCA	[16]
	AF1R: ATTAGTGCAGCATTCGCTC	
	AF2F: GCTGCAGGAGCATTTAGTG	
	AF2R: TATCGGCAGGAGCATCAA	
	AF3F: GGTGGTGGTAACGTAATC	
	AF3R: CGTCAGTTATTGTAACGGC	
	AF4F: GGAACAGTTGCAGAAATCAA	
	AF4R: CTGCTACATTACTCCCAATA	
ompB	120-2113: CGATGCTAACGTAGGTTCTT	[77]
	120-2988: CCGGCTATACCGCCTGTAGT	
	120-2788: AAACAATAATCAAGGTACTGT	
	120-3599: TACTTCCGGTTACAGCAAAGT	
	120-3462: CCACAGGAACTACAACCATT	
	120-4879: TTAGAAGTTTACACGGACTTTT	
sca4	D1f: ATGAGTAAAGACGGTAACCT	[78]
	D1390r: CTTGCTTTTCAGCAATATCAC	
	D1219f: CCAAATCTTCTTAATACAGC	
	D2482r: CTATAACAGGATTAACAGCG	
	D2338f: GATGCAGCGAGTGAGGCAGC	
	D3069r: TCAGCGTTGTGGAGGGGAAG	
groEL	RF: GTTGAAGTWGTCAAAGG	[75]
	RR:TTTTTCTTTWTCATAATC	

SFG Rickettsiae, but is not possessed by all representatives of the genus *Rickettsia* [77]. Roux and Raoult [77] were able to determine *ompB* sequences in all *Rickettsia* spp. investigated except for *R. bellii* and *R. canadensis.*

At present, *sca1* is the only gene allowing amplification, sequencing, and thus identification of *Rickettsia* spp. belonging to the SFG, TG, and the so-called AG [36]. A duplex PCR assay based on *groEL* has been designed for the identification of both *O. tsutsugamushi* and SFG Rickettsiae [80]. In addition, PCR-coupled restriction fragment length polymorphism (PCR-RFLP) has been proven to be sensitive and practical when applied to *sca4*, *ompB*, and *ompA*. Recently, PCR-RFLP based on *sca4* has been successfully applied for epidemiological studies including tick samples [81]. Finally, identification of rickettsial isolates at species level and further differentiation of strains within selected *Rickettsia* spp. has been achieved by multi-spacer typing based on three intergenic spacers: *dksA-xerC*, *mppA-purC*, and *rpmE*-tRNA^{fMet} [82]. Recent advances in real-time PCR allow for increased sensitivity and speed in the detection of Rickettsiae, enabling epidemiological investigations in both ticks and for human diagnostics based on a pan-*Rickettsia* PCR protocol for amplification of the *gltA* gene using primer PanRick_2_for 5'-ATAGGACAACCGTTTATTT-3', PanRick_2_rev 5'-CAAACATCATATGCAGAAA-3', and probe PanRick_3_taq 5'-FAM-CCTGATAATTCGTTAGATTTTACCG-TMR-3' [83].

8.5 Pathogenesis

Pathogenesis of the Rickettsiae is dependant upon several stages, including its arthropod interactions and those with the host endothelial cells such as entry, phagosomal escape, subsequent actin-based motility, and consequent spread to adjacent cells, cumulating in injury of the host. The best studied and most notorious members of this genus with regard to infective potential in vertebrate hosts are *R. prowazekii* (the cause of epidemic typhus), *R. typhi* (flea-borne murine typhus), and *R. felis* (a clinically similar flea-borne infection). Some of the best studied tick-borne SFG Rickettsiae are *R. akari* (the cause of rickettsial pox), *R. rickettsii* (RMSF), *R. conorii* (Boutonneuse fever), and *R. sibirica* (Asian tick typhus).

The genes determining the pathogenic mechanisms of rickettsial diseases may not be fully determined. About 42 genes have been identified to be involved in the virulence of rickettsial pathogens [84, 85]. Some genes affecting the pathogenesis of rickettsial diseases are given in Table 8.2, however the precise mechanisms of causing host damage remain poorly understood.

Rickettsiae introduced through the skin spread via the lymphatics and blood vessels, where the Rickettsiae display a distinct preference for the endothelial cells lining small blood vessels [86]. The tropism for endothelial cells is poorly understood. The main mechanism through which Rickettsiae asserts their pathogenic activity appears to be mediated through an increased microvascular permeability, with resulting leakage of blood into the tissue spaces. Generalized vascular injury follows, resulting in the activation of clotting factors, extravasation of fluid, edema, hypovolemia, hypotension, and hypoalbuminemia. These effects can accumulate, causing widespread vascular dysfunction, cerebral, and non-cardiogenic pulmonary edema. Several mechanisms are thought to result in reactive oxygen species (ROS) and down-regulation of the enzymes involved in protection against oxidative injury [66, 87]. Furthermore, phospholipase A of both *R. prowazeki* and *R. conorii* are cytotoxic for vero cells *in vitro*. It is thought to mediate entry into the host cell and cause cell injury [88]. Although the role of phospholipase A may not be fully

Table 8.2 Some rickettsial genes involved in pathogenesis [85].

Rickettsial gene	Product	Hypothesized function
Pat1	Patatin B1 precursor	Membrane phospholipase a host cell escape
tlyA	Hemolysin A	Membrane traversal of host cell membrane
thyC	Hemolysin C	Phagosomal membrane escape
pld	Phospholipase D	Phagosomal membrane escape
invA	Dinucleoside polyphosphate hydrolase	Hydrolysis of toxic dinucleoside pholyphosphates to ATP
coxABC	Cytochrome oxidase	Aerobic respiration
cydAB	Cytochrome d oxidase	Aerobic respiration (low O_2 concentrations)
sodB	Superoxide dismutase	Neutralization of reactive oxygen species
Lipopolysaccharide synthesis genes	Lipopolysaccharide	Endotoxin-mediated inflammation
Sca genes	Surface cell antigens (except *sca4* – intracellular)	Autotransporter outer membrane proteins
ompA	Outer membrane protein A	SFG rickettsial attachment protein
ompB	Outer membrane protein B	Rickettsial attachment protein
virB4, virB6, virB7, virB8, virB9, virBro, virB11	Type IV secretion system	Transport of rickettsial proteins/DNA
rickA	Actin tail polymerization genc	Actin-mediated polymerization

determined, it is hypothesized that it mediates escape from phagosomal vacuoles of the host cell [50].

Entry of Rickettsiae into the host cell is an active process requiring the recruitment of Ku70 (a component of DNA-dependent protein kinase present in cholesterol-rich lipid rafts) to the plasma membrane whereby it is able to interact with OmpB on the rickettsial cell surface [85]. This serves as a trigger resulting in a biological cascade including GTPase, Cdc42, phosphoinositide 3-kinase, protein tyrosine kinase (src family), and tyrosine phosphorylation of focal adhesion kinase (FAK), resulting in rapidly induced phagocytosis of Rickettsiae to within the intracellular phagosome. Rapid escape into the cytoplasm occurs prior to phagolysosomal fusion. This process is mediated by the up-regulation of genes encoding for membranes with membranolytic activity like *tlyC* (hemolysin) and *pld* (phospholipase D) [89].

Once the Rickettsiae escape the confines of the phagosome and enter the cytoplasm, they are able to harness the host cell's actin polymerization machinery to propel the Rickettsiae through the cytosol, facilitating the infection of adjacent cells. The RickA associated products play an essential role in this process through binding actin monomers and activating the Arp2/3 complex essential for the polymerization of "actin tails." *R. prowazekii* is unable to move intracellularly by using actin polymerization. Neither the Madrid E nor Breinl strains exhibit actin-associated motility. In contrast, *R. typhi* does display polar F-actin projections, but of a morphology different from that associated with SFG organisms. F-actin tails are very short and hook-shaped, never exceeding a few micrometers in length [70].

The infection foci enlarge as the Rickettsiae spread from cell to cell, forming a continuous network of infected endothelial cells in the microvasculature of the dermis, brain, lungs, liver, and other visceral organs and tissues [90].

8.5.1
Clinical and Pathological Findings

The route of entry is often mediated through the bite of an infected tick. This is usually followed by the development of a skin lesion known as an *eschar*. This may not always be obvious, particularly if covered by hair. For those *Rickettsia* spp. using other vectors, for example, flea-borne *Rickettsia* spp., these are transmitted through the contamination of irritated, abraded skin, and mucosal surfaces by infected vector feces during or after blood feeding [29]. The bacterium is then hematogenously spread and ultimately invades endothelial cells [25]. The transmission can also occur via the inhalation of aerosolized fecal particles. A similar scenario is utilized by the louse-borne Rickettsiae.

Of the 25 fully validated rickettsial species, 17 are known to have pathogenic potential [91]. Pathogenicity varies significantly between these species as does the extent of genomic decay found between these obligate intracellular pathogens. It has been hypothesized that those species associated with greater severity are those in which genetic decay is more advanced, potentially resulting from a loss of regulatory genes [91].

Varying manifestations of rickettsial disease continue to be described since its discovery. The spectrum of rickettsial infections differs from mild to severe fatal illness. The incubation period is 2–14 and 2–16 days for *R. rickettsii* and *R. conorii*, respectively, after the tick bite [12, 92].

Generally, the initial signs and symptoms of rickettsial illnesses in humans include: fever, headache, myalgia, chills, sweats, and cutaneous eruptions (eschars) [15, 86, 90]. Other reported symptoms include: lymphadenopathy [10], neurological signs [33], nausea, vomiting, and epistaxis. Neurologic signs noted in an epidemic typhus outbreak in Burundi included: mental confusion, seisures, delirium, and coma [93].

R. typhi is one of the leading causes of rickettsioses in the world and its minimum infectious dose is less than 10 organisms. Following an incubation of 6–14 days, those infected develop fever, headache, and myalgia, followed by a disseminated, multisystem disease, affecting the brain, lung, liver, kidney, and heart endothelia, lymphohistocytic vasculitis of the central nervous system, diffuse alveolar damage and hemorrhage, interstitial pneumonia, pulmonary edema, interstitial myocarditis and nephritis, portal triaditis, and cutaneous, mucosal, and serosal hemorrhages [19, 25, 82]. The nonspecificity and non-uniformity of symptoms and the lack of specific diagnostic tests that are effective during the acute stage of the illness often lead to misdiagnosis and a delay in appropriate treatment. Without the specific treatment, 99% of those infected clear the disease within weeks, making a proper accounting of *R. typhi* infection difficult [63]. The mortality rate for all ages is ~2% and increases with patient's age [39].

Epidemic typhus *R. prowazekii* usually manifests after an incubation period of 10–14 days, with high fever, headaches, and skin rash in patients with body lice or in persons who are living in crowded, cold, and unhygienic circumstances. Cough, pneumonia, diarrhea, splenomegaly, and conjunctivitis may also be observed. The disease is fatal in 10–30% of patients [50]. *R. prowazekii* is able to persist as a latent infection within human hosts, with potential for recrudescent illness called the *Brill–Zinsser disease*. The symptoms of recrudescent typhus are less pronounced than the initial infection and the mortality rate of Brill–Zinsser disease is less than 1% [39], but importantly, recrudescent typhus can serve as a long-term source of *R. prowazekii*. The skin rash is rarely seen. *R. prowazekii* has been isolated from sick dogs and humans in Northern Carolina. Unsurprisingly, these isolates infecting both dogs and people displayed high homology [94].

A rash, though considered the classic distinguishing characteristic of some rickettsial diseases like RMSF, may not appear until three days or more into the illness. The rash may not appear at all in a significantly great proportion of RMSF cases, or some patients may have a few cutaneous lesions that are only detectable after a meticulous examination [95]. About 50% of the patients suffering from murine typhus may develop a nonspecific rash [6].

A similar clinical picture has been noticed in tick-borne lymphadenopathy (TIBOLA), also known as *Dermacentor*-borne necrosis erythema and lymphadenopathy (DEBONEL). TIBOLA/DEBONEL is an emerging tick-borne rickettsiosis caused by *R. slovaca* and *R. raoultii*. An eschar, alopecia, and cervical lymphadenopathy are some peculiar clinical signs associated with the latter [22].

Rickettsia sibirica mongolotimonae has been associated with a specific lymphagitis that extend from the inoculation eschar to the draining lymph node [20].

After inoculation of the pathogen, attachment, spread, and growth, the rash develops in several different stages. The damage in the tissue begins as erythematous macules measuring 1–5 mm that frequently start on the ankles and wrists. Thence they spread to trunk, palms, and soles. The macules consist of areas of vasodilation that blanch on pressure and later become maculopapular because of extravasations of fluid from the damaged vessels. Severe illnesses lead to more pronounced endothelial destruction resulting in pinpoint areas of hemorrhage (petechiae) at the center of the maculopapule [90]. The events described herein may occur in all affected tissues, including the brain and lungs, which may be catastrophic.

Increased permeability of the vascular endothelium due to damage to the vascular lining may lead to changes in endothelial function. Pathophysiologic events include altered blood chemistry and hematological picture. Serum biochemistry may reveal hyponatremia, thrombocytopenia, increased alkaline phosphatase, lactate dehydrogenase, and serum glutaminoxalo acetic transaminase [12]. Similar changes have been documented in dogs suffering from RMSF.

Involvement of the pulmonary microvascular endothelium may result in noncardiogenic edema and interstitial pneumonia which are associated with adult respiratory distress. Congestive heart failure may develop, particularly in older patients, and is manifested as arrhythmias. Neurological signs occurring as a result of encephalitis may be manifested as lethargy, confusion, seizures, other neurological signs may vary from cranial nerve palsies, hemiphlegia, and paraplegia to complete paralysis. Dogs suffering from RMSF manifest neurological signs too. These signs may be a result of encephalitis and edema. These neurological signs together with adult respiratory distress are the major determinant of a fatal outcome [90]. *R. felis* has been associated with neurological signs and severe consequences [28].

Increased vascular permeability may lead to acute renal failure as a result of reduced glomerular filtration rate and hypovolemia. Acute tubular necrosis occurs as sequelae to prerenal azotemia and severe hypotensive shock. Focal hepatic necrosis has been documented which leads to an elevated hepatic enzyme concentration. There is involvement of skeletal muscles manifested as myalgia as a result of skeletal muscle cell necrosis. This is associated with elevated serum creatinine kinase concentration [90].

Major complications and long-term health problems associated with rickettsial illnesses (with more emphasis on RMSF) include renal involvement, paralysis of the lower extremities, gangrene formation that may require amputation of limbs, loss of bladder and bowel control, hearing loss, peripheral neuropathy, and cerebral, vestibular, and motor dysfunction [33, 96]. The mortality rate can be as high as 20–25% for some rickettsial diseases if left untreated [6]. Risk factors associated with mortality in patients with RMSF or MSF include: being more than 60 years, male, absence of a headache, lack of history of a tick bite (thus often a delayed treatment), and gastro-intestinal symptoms [12].

Some rickettsial diseases are peculiar in their own way. For instance the TIBOLA/DEBONEL caused by *R. slovaca* and *R. raoultii* have been shown to occur more in women and children. Furthermore it is not known why *Dermacentor marginatus* and *D. reticularis* prefer the scalp except for speculation that they may infest the scalp of women and children because of the long hair maintained by the latter [97].

8.6 Epidemiology

Our epidemiological knowledge of the Rickettsiae is expanding, especially with the recent discovery of several newly recognized species. It was held that the main vectors for these agents were lice, fleas, and hard ticks, with particular species transmitted by each. *R. prowazekii*, the cause of epidemic typhus, is classically transmitted through clothing lice (see Figure 8.1). These once had a worldwide distribution, but are now restricted to areas of poverty and typically show resurgence during political upheavals. The mechanisms of transmission include crushing of the louse or its feces into bite wounds [44, 55]. Infection of the louse has a dramatic deleterious effect upon the louse with it showing significant reduction in its lifespan associated with disruption to the gut epithelium and associated leakage of blood into the body cavity giving the louse a red appearance [44]. Studies of the survival of *R. prowazekii* in louse feces have shown viability over several weeks [44]. More recently, *R. prowazekii* has been detected both in African ticks and in flying squirrels in the United States [40, 98]. Indeed, cases of sylavatic infection have been reported following contact with flying squirrels (and possibly their fleas) [98].

The flea-borne *R. typhi* has been shown to have potential for infection of lice, but as seen with *R. prowazekii*, this is rapidly followed by leakage of blood and death of this "red louse" [55]. Although not a typical vector for this species, and resulting in reduction in the life-span of this vector, the organism remained viable in excess of 80 days in louse feces [55]. Similar studies have also shown that lice could potentially serve as vectors for other typically tick-borne species, namely *R. rickettsia* and *R. conorii* [99].

Rickettsiae are associated with arthropods, which may act as vectors, reservoirs, and/or amplifiers in the life cycles of different Rickettsiae. Rickettsiae infect and multiply in almost all organs of their invertebrate hosts. Arthropods serving as vectors for the transmission of Rickettsiae include ticks, mites, lice, and fleas [101]. According to currents concepts, Rickettsiae infect and multiply in almost all organs of their invertebrate hosts. In ticks oocytes of an adult female tick may be infected and Rickettsiae subsequently transmitted transovarially to at least some of its offspring [102]. Infected offspring will excrete the Rickettsiae during their whole life and also from one to the next stage (transstadial transmission). Ticks transmit Rickettsiae with salivary glands directly into vertebrates during blood-feeding. As arthropods are essential for the life cycles of Rickettsiae their geographic distribution and the epidemiology are closely associated with the geographic distribution and the ecologic conditions of their vectors.

Single rickettsial species can have a strong association to one single vector (e.g., *R. conorii* and the brown dog tick, *Rhipicephalus sanguineus*) [47]. Conversely, a single species can also be transmitted by several different species (e.g., *R. rickettsii* in North America by *Dermacentor variabilis* and *Dermacentor andersoni*) [30]. One single tick of one species may even contain several different rickettsial species. This may have implications for the transmission of Rickettsiae to vertebrates. The co-infection of *Dermacentor andersonii* with *R. peacockii* and *R. rickettsii* leads to failure of transovarial transmission of *R. rickettsii* and its subsequent elimination from areas where *R. peacockii* is prevalent in ticks [41]. For European the epidemiological picture is less clear, but it is known that for example *R. helvetica* and *R. monacensis* simultaneously occur in *Ixodes ricinus* ticks in one area in southern Germany (Dobler, unpublished observation).

From a historical perspective, epidemic typhus (louse-borne typhus) caused by *R. prowazekii* is the most important rickettsiosis. During Greek and Roman times it is believed to have caused numerous outbreaks and deaths [1]. It is now known that a major part of Napoleon's army during the Russian war died of louse-transmitted diseases, and prominent among them was epidemic typhus. Originally, epidemic typhus was distributed worldwide. Now it is limited to areas of poverty or remote mountainous areas of the world (e.g., Andean mountains, Ethiopian highlands). It reappears during wars such as that in Burundi-Rwanda and also sporadically in Algeria [103]. Epidemic typhus is the only for of rickettsiosis known to become chronic in humans. After decades, a recrudescence of typhus can occur in patients with the potential of transmission to lice and the initiation of a new epidemic occurrence of typhus. The detection of a natural reservoir of *R. prowazekii* in North American squirrels (*Glaucomys volans*) implies another mode of persistence in nature and a potential for re-occurrence of epidemic typhus under favorable circumstances for the pathogen [104, 105].

The second rickettsiosis with worldwide distribution is murine typhus (endemic typhus). Caused by *R. typhi* this infection generally manifests as a milder form of typhus. *R. typhi* is transmitted by fleas, mainly by *Xenopsylla cheopis*, the rat flea, but also via cat fleas *Ctenocephalides felis* or the rat louse *Polyplax spinulosa* [106–108]. It has a reservoir in rodents (*Rattus rattus, R. norvegicus,* Mus *musculus*), but also from opossums (*Didelphis virginiana*) and domestic cats. The classical transmission cycle for murine typhus is rat–flea–rat and accidentally rat–flea–man. However, the classical cycle could be replaced by Virginia Opossum (*Didelphis virginiana*) and the cat flea cycle [108, 109]. *R. typhi* is maintained transovarially in fleas where it multiplies in the epithelium of the flea midgut. Fleas are capable of multiple feeding, and thus potentially transmitting the *Rickettsia* to several hosts.

Murine typhus is especially prevalent in coastal areas and harbors with high rat populations. Human cases of murine typhus are sporadically diagnosed in travellers returning from tropical and subtropical countries and should therefore be included within the differential diagnosis of febrile illness.

The third rickettsiosis that occurs worldwide is caused *R. felis*, the agent of flea-borne rickettsiosis, and, although transmitted by fleas (mainly *Ctenocephalides felis*), phylogenetically it has features of both the SFG that includes rickettsioses mainly transmitted by ticks and the TG. Subsequently whole genome analysis has suggested that this species falls within a TRG that falls between both the above groups. Recently, *R. felis* was also described in ticks, which suggests also a natural transmission cycles exist (or existed) and *R. felis* spilled over into fleas. Its global distribution could be attributed with that of rats and/or their fleas around the world.

Rickettsia akari is the only *Rickettsia* of medical importance in humans that is transmitted by mites. The rodent-associated mite (*Liponyssoides sanguineus*), naturally parasitises several rodent species, including the house mouse (*Mus musculus*) [110]. *Rickettsia akari* is the etiologic agent of rickettsialpox. Sporadic cases of rickettsialpox seem to occur worldwide, again probably reflecting the wide distribution by the house mouse around the world. Phylogenetically, *R. akari* is most closely related to *R. felis* and *R. australis* [39, 79].

In contrast to the rickettsioses above that occur worldwide, the tick-borne SFG of rickettsioses are only endemic to limited global areas, probably driven by preferred vector habitats (Table 8.3) [10]. These range from whole continents (e.g., *Rickettsia rickettsii* in North and South America) to little islands (*Rickettsia honei* on Flinders Island at the Australian coast). This is likely to arise from the limited

Table 8.3 Characteristics of Rickettsiae with human medical importance.

Rickettsial species	Disease	Vector	Geographic distribution
R. rickettsii	Rocky mountain spotted fever, Brazilian spotted fever	*Dermacentor andersoni*, *D. variabilis*	North America, Central America, and South America
R. conorii ssp. *conorii*	Mediterranean spotted fever;	*Rhipicephalus sanguineus*	Southern Europe, near Asia, central Asia, India, and northern Africa
R. conorii ssp. *israelensis*	Israelian spotted fever,		
R. conorii ssp. *caspia*	Astrakhan spotted fever;		
R. conorii ssp. *indica*	Indian spotted fever		
R. africae	African tick bite fever	*Amblyomma hebraeum*, *A. variegatum*	Sub-Saharan Africa, and Caribbean islands
R. helvetica	Aneruptive tick bite fever	*Ixodes* spp.	Europe and Asia
R. marpioni	Australian spotted fever	*Hyalomma* spp., *Ixodes* spp.	Australia
R. heilongjiangensis	Far eastern spotted fever	*Dermacentor* spp., *Haemaphysalis* spp.	Far eastern Russia and northern China
R. honei	Flinders Island spotted fever; Thai spotted fever	*Aponomma* spp., *Rhipicephalus* spp.	Australia and Thailand
R. sibirica ssp. *mongolotimonae*	Tick-borne lymphangitis (TIBONA)	*Hyalomma* spp.	Southern Europa, Asia, and Africa
R. parkeri	Macular fever	*Amblyomma* spp.	North America and South America
R. sibirica	Asian tick bite fever	*Haemaphysalis* spp., *Dermacentor* spp.	Russia, China, and Mongolia
R. japonica	Japanese spotted fever	*Hyalomma* spp., *Dermacentor* spp., *Ixodes* spp.	Japan
R. australis	Queensland spotted fever	*Ixodes* spp.	Australia and Tasmania
R. monacensis	Tick bite fever	*Ixodes ricinus*	Europe
R. massiliae	Tick bite fever	*Rhipicephalus* spp.	Europe
R. slovaca	TIBOLA	*Dermacentor* spp.	Europe and Asia
R. felis	Flea-borne spotted fever	*Ctenocephalides felis* (also some tick species)	Worldwide
R. typhi	Murine typhus	*Xenopsylla cheopis*, *C. felis*	Worldwide
R. prowazekii	Epidemic typhus	*Pediculus humanus* (lice), fleas of flying squirrel (*Glaucomys volans volans*)	Worldwide (United States)
R. akari	Rickettsialpox	*Liponyssoides sanguinus*	Worldwide

distribution range of tick species. However, *R. africae*, the etiologic agent of ATBF, that mainly occurs in sub-Saharan Africa, has been detected on some Caribbean islands suggesting that it was imported with the slave trade from Africa [111]. For many of the tick-transmitted Rickettsiae, only limited data on their distribution and pathogenic potential are available so far. Almost every year new candidates of rickettsioses or new areas of distribution for known *Rickettsia* species are reported. The spotted fever Rickettsiae are among the most important emerging infections. A list comprising some of the major rickettsial species, their pathogenic potential for humans and some epidemiological characteristics are listed in Table 8.3. A suitable vaccine is not available [100].

8.7
Conclusions

In this chapter we have reviewed our current understanding of this intriguing group of organisms, the Rickettsiae. Although we still have significant knowledge gaps, our understanding has advanced significantly in recent years, in part through the inclusion of *R. prowazekii* as a potential warfare agent (classified on the B list of bioterrorism agents by the Centers for Diseases Control and Prevention, Atlanta, Ga., USA). Likewise, it is on the list of high-risk biological agents in all European Union countries.

Once descendants of free-living organisms, these bacteria now have some of the smallest bacterial genomes and provide an insight into the adaptations to an obligate intracellular lifestyle. As a group, they are reliant upon their arthropod interactions, providing both a reservoir and a means of transmission. *R. prowazekii* is acquired and maintained via horizontal transmission involving both humans and human body lice. In contrast, flea-borne rickettsiosis is acquired and maintained via both horizontal and vertical transmissions.

Huge variation is seen clinically, ranging from the high rate of mortality from epidemic typhus (reaching 10–30%) through to the relatively benign murine typhus. Many newly detected species remain of uncertain clinical significance.

Despite the recent diagnostic advances, diagnosing rickettsial diseases remains challenging. The most widely used serological assays are plagued by intensive serological cross-reactions between species. Detailed proteomic analysis coupled with improved molecular diagnostics will enable us to develop better future diagnostics enabling rapid and precise diagnosis and clinical management of cases.

Acknowledgments

Authors S.P. and R.T. thank the Scientific Grant Agency of Ministry of Education of Slovak Republic for grants No. 2/0065/09 and 2/0127/09 Slovak Academy of Sciences and thank the for their support.

References

1. Raoult, D., Dutour, O., Houhamdi, L., Jankauskas, R., Fournier, P.E., Ardagna, Y., Drancourt, M., Signoli, M., La, V.D., Macia, Y. *et al.* (2006) Evidence for louse-transmitted diseases in soldiers of napoleon's grand army in vilnius. *J. Infect. Dis.*, **193**, 112–120.
2. Gillespie, J., Williams, K., Shukla, M., Snyder, E., Nordberg, E., Ceraul, S., Dharmanolla, C., Rainey, D., Soneja, J., Shallom, J. *et al.* (2008) Rickettsia phylogenomics: unwinding the intricacies of obligate intracellular life. *PLoS ONE*, **3**, e2018.
3. Gottlieb, Y., Ghanim, M., Chiel, E., Gerling, D., Portnoy, V., Steinberg, S., Tzuri, G., Horowitz, A.R., Belausov, E., Mozes-Daube, N. *et al.* (2006) Identification and localization of a rickettsia sp. In bemisia tabaci (homoptera: Aleyrodidae). *Appl. Environ. Microbiol.*, **72**, 3646–3652.
4. Perotti, M.A., Clarke, H.K., Turner, B.D., and Braig, H.R. (2006) Rickettsia as obligate and mycetomic bacteria. *FASEB.J.*, **20**, 2372–2374. doi: 10.1096/fj.06-5870fje
5. von der Schulenburg, J.H.G., Habig, M., Sloggett, J.J., Webberley, K.M., Bertrand, D., Hurst, G.D.D., and Majerus, M.E.N. (2001) Incidence of male-killing rickettsia spp. ({alpha}-proteobacteria) in the ten-spot ladybird beetle adalia decempunctata l. (coleoptera: Coccinellidae). *Appl. Environ. Microbiol.*, **67**, 270–277.
6. Raoult, D. and Roux, V. (1997) Rickettsioses as paradigms of new or emerging infectious diseases. *Clin. Microbiol. Rev.*, **10**, 694–719.
7. Hechemy, K.E., Avsic-Zupanc, T., Childs, J.E., and Raoult, D.A. (2003) Rickettsiology: Present and future directions: Preface. *Ann. N.Y. Acad. Sci.*, **990**, 1–xvii.
8. Raoult, D., Fournier, P., Eremeeva, M., Graves, S., Kelly, P., Oteo, J., Sekeyova, Z., Tamura, A., Tarasevich, I., and Zhang, L. (2005) Naming of Rickettsiae and rickettsial diseases. *Ann. N. Y. Acad. Sci.*, **1063**, 1–12.
9. Letaief, A. (2006) Epidemiology of rickettsioses in North Africa. *Ann. N. Y. Acad. Sci.*, **1078**, 34–41.
10. Parola, P., Paddock, C.D., and Raoult, D. (2005) Tick-borne rickettsioses around the world: Emerging diseases challenging old concepts. *Clin. Microbiol. Rev.*, **18**, 719–756.
11. Blanco, J. and Oteo, J. (2006) Rickettsiosis in Europe. *Ann. N. Y. Acad. Sci.*, **1078**, 26–33.
12. Raoult, D., Weiller, P.J., Chagnon, A., Chaudet, H., Gallais, H., and Casanova, P. (1986) Mediterranean spotted fever: clinical, laboratory and epidemiological features of 199 cases. *Am.J. Trop. Med. Hyg.*, **35**, 845–850.
13. Bitam, I., Parola, P., De La Cruz, K.D., Matsumoto, K., Baziz, B., Rolain, J.-M., Belkaid, M., and Raoult, D. (2006) First molecular detection of rickettsia felis in fleas from algeria. *Am.J. Trop. Med. Hyg.*, **74**, 532–535.
14. Raoult, D. and Parola, P. (2008) Rocky mountain spotted fever in the USA: a benign disease or a common diagnostic error? *Lancet Infect. Dis.*, **8**, 587–589.
15. Whitman, T.J., Richards, A.L., Paddock, C.D., Tamminga, C.L., Sniezek, P.J., Jiang, J., Byers, D.K., and Sanders, J.W. (2007) Rickettsia parkeri infection after tick bite, virginia. *Emerg. Infect. Dis.*, **13**, 153–157.
16. Raoult, D., Fournier, P.E., Fenollar, F., Jensenius, M., Prioe, T., de Pina, J.J., Caruso, G., Jones, N., Laferl, H., Rosenblatt, J.E. *et al.* (2001) Rickettsia africae, a tick-borne pathogen in travelers to sub-saharan africa. *N. Engl.J. Med.*, **344**, 1504–1510.
17. Parola, P., Vestris, G., Martinez, D., Brochier, B., Roux, V., and Raoult, D. (1999) Tick-borne rickettiosis in guadeloupe, the French West Indies: Isolation of rickettsia africae from amblyomma variegatum ticks and serosurvey in humans, cattle, and goats. *Am.J. Trop. Med. Hyg.*, **60**, 888–893.
18. Fournier, P., Grunnenberger, F., Jaulhac, B., Gastinger, G., and Raoult, D. (2000) Evidence of rickettsia helvetica

infection in humans, eastern France. *Emerg. Infect. Dis.*, **6**, 389–392.
19. Fournier, P.E., Tissot-Dupont, H., Gallais, H., and Raoult, D. (2000) Rickettsia mongolotimonae: a rare pathogen in france. *Emerg. Infect. Dis.*, **6**, 290–292.
20. Parola, P. (2006) Rickettsioses in sub-saharan africa. *Ann. N. Y. Acad. Sci.*, **1078**, 42–47.
21. Mura, A., Masala, G., Tola, S., Satta, G., Fois, F., Piras, P., Rolain, J., Raoult, D., and Parola, P. (2008) First direct detection of rickettsial pathogens and a new rickettsia, 'candidatus rickettsia barbariae', in ticks from sardinia, Italy. *Clin. Microbiol. Infect.*, **14**, 1028–1033.
22. Parola, P., Rovery, C., Rolain, J., Brouqui, P., Davoust, B., and Raoult, D. (2009) Rickettsia slovaca and r. Raoultii in tick-borne rickettsioses. *Emerg. Infect. Dis.*, **15**, 1105–1108.
23. Duh, D., Punda-Polic, V., Avsic-Zupanc, T., Bouyer, D., Walker, D.H., Popov, V.L., Jelovsek, M., Gracner, M., Trilar, T., Bradaric, N. *et al.* (2010) Rickettsia hoogstraalii sp. Nov., isolated from hard- and soft-bodied ticks. *Int. J. Syst. Evol. Microbiol.*, **60**, 977–984.
24. Matsumoto, K., Parola, P., Rolain, J., Jeffery, K., and Raoult, D. (2007) Detection of "Rickettsia sp. Strain uilenbergi" And "Rickettsia sp. Strain davousti" In amblyomma tholloni ticks from elephants in africa. *BMC Microbiol.*, **7**, 74.
25. Bouyer, D., Stenos, J., Crocquet-Valdes, P., Moron, C., Popov, V., Zavala-Velazquez, J., Foil, L., Stothard, D., Azad, A., and Walker, D. (2001) Rickettsia felis: molecular characterization of a new member of the spotted fever group. *Int. J. Syst. Evol. Microbiol.*, **51**, 339–347.
26. Perez-Osorio, C., Zavala-Velazquez, J., Arias Leon, J., and Zavala-Castro, J. (2008) Rickettsia felis as emergent global threat for humans. *Emerg. Infect. Dis.*, **14**, 1019–1023.
27. Richards, A., et al. (2010) Human infections with rickettsia felis, kenya. *Emerg. Infect. Dis.*, **16**.
28. Zavala-Castro, J., Zavala-Velázquez, J., Walker, D., Pérez-Osorio, J., and Peniche-Lara, G. (2009) Severe human infection with rickettsia felis associated with hepatitis in yucatan, Mexico. *Int. J. Med. Microbiol.*, **299**, 529–533.
29. Azad, A. and Radulovic, S. (2003) Pathogenic Rickettsiae as bioterrorism agents. *Ann. N.Y. Acad. Sci.*, **990**, 734–738.
30. Azad, A.F. (2007) Pathogenic Rickettsiae as bioterrorism agents. *Clin. Infect. Dis.*, **45**, S52–S55.
31. Baldridge, G.D., Burkhardt, N.Y., Simser, J.A., Kurtti, T.J., and Munderloh, U.G. (2004) Sequence and expression analysis of the ompa gene of rickettsia peacockii, an endosymbiont of the rocky mountain wood tick, dermacentor andersoni. *Appl. Environ. Microbiol.*, **70**, 6628–6636.
32. Balraj, P., Karkouri, K.E., Vestris, G., Espinosa, L., Raoult, D., and Renesto, P. (2008) Ricka expression is not sufficient to promote actin-based motility of rickettsia raoultii. *PLoS ONE*, **3**, e2582.
33. Aliaga, L., Sanchez-Blazquez, P., Rodriguez-Granger, J., Sampedro, A., Orozco, M., and Pastor, J. (2009) Mediterranean spotted fever with encephalitis. *J. Med. Microbiol.*, **58**, 521–525.
34. Roux, V., Rydkina, E., Eremeeva, M., and Raoult, D. (1997) Citrate synthase gene comparison, a new tool for phylogenetic analysis, and its application for the Rickettsiae. *Int. J. Syst. Bacteriol.*, **47**, 252–261.
35. Fournier, P., Roux, V., and Raoult, D. (1998) Phylogenetic analysis of spotted fever group Rickettsiae by study of the outer surface protein rompa. *Int. J. Syst. Bacteriol.*, **48**, 839–849.
36. Ngwamidiba, M., Blanc, G., Raoult, D., and Fournier, P. (2006) Sca1, a previously undescribed paralog from autotransporter protein-encoding genes in rickettsia species. *BMC Microbiol.*, **6**, 12.
37. Vitorino, L., Chelo, I.M., Bacellar, F., and Ze-Ze, L. (2007) Rickettsiae phylogeny: a multigenic approach. *Microbiology*, **153**, 160–168.

38. Merhej, V., El Karkouri, K., and Raoult, D. (2009) Whole genome-based phylogenetic analysis of Rickettsiae. *Clin. Microbiol. Infect.*, **15**, 336–337.
39. Gillespie, J., Beier, M., Rahman, M., Ammerman, N., Shallom, J., Purkayastha, A., Sobral, B., and Azad, A. (2007) Plasmids and rickettsial evolution: Insight from rickettsia felis. *PLoS ONE*, **2**, e266.
40. Foley, J.E., Nieto, N.C., Clueit, S.B., Foley, P., Nicholson, W.N., and Brown, R.N. (2007) Survey for zoonotic rickettsial pathogens in northern flying squirrels, glaucomys sabrinus, in california. *J. Wildl. Dis.*, **43**, 684–689.
41. Kurtti, T., Simser, J., Baldridge, G., Palmer, A., and Munderloh, U. (2005) Factors influencing in vitro infectivity and growth of rickettsia peacockii (rickettsiales: Rickettsiaceae), an endosymbiont of the rocky mountain wood tick, dermacentor andersoni (acari, ixodidae). *J. Invertebr. Pathol.*, **90**, 177–186.
42. Mattila, J., Burkhardt, N., Hutcheson, H., Munderloh, U., and Kurtti, T. (2007) Isolation of cell lines and a rickettsial endosymbiont from the soft tick carios capensis (acari: Argasidae: Ornithodorinae). *J. Med. Entomol.*, **44**, 1091–1101.
43. Emelyanov, V. (2001) Rickettsiaceae, rickettsia-like endosymbionts, and the origin of mitochondria. *Biosci. Rep.*, **21**, 1–17.
44. Houhamdi, L., Fournier, P., Fang, R., Lepidi, H., and Raoult, D. (2002) An experimental model of human body louse infection with rickettsia prowazekii. *J. Infect. Dis.*, **186**, 1639–1646.
45. Weisburg, W.G., Dobson, M.E., Samuel, J.E., Dasch, G.A., Mallavia, L.P., Baca, O., Mandelco, L., Sechrest, J.E., Weiss, E., and Woese, C.R. (1989) Phylogenetic diversity of the Rickettsiae. *J. Bacteriol.*, **171**, 4202–4206.
46. Eremeeva, M., Madan, A., Shaw, C., Tang, K., and Dasch, G. (2005) New perspectives on rickettsial evolution from new genome sequences of rickettsia, particularly r. Canadensis, and orientia tsutsugamushi. *Ann. N. Y. Acad. Sci.*, **1063**, 47–63.
47. Weinert, L., Werren, J., Aebi, A., Stone, G., and Jiggins, F. (2009) Evolution and diversity of rickettsia bacteria. *BMC Biol.*, **7**, 6.
48. Raoult, D., Ndihokubwayo, J.B., Tissot-Dupont, H., Roux, V., Faugere, B., Abegbinni, R., and Birtles, R.J. (1998) Outbreak of epidemic typhus associated with trench fever in burundi. *Lancet*, **352**, 353–358.
49. Lutwick, L.I. (2001) Brill-zinsser disease. *Lancet*, **357**, 119–1200.
50. Driskell, L.O., Yu, X.-j., Zhang, L., Liu, Y., Popov, V.L., Walker, D.H., Tucker, A.M., and Wood, D.O. (2009) Directed mutagenesis of the rickettsia prowazekii pld gene encoding phospholipase D. *Infect. Immun.*, **77**, 3244–3248.
51. Medina-Sanchez, A., Bouyer, D., Alcantara-Rodrigu, V., Mafra, C., Zavala-Castro, J., Whitworth, T., Popov, V., Fernandez-Salas, I., and Walker, D. (2005) Detection of a typhus group rickettsia in amblyomma ticks in the state of nuevo leon, mexico. *Ann. N. Y. Acad. Sci.*, **1063**, 327–332.
52. Bechah, Y., Capo, C., Mege, J., and Raoult, D. (2008) Epidemic typhus. *Lancet Infect. Dis.*, **8**, 417–426.
53. Lledo, L., Gegundez, M., Medina, J., Gonzalez, J., Alamo, R., and Saz, J. (2005) Epidemiological study of rickettsia typhi infection in two provinces of the north of Spain: analysis of sera from the general population and sheep. *Vector Borne Zoonotic Dis.*, **5**, 157–161.
54. Traub, R. and Wisseman, C. (1978) The ecology of murine typhus-a critical review. *Trop. Dis. Bull.*, **75**, 237–317.
55. Houhamdi, L., Fournier, P.-E., Fang, R., and Raoult, D. (2003) An experimental model of human body louse infection with rickettsia typhi. *Ann. N.Y. Acad. Sci.*, **990**, 617–627.
56. Jensenius, M., Fournier, P., Kelly, P., Myrvang, B., and Raoult, D. (2003) African tick bite fever. *Lancet Infect. Dis.*, **3**, 557–564.
57. Rovery, C., Brouqui, P., and Raoult, D. (2008) Questions on Mediterranean spotted fever a century after its discovery. *Emerg. Infect. Dis.*, **14**, 1360–1367.
58. Zhu, Y., Fournier, P., Eremeeva, M., and Raoult, D. (2005) Proposal to create

subspecies of rickettsia conorii based on multi-locus sequence typing and an emended description of rickettsia conorii. *BMC Microbiol.*, **5**, 11.

59. Paddock, C.D., Koss, T., Eremeeva, M.E., Dasch, G.A., Zaki, S.R., and Sumner, J.W. (2006) Isolation of rickettsia akari from eschars of patients with rickettsial pox. *Am. J. Trop. Med. Hyg.*, **75**, 732–738.
60. Gillespie, J.J., Ammerman, N.C., Beier-Sexton, M., Sobral, B.S., and Azad, A.F. (2009) Louse- and flea-borne rickettsioses: Biological and genomic analyses. *Vet. Res.*, **40**, in press.
61. Oliveira, K.A., Oliveira, L.S., Dias, C.C.A., Silva, A. Jr., Almeida, M.R., Almada, G., Bouyer, D.H., Galvão, M.A.M., and Mafra, C.L. (2008) Molecular identification of rickettsia felis in ticks and fleas from an endemic area for brazilian spotted fever. *Mem. Inst. Oswaldo Cruz*, **103**, 191–194.
62. Ogata, H., Renesto, P., Audic, S., Robert, C., Blanc, G., Fournier, P., Parinello, H., Claverie, J., and Raoult, D. (2005) The genome sequence of rickettsia felis identifies the first putative conjugative plasmid in an obligate intracellular parasite. *PLoS Biol.*, **3**, e248.
63. Gillespie, J., Ammerman, N., Dreher-Lesnick, S., Rahman, M., Worley, M., Setubal, J., Sobral, B., and Azad, A. (2009) An anomalous type iv secretion system in rickettsia is evolutionarily conserved. *PLoS ONE*, **4**, e4833.
64. Ogata, H., La Scola, B., Audic, S., Renesto, P., Blanc, G., Robert, C., Fournier, P., Claverie, J., and Raoult, D. (2006) Genome sequence of rickettsia bellii illuminates the role of amoebae in gene exchanges between intracellular pathogens. *PLoS Genet.*, **2**, e76.
65. Blanc, G., Ogata, H., Robert, C., Audic, S., Suhre, K., Vestris, G., Claverie, J., and Raoult, D. (2007) Reductive genome evolution from the mother of rickettsia. *PLoS Genet.*, **3**, e14.
66. Eremeeva, M.E. and Silverman, D.J. (1998) Rickettsia rickettsii infection of the ea.Hy 926 endothelial cell line: Morphological response to infection and evidence for oxidative injury. *Microbiology*, **144**, 2037–2048.
67. Fenolla, F., Fournier, P.-E., and Raoult, D. (2007) in *Rickettsial Diseases* (eds D. Raoult and P. Parola), Informa Health Care, New York, London, pp. 315–330.
68. La Scola, B. and Raoult, D. (1997) Laboratory diagnosis of rickettsioses: current approaches to diagnosis of old and new rickettsial diseases. *J. Clin. Microbiol.*, **35**, 2715–2727.
69. Walker, D.H. and Bouyer, D.H. (2007) in *Manual of Clinical Microbiology* (eds P.R. Murray, E.J. Baron, J. Jorgensen, M. Pfaller, and M.L. Landry), John Wiley & Sons, Inc., pp. 1036–1045.
70. Dobler, G. and Wölfel, R. (2009) Typhus and other rickettsioses – emerging infections in Germany. *Dtsch. Arztebl. Int.*, **106**, 348–354.
71. La Scola, B. and Raoult, D. (1996) Diagnosis if mediterranean spotted fever by cultivation of rickettsia conorii from blood and skin samples using centrifugation-shell vial technique and by detection of R. conorii in circulating endothelial cells: A 6-year follow-up. *J. Clin. Microbiol.*, **34**, 2722–2727.
72. Brouqui, P., Bacellar, F., Baranton, G., Birtles, R., Bjoersdorff, A., Blanco, J., Caruso, G., Cinco, M., Fournier, P., Francavilla, E. *et al.* (2004) Guidelines for the diagnosis of tick-borne bacterial diseases in europe. *Clin. Microbiol. Infect.*, **10**, 1108–1132.
73. Marrero, M. and Raoult, D. (1989) Centrifugation-shell vial technique for rapid detection of mediterranean spotted fever rickettsia in blood culture. *Am. J. Trop. Med. Hyg.*, **40**, 197–199.
74. Quesada, M., Sanfeliu, I., Cardenosa, N., and Segura, F. (2006) Ten years' experience of isolation of rickettsia spp. From blood samples using the shell-vial cell culture assay. *Ann. N. Y. Acad. Sci.*, **1078**, 578–581.
75. Lee, J.-H., Park, H.-S., Jang, W.-J., Koh, S.-E., Kim, J.-M., Shim, S.-K., Park, M.-Y., Kim, Y.-W., Kim, B.-J., Kook, Y.-H. *et al.* (2003) Differentiation of Rickettsiae by groel gene analysis. *J. Clin. Microbiol.*, **41**, 2952–2960.
76. Nilsson, K., Jaenson, T., Uhnoo, I., Lindquist, O., Pettersson, B., Uhlen, M.,

Friman, G., and Pahlson, C. (1997) Characterization of a spotted fever group rickettsia from ixodes ricinus ticks in sweden. *J. Clin. Microbiol.*, **35**, 243–247.

77. Roux, V. and Raoult, D. (2000) Phylogenetic analysis of members of the genus rickettsia using the gene encoding the outer-membrane protein rompb (ompb). *Int. J. Syst. Evol. Microbiol.*, **50**, 1449–1455.
78. Sekeyova, Z., Roux, V., and Raoult, D. (2001) Phylogeny of rickettsia spp. Inferred by comparing sequences of 'gene d', which encodes an intracytoplasmic protein. *Int. J. Syst. Evol. Microbiol.*, **51**, 1353–1360.
79. Roux, V. and Raoult, D. (1995) Phylogenetic analysis of the genus rickettsia by 16s rDNA sequencing. *Res. Microbiol.*, **146**, 385–396.
80. Park, H., Lee, J., Jeong, E., Kim, J., Hong, S., Park, T., Kim, T., Jang, W., Park, K., Kim, B. *et al.* (2005) Rapid and simple identification of orientia tsutsugamushi from other group Rickettsiae by duplex PCR assay using groel gene. *Microbiol. Immunol.*, **49**, 545–549.
81. Matsumoto, K. and Inokuma, H. (2009) Identification of spotted fever group rickettsia species by polymerase chain reaction-restriction fragment length polymorphism analysis of the sca4 gene. *Vector Borne Zoonotic Dis.*, **9**, 747–749.
82. Fournier, P.-E. and Raoult, D. (2007) Identification of rickettsial isolates at the species level using multi-spacer typing. *BMC Microbiol.*, **7**, 72.
83. Wölfel, R., Essbauer, S., and Dobler, G. (2008) Diagnostics of tick-borne rickettsioses in germany: a modern concept for a neglected disease. *Int. J. Med. Microbiol.*, **298**, 368–374.
84. Walker, D. and Yu, X. (2005) Progress in rickettsial genome analysis from pioneering of rickettsia prowazekii to the recent rickettsia typhi. *Ann. N. Y. Acad. Sci.*, **1063**, 13–25.
85. Walker, D.H. and Ismail, N. (2008) Emerging and re-emerging rickettsioses: endothelial cell infection and early disease events. *Nat. Rev. Microbiol.*, **6**, 375–386.
86. Walker, D. (2007) Rickettsiae and rickettsial infections: the current state of knowledge. *Clin. Infect. Dis.*, **1** (Suppl. 45), S39–S44.
87. Woods, M. and Olano, J. (2008) Host defenses to rickettsia rickettsii infection contribute to increased microvascular permeability in human cerebral endothelial cells. *J. Clin. Immunol.*, **28**, 174–185.
88. Walker, D., Feng, H., and Popov, V. (2001) Rickettsial phospholipase a2 as a pathogenic mechanism in a model of cell injury by typhus and spotted fever group Rickettsiae. *Am. J. Trop. Med. Hyg.*, **65**, 936–942.
89. Olano, J. (2005) Rickettsial infections. *Ann. N. Y. Acad. Sci.*, **1063**, 187–196.
90. Walker, D., Valbuena, G., and Olano, J. (2003) Pathogenic mechanisms of diseases caused by rickettsia. *Ann. N. Y. Acad. Sci.*, **990**, 1–11.
91. Fournier, P.-E., El Karkouri, K., Leroy, Q., Robert, C., Giumelli, B., Renesto, P., Socolovschi, C., Parola, P., Audic, Sp., and Raoult, D. (2009) *Analysis of the Rickettsia Africaegenome Reveals that Virulence Acquisition in Rickettsiaspecies may be Explained by Genome Reduction*, BioMed Central Ltd..
92. Burgdorfer, W. (1977) Tick-borne diseases in the united states: rocky mountain spotted fever and colorado tick fever. A review. *Acta Trop.*, **34**, 103–126.
93. Raoult, D., Roux, V., Ndihokubwayo, J., Bise, G., Baudon, D., Marte, G., and Birtles, R. (1997) Jail fever (epidemic typhus) outbreak in burundi. *Emerg. Infect. Dis.*, **3**, 357–360.
94. Kidd, L., Hegarty, B., Sexton, D., and Breitschwerdt, E. (2006) Molecular characterization of rickettsia rickettsii infecting dogs and people in North Carolina. *Ann. N. Y. Acad. Sci.*, **1078**, 400–409.
95. Masters, E.J., Olson, G.S., Weiner, S.J., and Paddock, C.D. (2003) Rocky mountain spotted fever: a clinician's dilemma. *Arch. Intern. Med.*, **163**, 769–774.
96. Tikare, N.V., Shahapur, P.R., Bidari, L.H., and Mantur, B.G. (2009) Rickettsial meningoencephalitis in a child–a case report. *J. Trop. Pediatr.*, **15**, 24–27
97. Parola, P., Labruna, M.B., and Raoult, D. (2009) Tick-borne

rickettsioses in America: unanswered questions and emerging diseases. *Curr. Infect. Dis. Rep.*, **11**, 40–50.

98. Chapman, A.S., Swerdlow, D.L., Dato, V.M., Anderson, A.D., Moodie, C.E., Marriott, C., Amman, B., Hennessey, M., Fox, P., Green, D.B. *et al.* (2009) Cluster of sylvatic epidemic typhus cases associated with flying squirrels, 2004–2006. *Emerg. Infect. Dis.*, doi: 10.3201/eid1507.081305

99. Houhamdi, L. and Raoult, D. (2006) Experimentally infected human body lice (pediculus humanus humanus) as vectors of rickettsia rickettsii and rickettsia conorii in a rabbit model. *Am. J. Trop. Med. Hyg.*, **74**, 521–525.

100. Walker, D. (2009) The realities of biodefense vaccines against rickettsia. *Vaccine*, **4** (Suppl. 27), D52–D55.

101. Azad, A. and Beard, C. (1998) Rickettsial pathogens and their arthropod vectors. *Emerg. Infect. Dis.*, **4**, 179–186.

102. Burgdorfer, W. and Brinton, L. (1975) Mechanisms of transovarial infection of spotted fever Rickettsiae in ticks. *Ann. N. Y. Acad. Sci.*, **266**, 61–72.

103. Houhamdi, L. and Raoult, D. (2007) in *Rickettsial Diseases* (eds D. Raoult and P. Parola), Informa Health Care, New York, London, pp. 51–61.

104. Bozeman, F., Masiello, S., Williams, M., and Elisberg, B. (1975) Epidemic typhus Rickettsiae isolated from flying squirrels. *Nature*, **255**, 545–547.

105. Chapman, A.S., Swerdlow, D.L., Dato, V.M., Anderson, A.D., Moodie, C.E., Marriott, C., Amman, B., Hennessey, M., Fox, P., Green, D.B. *et al.* (2009) Cluster of sylvatic epidemic typhus cases associated with flying squirrels, 2004–2006. *Emerg. Infect. Dis.*, **15**, 1005–1011.

106. Civen, R. and Ngo, V. (2008) Murine typhus: an unrecognized suburban vectorborne disease. *Clin. Infect. Dis.*, **46**, 913–918.

107. Mouffok, N., Parola, P., and Raoult, D. (2008) Murine typhus, algeria. *Emerg. Infect. Dis.*, **14**, 676–678.

108. Schriefer, M., Sacci, J., Taylor, J., Higgins, J., and Azad, A. (1994) Murine typhus: updated roles of multiple urban components and a second typhuslike rickettsia. *J. Med. Entomol.*, **31**, 681–685.

109. Williams, S.G., Sacci, J.B., Schriefer, M.E., Andersen, E.M., Fujioka, K.K., Sorvillo, F.J., Barr, A.R., and Azad, A.F. Jr. (1992) Typhus and typhuslike Rickettsiae associated with opossums and their fleas in Los Angeles County, California. *J. Clin. Microbiol.*, **30**, 1758–1762.

110. Paddock, C.D. and Eremeeva, M.E. (2007) in *Rickettsial Diseases* (eds D. Raoult and P. Parola), Informa Health Care, New York, London, pp. 63–86.

111. Parola, P., Attali, J., and Raoult, D. (2003) First detection of rickettsia africae on martinique, in the french west indies. *Ann. Trop. Med. Parasitol.*, **97**, 535–537.

9
Mycobacterium tuberculosis: Tuberculosis

Stefan Panaiotov, Massimo Amicosante, Marc Govaerts, Patrick Butaye, Elizabeta Bachiyska, Nadia Brankova, and Victoria Levterova

9.1
Introduction

The genus *Mycobacterium*, belonging to the Actinobacteria, is closely related to *Corynebacterium*, *Nocardia*, *Propionibacterium*, and *Rhodococcus*. It consists of over 130 species of which some are the cause of the most serious infectious diseases in animals and humans, such as tuberculosis and leprosy. *Mycobacterium tuberculosis (M. tuberculosis)* causes over 9 million new cases of human tuberculosis and 1.7 million of deaths a year [1]; whereas *M. leprae*, a leading cause of death over the past millennia, is still responsible for over 250 000 new cases of leprosy worldwide annually. In contrast, most nontuberculosis mycobacteria (NTM) are opportunistic pathogens, only causing severe disease among immune deficient individuals [2].

Mycobacteria are globally spread, typically living in a variety of environments, including water (such as tap water, even after treatment with chlorine) and food sources. They are aerobic and nonmotile bacteria that are characterized by their acid-alcohol fast staining properties. They typically show up as straight or slightly curved rods 0.2–0.6 μm wide and 1.0–10 μm long. Mycobacteria do not contain endospores or capsules. From a staining point of view mycobacteria are usually considered weakly Gram-positive (i.e., they partially retain the crystal violet stain). However, the general structure of the cell wall is very different from the Gram-positive and Gram-negative bacteria. All mycobacterial species share a characteristic cell wall, thicker than in many other bacteria, which is hydrophobic, waxy, and rich in mycolic acids. The cell wall consists of a hydrophobic mycolate layer and a peptidoglycan layer held together by a polysaccharide, arabinogalactan, and a lipopolysaccharide, lipoarabinomannan. The cell wall makes a substantial contribution to the hardiness of this genus and is at the basis of its acid-alcohol fast stain [3].

Many *Mycobacterium* species adapt readily to growth on very simple substrates, using ammonia or amino acids as nitrogen sources and glycerol as a carbon source in the presence of minerals. Optimum growth temperatures vary widely according to species and range from 25 to over 50 °C. However, some species can be very

BSL3 and BSL4 Agents: Epidemiology, Microbiology, and Practical Guidelines, First Edition.
Edited by Mandy C. Elschner, Sally J. Cutler, Manfred Weidmann, and Patrick Butaye.

difficult to culture, sometimes taking over nine months to develop in culture. Further, some species also have extreme long reproductive cycles – *M. leprae* may take more than 20 days to proceed through one division cycle – making laboratory culture a slow, nearly impossible process to observe [3]. A natural division is drawn between slow- and rapid-growing species. Mycobacteria that form clearly visible colonies within seven days of culturing are named rapid growers, while those requiring longer periods are named slow growers [3]. Another division is made based on their pigment production. Some mycobacteria produce carotenoid pigments without light. Others require photoactivation for pigment production. This is at the basis of Runyon's landmark phenotypical classification of the *Mycobacterium* genus, resulting in four groups [3]:

- **Photochromogens (group I)**: Produce nonpigmented colonies when grown in the dark and pigmented colonies when grown under light (examples: *M. kansasii, M. marinum, M. simiae*).
- **Scotochromogens (group II)**: Always produce deep yellow to orange colonies (examples: *M. scrofulaceum, M. gordonae, M. xenopi, M. szulgai*).
- **Nonchromogens (groups III and IV)**: Produce nonpigmented colonies under all conditions (examples: *M. tuberculosis/bovis, M. avium/intracellulare, M. ulcerans*), or have only a pale yellow, buff, or tan pigment that does not intensify after light exposure (examples: *M. fortuitum, M. chelonae*).

From a medical point of view, Mycobacteria are classified into several groups for the purpose of diagnosis and treatment:

- *M. tuberculosis* complex, which can cause tuberculosis, includes: *M. tuberculosis, M. bovis, M. africanum, M. microti, M. canettii, M. caprae, M. pinnipedii*, and the *oryx*, and *dassie bacilli*.
- *M. leprae*, which causes Hansen's disease or leprosy.
- NTM are all the other mycobacteria, which can cause pulmonary disease resembling tuberculosis, lymphadenitis, skin, soft tissue and skeletal infections, catheter-related bloodstream infections, and disseminated disease. Most NTM are accidental pathogens causing only severe disease in immune compromised individuals, yet of growing incidence [4, 5]. Infections with NTM or with the bacille Calmette–Guérin (BCG) attenuated TB vaccine strain, are rare and grouped under the name of Mendelian syndrome. However, NTM can also in rare cases cause disease in otherwise healthy individuals [6].

9.2 Diagnostic Microbiology of Mycobacteria

The contribution of the microbiology laboratory to the diagnosis and management of tuberculosis and other NTM infections involves the detection and isolation of mycobacteria, the identification of the mycobacterial species, and susceptibility testing.

A first step for the successful isolation of mycobacteria is the preparation of the clinical sample (i.e., digestion and decontamination of specimens) as most clinical specimens contain an abundance of nonmycobacterial organisms, who can quickly overgrow the generally more slowly reproducing mycobacteria (18–24 h generation time in culture). It is also necessary to liquefy organic samples (tissue, serum, and other material) so that decontaminating agents may kill undesirable microbes, and surviving mycobacteria may gain access to the nutrients. Since mycobacteria are more refractory to harsh chemicals than are most other microorganisms, chemical decontamination procedures can be successfully applied [3].

Most mucin-rich respiratory specimens are homogenized with a mucolytic agent (such as *N*-acetyl-L-cysteine) and a decontaminant (such as a 1–2% sodium hydroxide solution, 8% sulfuric acid, or 5% oxalic acid). The mildest decontamination procedure that provides sufficient control of the contaminants without killing the mycobacteria is likely to yield the best results [7]. However, even under the most optimal conditions, these procedures kill all but 10–20% of the mycobacteria in a specimen, lowering the diagnostic sensitivity [8, 9]. Even then, 2–5% of the sputum samples inoculated on solid Löwenstein–Jensen medium are contaminated. However, if fewer than 2% of the specimens yield contaminations, it is supposed that the decontamination process may be too harsh, killing too many of the possibly present mycobacteria. If more than 5% of the cultures are contaminated, the decontamination process is inadequate [10]. Tissues may be ground in homogenizers and likewise decontaminated. Specimens collected from normally sterile sites may be homogenized and placed directly into the culture medium.

9.3
Staining and Microscopic Examination

The microscopic detection of acid-fast bacilli (AFB) in smears is the first bacteriological evidence for the presence of mycobacteria in a clinical specimen. It is the easiest and quickest procedure, and it provides the physician with a preliminary confirmation of a diagnostic suspicion [3]. Because it gives a quantitative estimation of the number of bacilli being excreted, the smear is also of clinical and epidemiologic importance in assessing the patient's infectiousness. Smears may be prepared directly from clinical specimens or from concentrated preparations. The acid-fast staining procedure is based on the ability of mycobacteria to retain a dye when treated with a mineral acid or an acid-alcohol solution. Two procedures are commonly used for acid-fast staining: the carbolfuchsin method, which includes the Ziehl–Neelsen and Kinyoun methods, and a fluorochrome procedure using auramine-O or auramine–rhodamine dyes [3]. Several quantitative studies have shown that there must be at least 5000–10 000 bacilli/ml of specimen to allow the detection of bacteria in stained smears [11], which is a low sensitivity. In contrast, 10–100 organisms are enough for a positive culture [12]. Concentration procedures in which a liquefied specimen is centrifuged and the sediment is used for staining increase the sensitivity of the test; thus, smears of concentrated material are preferred. Negative smears, however, do not preclude tuberculosis or mycobacterial disease. Various studies have indicated that only 50–80% of patients with pulmonary tuberculosis have positive sputum smears.

Acid-fast bacteria seen on a smear may represent either *M. tuberculosis* or NTM. The percentage of specimens shown positive by smear but negative in culture is less than 1%, indicating a good specificity [13]. Most smear-positive, culture-negative specimens are seen in patients who are taking anti-mycobacterial therapy. Laboratory errors, prolonged specimen decontamination, shortened incubation times of culture, cross-contamination of smears, and water or stains contaminated with acid-fast organisms can result in an augmentation of smear-positive, culture-negative specimens [7].

9.4
Cultivation of Mycobacteria

All clinical specimens suspected of containing mycobacteria should be cultured for four reasons:

1) Culture is much more sensitive than microscopy, being able to detect as few as 10 bacteria/ml of material [12].
2) Growth of the organisms is necessary for species identification.
3) Drug susceptibility testing requires live cultures.
4) Genotyping of cultured organisms may be useful to identify epidemiological links between patients or to detect laboratory cross-contamination.

In general, the sensitivity of culture is 80–85%, with a specificity of approximately 98% [14, 15]. Although the diagnosis of tuberculosis disease is mainly bacteriologic in adults, it remains based on epidemiologic data for children [16]. For example, from 1985 to 1988 in the United States, 90% of tuberculosis cases in adults were bacteriologically confirmed, while only 28% in children [17]. Also in HIV-infected pediatric tuberculosis cases, the sensitivity of culture appears to be much higher, possibly because of the dissemination of the organism or higher bacterial burdens.

Three different types of traditional culture media are available: egg-based (Löwenstein–Jensen), agar-based (Middlebrook 7H10 or 7H11 medium), and liquid (Middlebrook 7H12 and other commercially available broths), and by the addition of antibiotics, the selectivity is guaranteed. The growth of mycobacteria on solid media shows a better sensitivity on egg-based media but the agar-based media give a faster diagnosis.

Growth in liquid media is faster than growth on solid media (detection of mycobacterial growth within one to three weeks compared with solid media, where growth takes three to eight weeks [14]). However, liquid media can be used for primary isolation of mycobacteria from nonsterile samples only if supplemented with an appropriate antibiotic cocktail. The increasing antibiotic resistance in all bacterial species complicates this selectivity. Nowadays, automated culture systems such as BACTEC 460 (Becton Dickinson Microbiology Systems, Sparks, Md., USA), mycobacterial growth indicator tube (MGIT) systems, Extra Sensing Power (ESP) Myco-ESPculture System II (Trek Diagnostic Systems, Inc., Westlake, Ohio, USA), and BacT/ALERT MB Susceptibility Kit (Organon Teknika, Durham, N.C., USA) use Middlebrook 7H12 media for the detection of mycobacteria which is based on radiometric or colorimetric systems. Automated liquid systems should be checked at least every two to three days for growth.

9.5 Identification of Mycobacteria from Culture

More than half of the mycobacterial species may be isolated from humans, as pathogens or saprophytes. A clear-cut distinction between pathogen and saprophyte is not always possible for the individual isolate. Isolation of a nontuberculous organism of potential clinical significance is not *ipso facto* evidence that the patient has a disease caused by the organism; conversely, all such isolates, which are usually clinically insignificant, should not be regarded as saprophytes. Each mycobacterial isolate, like each patient, must be evaluated individually [18].

Historically, *M. tuberculosis* could readily be identified by: (i) its rough, nonpigmented, corded colonies on oleic acid–albumin agars, (ii) a positive niacin test, (iii) generally weak catalase activity, that is lost completely by heating to 68 °C, and (iv) a positive nitrate reduction test. Observation of mycobacterial colonial morphology remains a valuable tool. Although the colonial morphologies of mycobacteria on various egg-based media may be quite similar, their appearance on Middlebrook 7H10 or 7H11 agar is distinctive [19].

Biochemical methods can distinguish mycobacterial species, but are time-consuming and laborious. Two identification procedures for *M. tuberculosis* have gained widespread use: nucleic acid hybridization and high performance liquid chromatography (HPLC). Nucleic acid hybridization uses molecular probes that can hybridize specifically with *M. tuberculosis* complex, *M. avium* complex, *M. kansasii*, and *M. gordonae* [20]. A commercial 16S probe-based reverse blot assay is available that allows the differentiation of 17 mycobacterial species or complexes (Inno-Lipa by Innogenetics). A competitor offers simultaneous differentiation of 5–16 mycobacterial species or complexes (GenoType®, Hain Lifescience). These assays can be completed within hours and have sensitivities and specificities approaching 100% when at least 10^5 organisms are present. This requirement is easily met when pure cultures are used but is rarely achieved with clinical specimens. Thus nucleic acid hybridization is typically used after the organisms are grown in culture.

HPLC is based on the observation that each *Mycobacterium* species synthesizes a unique set of mycolic acids, β-hydroxy-α-fatty acids that are components of the cell wall [21]. HPLC can produce a pattern that reliably identifies and distinguishes about 50 mycobacterial species. It can be performed in a few hours but requires pure cultures. One major drawback is that it cannot differentiate *M. tuberculosis* from *M. bovis*, although it can differentiate *M. bovis* BCG from *M. tuberculosis* complex.

The initial equipment cost for HPLC limits its availability. Newer developed techniques for species identification, such as spoligotyping, polymerase chain reaction restriction analysis (PRA), and DNA sequencing are nowadays common tools for the reference laboratories. Although all these newly developed techniques form the basis for species identification, they have found their greatest use in molecular epidemiology analysis of *M. tuberculosis*. The details of the molecular epidemiology analysis is largely provided later in this chapter, particularly for the spoligotyping and its related techniques (see Section 9.8). Regarding the DNA sequencing, although at present with pirosequencing approach it is possible to type a whole genome within one day and have the full characterization of the bacteria under evaluation, for mycobacteria the largest use for identification has been done with the sequencing of the 16SRNA gene. This technique is quite robust, allowing rapid and precise species identification. However, due to the relatively high costs, it has not entered into general practice, only in reference laboratories.

9.6 Identification of Mycobacteria Directly from Clinical Specimens

Because of their ability to detect low numbers of *M. tuberculosis* DNA, nucleic acid amplification tests (NAATs) have boosted diagnostic capacities, are capable of reducing the time to diagnosis of patients with suspected tuberculosis from weeks to hours, and are taking hold in clinical practice [22, 23]. The NAAT have been used to:

1) Diagnose tuberculosis rapidly by identifying DNA or RNA from *M. tuberculosis* in clinical samples;
2) Determine rapidly whether acid-fast microorganisms identified by microscopic examination in clinical specimens are *M. tuberculosis* or atypical mycobacteria;
3) Identify the presence of genetic modifications known to be associated with resistance to some anti-mycobacterial agents.

However, it is clear that NAAT are not reliable enough to replace conventional diagnostic methods. Both inherent test characteristics and errors in testing procedures may account for NAAT inaccuracy [23]. One of the key factors for NAAT false negatives is the local density of mycobacteria in a specimen, since it can result in the absence of nucleic acids in the small sample generally used for testing. Also the presence of inhibitory enzymes in respiratory secretions may inhibit amplification reactions. The latter may account for 3–25% of false negative NAAT results. However, false positives mainly arise from contamination [23].

In the attempt to minimize these defects, the industry developed automated commercial systems which were made more robust by means of the use of standardized procedures and reagents for sample processing, amplification, and detection. These tests, allowing different steps of the process to take place in a single sealed tube, should render commercial NAAT less prone to contamination. At the same time, other measures such as the choice of a multi copy target sequence, the use of larger sample volumes, and the introduction of internal amplification controls to detect inhibitors, have been adopted for cutting down on false negatives. However, notwithstanding this, a considerable variability of diagnostic accuracy of commercial NAAT is still apparent [24]. Therefore the United States Centers for Disease Control (CDC) indicates that, if commercial NAAT is used, microscopy has to be performed in a parallel test to improve diagnostic certainty, and culture should be performed as confirmation. A positive pulmonary TB is considered when microscopy and NAAT are positive, and when microscopy is negative but with two subsequent positive NAAT results. In the case of at least two negative, inhibitor-free NAAT results, a microscopy positive smear is indicative for a NTM disease, while a microscopy negative smear can be presumed as negative for mycobacterial disease [25]. The CDC recommendations are limited to the two FDA-approved commercial NAATs [25]. Furthermore, the CDC strongly advises physicians to rely upon clinical judgment in the interpretation of the NAAT laboratory data [24, 25].

9.7 Immunological Tests for the Diagnosis of *Mycobacterium tuberculosis* Infection

Mycobacterium tuberculosis, as well as other mycobacteria, elicits a strong immune response upon infection by stimulating both CD4+ and CD8+ T-cells as well as other cells of the immune system, determining a strong type-1 proinflammatory-like response dominated by interferon (IFN)-gamma and tumor necrosis factor (TNF)-alpha [26]. The overall response is at the basis of the so-called delayed type hypersensitivity (DTH) caused by *M. tuberculosis* antigens [27]. This phenomenon has been used for more than a century for the identification of subjects infected by *M. tuberculosis* by using the tuberculin skin test (TST), which attempts to measure cell-mediated immunity in the form of a DTH response to the most commonly used purified protein derivative (PPD) of tuberculin [27]. However, the test is affected by many limitations, not least the fact of a high rate of positivity consequent to vaccination with *M. bovis* BCG [27].

In the last decade, extensive studies have shown that immunodominant antigens, such as the 6-kDa early secretory antigenic target (ESAT-6) and its homologs, are highly suitable for detecting infection. There is no cross-reaction with the BCG vaccine, since these antigens are absent in the BCG vaccine strains. By screening eluted fractions of antigens from *M. tuberculosis* and *M. bovis* culture filtrates for recognition by T cells from infected humans and cattle, respectively, Andersen and coworkers identified several low molecular mass antigens that are major targets of cellular mediated immune responses [28]. Subtractive DNA hybridization of pathogenic *M. bovis* and BCG [29] and comparative genome-wide DNA microarray analysis of *M. tuberculosis* H37Rv and BCG [30] identified several regions of difference (RD), designated RD1 to RD16, between *M. tuberculosis* and *M. bovis*. All represent segments that have been deleted from the *M. bovis* genome. RD1 was lost early during the process of *M. bovis* BCG attenuation and is therefore missing in all the daughter strains known today [29]. This region has been the subject of detailed studies and a number of antigens have recently been characterized as candidate antigens for diagnostic and vaccine development [31, 32]. Antigens, such as ESAT-6 and culture filtrate protein 10 (CFP-10), are located in this region and have already shown great potential for tuberculosis diagnosis [33–36].

Based on these studies, one of the most significant developments in the diagnostic armamentarium for tuberculosis in the last 100 years seems to be the assays based on IFN-γ determination [interferon gamma release assays (IGRAs)]. The assays stem from the principle that T-cells of sensitized individuals produce IFN-gamma when they re-encounter the antigens of *M. tuberculosis* [37], and this was initially developed to screen cattle for bovine tuberculosis under free ranching conditions [38]. IGRAs' clearest advantage is increased specificity for detection of *M. tuberculosis* infection thanks to their utilization of *M. tuberculosis*-specific antigens encoded in RD1, a genomic segment absent from the BCG vaccine and most environmental mycobacteria. Recent evaluations showed that IFN-γ assays that use *M. tuberculosis* RD1 antigens, such as ESAT6 and CFP10, may have advantages over tuberculin skin testing [4, 34, 35].

The IFN-gamma assays (IGRA) that are now commercially available are: the original QuantiFERON-TB and its enhanced versions QuantiFERON-TB Gold and QuantiFERON-TB Gold in-tube assays (Cellestis International, Carnegie, Australia), the enzyme-linked immunospot (ELISPOT) T SPOT-TB assay (Oxford Immunotec, Oxford, UK), and various veterinary specialties (Bovigam®, Cervigam®, Primagam®; Prionics, Schlieren-Zurich, Switzerland).

IGRAs have been successfully applied in both active and latent tuberculosis infection (LTBI), pulmonary and extrapulmonary tuberculosis, patients with MDR tuberculosis, and in immunocompromised individuals with HIV/AIDS. IGRAs could be successfully used in patients with transplantation, immune suppressive therapies, anti-TNF-alpha treatment, cancers, chronic renal failure, and diabetes. Also, the use of IGRAs is successful in children of any age, epidemiological studies (contact tracing, testing of health care workers), monitoring of anti-tuberculous therapy, and discrimination between BCG vaccination and *M. tuberculosis* (MTB) infection.

In immunocompromised hosts, all available data should be used to demonstrate or exclude latent infection with *M. tuberculosis*. The risk of developing active tuberculosis is different between various immunocompromising conditions. However, screening for latent infection with *M. tuberculosis* in immunocompromised patients is carried out irrespective of the type of immunosuppression, because the risk of developing active tuberculosis is probably higher compared with that of immunocompetent individuals. Sensitivity of the TST is limited in immunocompromised individuals and specificity is limited because of cross-reactivity due to prior infection with environmental mycobacteria or BCG vaccination. IGRAs have a higher specificity in populations with a high prevalence of BCG vaccination compared with TST [4, 35, 37].

Tuberculosis contacts potentially benefiting from preventive therapy should be identified hierarchically according to their likelihood of having become infected by a putative source and by the presence of potentially aggravating risk factors. IGRAs may be superior to TST in identifying contacts at risk of developing tuberculosis.

Children are more likely to develop tuberculosis than adults after exposure to an active tuberculosis case, hence contact screening and chemoprophylaxis are particularly important. In children with a high risk of infection (especially young children aged <5 years and immunocompromised children) an IGRA should be performed in addition to the TST to increase sensitivity. If either tests give a positive result, this may be interpreted as supportive evidence of infection, and the children should be offered preventive chemotherapy.

Children aged ≥ 5 years with exposure to sputum smear-positive tuberculosis should also be screened and a positive TST confirmed by IGRA, where available. In cases in which the treating pediatrician opts not to provide preventive therapy to TST-positive but IGRA-negative children, surveillance for a minimum of 12–24 months is indicated for observation and to collect outcome data, until the positive and negative predictive value of IGRA are better established in the setting of pediatric tuberculosis.

In a nutshell, the use of IGRAs' positive predictive value for the development of active tuberculosis it is likely to be equal or better than that of the TST for immunocompetent individuals.

9.8 Molecular Epidemiology of Tuberculosis

Molecular epidemiology is the integration of molecular techniques to track specific strains of pathogens with conventional epidemiologic approaches to understanding the distribution of disease in populations.

Phenotypic markers that distinguish between strains of *Mycobacterium tuberculosis* such as unusual antibiotic resistance patterns and mycobacterial phage susceptibility, have long been employed in the investigation of TB outbreaks. These investigations have provided much of the understanding of the transmission and pathogenesis of tuberculosis upon which the current approach to disease control is based. Two decades ago, these phenotypic markers started to be replaced by molecular markers. The modern methods for genotyping of *M. tuberculosis* were developed during the 1990s. Genotyping aims to solve several practical questions: (i) strain transmission from one patient to another, (ii) reinfection with the same strain, (iii) laboratory cross contamination, (iv) identification of geographically dominant *M. tuberculosis* genotypes, (v) development and implementation of specific targeted programs for the control of dominant genotypes of *M. tuberculosis*, (vi) development of new vaccines based on the dominant genotype in certain geographic areas (e.g., Beijing genotype in Asia), (vii) identification of subspecies of the *M. tuberculosis* complex, (viii) identification of strains with high pathogenicity and transmissibility, and so on. Here we describe the genetic elements of *M. tuberculosis* that may be exploited as strain-specific markers, the strain typing methods that are based on these elements and some of the DNA fingerprinting results obtained to date,

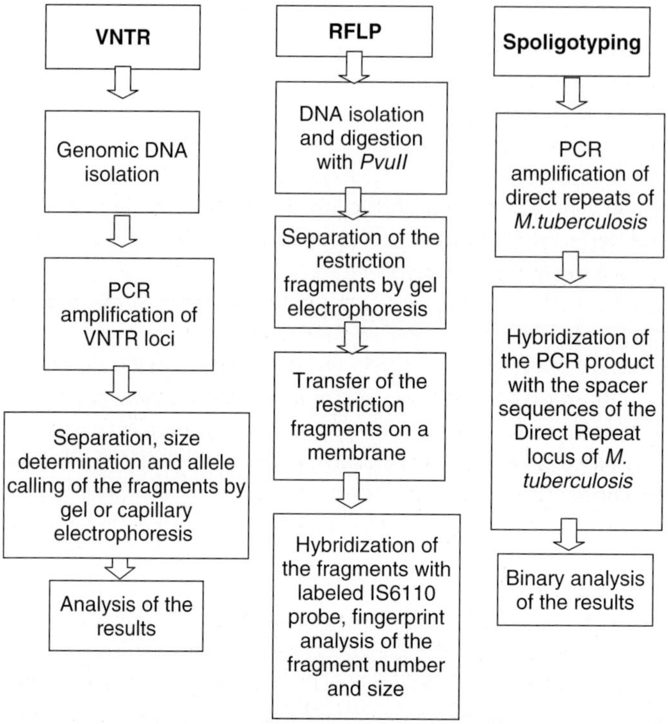

Figure 9.1 Schematic description of the major steps of the three most diffused *M. tuberculosis* molecular typing methods.

and speculate on future directions in this field. First the theoretical principles of genotyping will be explained and put into view of the reference methods applied for typing of *M. tuberculosis* and the goals of molecular epidemiology in controlling tuberculosis.

Over the past 15 years several methods for typing of *M. tuberculosis* were developed. Only three of them were established as reference methods and are presently widely applied (Figure 9.1). In the early 1990s, the restriction fragment length polymorphism (RFLP) technique was first developed. It is based on the restriction/hybridization analysis of IS6110, an insertion element specific for *M. tuberculosis*. At the end of 1990s, the so-called "spoligotyping" technology was developed. At the beginning of this century the methods for typing *M. tuberculosis* were complemented by the analysis of variable number of tandem repeats (VNTRs). These three methods were approved by the international scientific community as reference methods. Applying these methods gives a much better picture of the *M. tuberculosis* genotypes worldwide.

Genotyping approaches may permit more precise targeting of conventional control measures, such as the characterization of specific populations that are at particular risk of developing tuberculosis as a result of reactivated latent disease.

A more complete understanding of the secondary transmission would allow identifying those patients who are particularly prone to spread tuberculosis and pinpointing which population would benefit from increased efforts of treatment. These approaches may also provide sufficient understanding of the specific environmental and social factors that influence the efficiency of tuberculosis control strategies.

Some of the most important bacterial pathogens of humans exhibit strikingly low DNA sequence diversity. On average, these organisms harbor one nucleotide difference every 2–28 kbp and are thus referred to as *genetically monomorphic* [5]. Examples of such bacteria pathogens with highly conserved genomes are *Yersinia pestis* (the etiologic agent of plague), *Salmonella enterica* serovar Typhi (typhoid fever), *Bacillus anthracis* (anthrax), and the three most important pathogenic mycobacteria: *Mycobacterium leprae* (leprosy), *Mycobacterium ulcerans* (Buruli ulcer), and *Mycobacterium tuberculosis* complex members. *M. tuberculosis* complex bacteria constitute a remarkably homogeneous group. This fact was revealed by the inability of multilocus enzyme electrophoresis to differentiate individual strains, the minimal DNA polymorphism in restriction fragments of randomly chosen chromosomal DNA fragments and the little discriminatory power of amplified fragment length polymorphism (AFLP) applying any restriction enzyme combination. Genetically monomorphic bacteria such as *M. tuberculosis* contain so little sequence diversity that sequencing a few gene fragments, as performed in multilocus sequence typing (MLST), yields little or no information. In contrast, various repetitive DNA elements that contribute to strain variation have been discovered in *M. tuberculosis*. Two of these are IS and short repetitive DNA sequences with no known function or phenotype. More recently, genomic analysis of genetic variation to understand evolutionary and phylogeographic patterns has reached single nucleotide polymorphism (SNP) analysis [39, 40].

In general, cases in low-incidence areas tend to comprise mostly reactivation cases, while those in high-incidence regions include both reactivation and recent transmission. A crucial aspect in understanding the dynamics of a TB epidemic is the ability to track the spread of specific strains in the population. Previously unresolved issues, such as population estimates of recent transmission and the ability to distinguish endogenous reactivation from exogenous reinfection, have been made possible by the use of a variety of molecular techniques [27].

Programs to control TB developed by the WHO, CDC, and ECDC oblige each country to develop a program to control tuberculosis, which includes strain typing of *M. tuberculosis*, storage, and exchange of results. Typing of strains from different geographic regions revealed the *M. tuberculosis* population structures, in countries with low and high frequency of incidence of tuberculosis. In several European countries – Netherlands, Denmark, and Norway – all *M. tuberculosis* isolates are typed. In other countries such as Italy, Germany, and France only drug-resistant isolates are routinely typed. WHO recommends that countries with a low morbidity incidence type all strains, while countries with medium or high morbidity incidences (Bulgaria, Romania) should perform typing on all MDR isolates.

9.9 Theoretical Principles of Typing

Establishing subspecies differences is known as "*bacterial typing*," this is, after the species has been identified. The methods used for typing are phenotypic or genetic. Depending on the species, we encounter more or less significant genotypic and phenotypic variability, which reflects the evolutionary divergence due to mutations, rearrangements, and horizontal genetic flow. Different typing methods are based on the premise that associated clonal isolates have common biochemical and genomic characteristics, through which they can be differentiated from the clonally independent isolates. The use of a particular characteristic for typing is associated with their stability at strain level and variability at species level [41–44].

9.10 Performance Criteria Applied in Selecting the Method for Molecular Typing of Microorganisms

There are several criteria for the selection of typing methods, depending on the aim of the research. This may be looking at the evolution of species going over to the determination of a single clonal lineage: the fingerprint of a strain. The most significant are:

- Reproducibility;
- Discriminatory power;
- Typing ability or typeability;
- Ease of implementation and interpretation of data.

9.10.1 Reproducibility

With reproducibility we mean the ability of a typing method to produce the same result when the same strain is tested repeatedly in a same or in a different laboratory. This requests a well standardized method. The reproducibility of the method can be influenced by both technical and biological variables. Reproducibility is essential for the establishment of reliable international public databases containing profiles of tested strains that can be compared and classified.

9.10.2 Discriminatory Power

Depending on the relative need for strain typing, a typing method must have a lower or higher discriminatory power, respectively. It is clear when tracing back infections, like is the case for mycobacterial infections, a high discriminatory power is requested. It is defined as the possibility of the method to differentiate unrelated strains. Therefore to determine the discriminatory power, the strains in a test collection need to include both epidemiologically related and epidemiologically unrelated strains. The calculation is based on the probability that two unrelated isolates would be classified as the same type.

The discriminative power can be calculated on the basis of the Simpson index of biodiversity, the discrimination index (DI):

$$DI = 1 - [1/N(N-1)] \sum_{j=1}^{S} n_j (n_j - 1)$$

where: DI is the probability of two randomly chosen isolates from a set of unrelated isolates to be differentiated as separate genotypes, N is the total number of isolates, $S$is the total number of genotypes, and n_j is the number of isolates appertaining to the n_jth type.

9.10.3
Typeability

Special attention should be paid to the selection of an appropriate set of test strains for evaluation of the typeability of a typing system. The test population should include isolates that are epidemiologically unrelated on the basis of detailed clinical and epidemiological data. Thus, studies evaluating the typeability of the typing system need to be done with collections of isolates from unrelated patients [44].

9.11
Genetic Elements in *M. tuberculosis* that Contribute to DNA Polymorphism: Current Methods Applied for Genotyping of *M. tuberculosis*

In Figure 9.1 a schematic description is given of the three reference methods for typing *M. tuberculosis*. The detailed description of the different methods is given underneath.

9.12
IS6110-RFLP Analysis

IS6110-RFLP analysis is considered the "gold standard" for the molecular epidemiology of *M. tuberculosis*. In 1993, van Embden and colleagues proposed a standardized method for performing IS6110-based Southern blot hybridization analysis for typing of *M. tuberculosis* strains. The recommendation was based on the use of a common restriction endonuclease *Pvu*II, cleaving IS6110 at a single asymmetric site and yielding chromosomal fragments of reasonable size which can be separated by simple agarose gel electrophoresis. The recommended hybridization probe is targeted at the 3' fragment of IS6110, whereby each IS6110 insertion site will yield a single hybridizing band. Patterns obtained can be computerized with a specialized software. IS6110 is very stable and has proven to be very adequate for the study of transmission. Applications of this method include molecular epidemiology, evolutionary and phylogeny studies, and detection of error/cross laboratory contamination. Disadvantages of this method are that the process is laborious, it has a slow turnaround time (30–40 days) due to the prerequisite of a high quantity of DNA of good quality, the method cannot be used to reliably type strains with low copy (<6) number of IS6110, and interlaboratory comparative analysis of RFLP patterns can be tedious. In some cases, IS6110-RFLP still remains the most discriminatory technique. It provided the best resolution for the analysis of the Beijing genotype strains. IS6110 transposition and/or deletion events are always unidirectional and can be considered a form of divergent evolution [45–47].

9.13
Spacer Oligonucleotide Typing – Spoligotyping

M. tuberculosis complex strains contain a distinct chromosomal region consisting of multiple 36-bp direct repeats (DRs) interspersed by unique spacer DNA sequences (35–41 bp). Spoligotyping is an amplification–hybridization based analysis method aimed at detecting the 43 interspersed spacer sequences in the genomic DR region of *M. tuberculosis* complex strains. Additional spacers in this region have been reported. Membranes spotted with the 43 synthetic oligonucleotides are hybridized with a labeled PCR-amplified DR locus of the tested strain, resulting in a pattern that can be detected by chemiluminescence. The results are highly reproducible, and the binary (present or absent) data generated can be easily interpreted and computerized and are amenable to good interlaboratory comparisons (Figure 9.2). There are several international spoligo databases. SpolDB4 is the most representative, where different spoligotypes (STs) identified worldwide are organized into large ST families. ST families are named on the basis of a common motif of deleted spacers. Recently, two new web-based programs "MIRU-VNTR*plus*" (mycobacterial interspersed repetitive units) and "SpotClust" have been developed. Spoligotyping is the simplest technique for *M. tuberculosis* strain genotyping. Commercial hybridization membranes are available for simultaneous analysis of 45 samples. An advantage of this method is that the analysis can be performed on nonviable microorganisms, that is, from microscopic slides. Although spoligotyping is less discriminatory than IS6110-RFLP and VNTR methods, it is the most popular and widely used typing method and is extensively used for phylogeographic studies (Table 9.1). However, the discriminatory power is sometimes poor, giving non-existing relatedness between strains: the same patterns may be found on distinct evolutionary branches due to the fact that the same spacer may be lost independently in different lineages. The limited discriminatory power of spoligotyping is primarily because it targets a single locus that accounts for less than 0.1% of

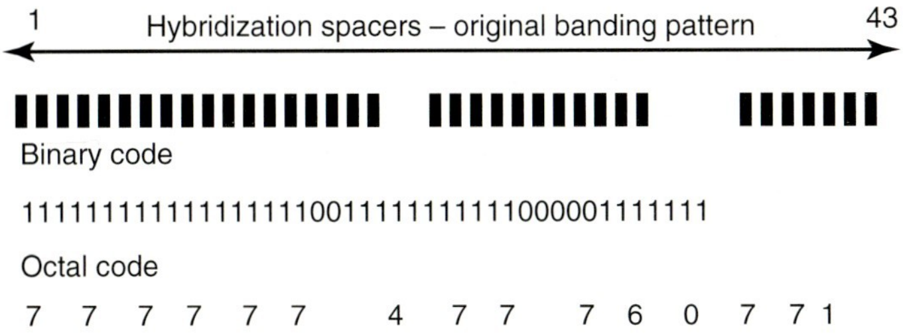

Figure 9.2 Spoligotyping. Binary and octal code designations. Hybridized spacers are represented as 1 (black blocks), absence of hybridization is 0. Designation can be represented as octal code with numbers from 0 to 7 by grouping the spacers in triplets.

the *M. tuberculosis* genome, while IS6110-based RFLP analysis and VNTR scan the whole genome more widely. The most recent evolution is the introduction of the Luminex system for a more automated analysis. There are attempts for transferring the method on a microchip in combination with the detection of drug resistance markers.

9.14 VNTR and MIRU Analysis

Since the *M. tuberculosis* complex genome is highly conserved compared to other bacterial pathogens, specific polymorphic genomic regions have been sought. Much like eukaryotic genomes, those of prokaryotes (including *M. tuberculosis*) are also characteristically punctuated by monomeric sequences repeated periodically (repeated units). Frothingham and Meeker-O'Connell [48] identified variable and informative polymorphic VNTR loci in *M. tuberculosis* complex strains and found 11 loci comprising five major polymorphic tandem repeats (MPTRs; A–E) and six exact tandem repeats (ETRs; A–F) ranging in size from 53 to 79 bp. Since then, additional VNTR loci have been reported. As such, Supply *et al.* [60] identified 41 VNTR of MIRU representing tandem repeats of 40–100 bp dispersed in the genome. VNTR genotyping basically relies on: (i) PCR amplification using primers specific for the flanking regions of the VNTRs and (ii) the determination of the sizes of the amplicons, after electrophoretic migration. As the length of the repeat units is known, these sizes reflect the number of the amplified VNTR copies. The final result is a numerical code, corresponding to the repeat number in each VNTR locus. Such numerical genotypes are simple for processing and are thus particularly convenient for both intra- and inter-laboratory comparative studies. In addition, compared to IS*6110*-RFLP, MIRU-VNTR typing has the advantages of being faster and appropriate for virtually all *M. tuberculosis* isolates, including strains that have few IS*6110* copies (GenoScreen Ltd., MIRU-VNTR typing manual). In 2006, Supply *et al.* [60] proposed a standardized protocol for 24 VNTR loci typing which

Table 9.1 Identified Bulgarian geographically specific spoligotypes and their prevalence worldwide (SpolDB4 database).

ST type	Spoligo octal designation / Phylogenetic lineage	Country prevalence of the ST spoligotype: MTB strain	Country	Prevalence (%)	Prevalence in SpolDB4 database: Number of ST genotype	Country	Database prevalence (%)	Worldwide distribution in SpolDB4 database (%)
ST125	000000007760731	196	Bulgaria	11.2	32/56	Bulgaria	57	0.08
		1213	Belgium	0.16	2/56	Belgium	3.5	
	LAM/S	684	Germany	0.3	2/56	Germany	3.5	
		226	Paraguay	0.9	2/56	Paraguay	3.5	
		240	Iran	0.8	2/56	Iran	3.5	
ST154	757777777760771	196	Bulgaria	5.1	11/59	Bulgaria	18.6	0.001
		172	Greece	2.3	4/59	Greece	6.7	
	T	841	Turkey	0.2	2/59	Turkey	3.3	
		13	Romania	Two strains	2/59	Romania	3.3	
		684	Germany	0.4	3/59	Germany	5	
		1994	Italy	0.5	10/59	Italy	18	
		110	Austria	7.2	8/59	Austria	13.5	
		1322	Argentina	0.2	3/59	Argentina	5	
ST878	777776777760571	196	Bulgaria	3	6/17	Bulgaria	35	0.0002
	X1 – VAR	841	Turkey	0.35	3/17	Turkey	17	
ST284	037637777760771	196	Bulgaria	6.1	12/107	Bulgaria	11	0.15
		110	Austria	3.6	4/107	Austria	3.7	
	T1	1213	Belgium	0.4	5/107	Belgium	4.6	
		172	Greece	2.3	4/107	Greece	3.7	
		841	Turkey	6	31/107	Turkey	29	
		240	Iran	2.5	6/107	Iran	5.6	
		1512	Saudi Arabia	1	15/107	Saudi Arabia	14	

is at present adopted as a reference method throughout the European Union. The discriminatory power of VNTR typing has been shown to be better than spoligotyping, and similar to IS6110-RFLP. Manual and automated analysis have been developed with fluorescently labeled primers and capillary electrophoresis or nondenaturing HPLC. The method is ideal for evolutionary and population genetic studies [49] since each locus has a different molecular clock [50].

9.15 Single Nucleotide Polymorphism

Another typing method that is gaining significance is single nucleotide polymorphism (SNP) typing. Informative single nucleotide genetic polymorphisms have provided researchers with markers to differentiate isolates as well as to study their phylogenetic relatedness. Studies have shown that the *M. tuberculosis* complex (i.e., *M. tuberculosis, M. bovis, M. microti, M. africanum, M. canettii,* and more recently, *M. pinnipedii, M. caprae*) genomes are highly conserved. Comparative sequence analysis of the 275-bp internal transcribed spacer (ITS) region which separates the 16S rRNA and the 23S rRNA is considered to be one of the most polymorphic regions in *Mycobacteria*. However, it revealed complete conservation between members of the *M. tuberculosis* complex [48]. Sequence analysis of 56 structural genes in several hundred phylogenetically and geographically diverse *M. tuberculosis* complex isolates suggested that allelic polymorphisms are extremely rare [51]. Similar results have been obtained by Comas *et al.* [52] who performed a multilocus sequence analysis (MLSA) on 108 *M. tuberculosis* complex (MTBC) strains in which they generated the complete coding sequences of 89 genes, corresponding to 70 kbp per strain. While the members of the *M. tuberculosis* complex display diverse phenotypic characteristics and host ranges, they represent an extreme example of interspecies genetic homogeneity, with an estimated rate of synonymous nucleotide polymorphisms of 0.01–0.03% [53] and no significant evidence for horizontal genetic transfer between genomes, unlike most other bacterial pathogens.

The disadvantage of the method is thus that it requires extensive genomic sequencing of multiple chromosome targets. Both specifically targeted nonsynonymous (ns)SNPs and synonymous (s)SNPs provide however useful genetic information that can be applied to differentiate *M. tuberculosis* strains. They address different biologic questions. sSNP mutations do not result in amino acid change and are not associated with selective pressure, hence they are ideal for population genetic studies. nsSNP mutations create an amino acid change and may be subject to selection pressure, thus they can be used to study drug resistance-determining genetic loci or the influence of other environmental, physical, or chemical stress factors. Nonsynonymous changes in drug resistance-determining genetic loci can result in phenotypic drug resistance. Accordingly, *M. tuberculosis* resistance to antituberculosis agents nearly always correlates with genetic alterations (nonsynonymous point mutations, small duplications, or deletions) in resistance-determining chromosomal regions [54]. Thus, SNPs have been extensively studied with the goal of developing a platform able to target multiple polymorphisms that are informative, such as phylogenetic grouping, drug resistance, virulence, and other epidemiologically significant markers.

Comparative genomic analysis of strains H37Rv and CDC1551 has revealed large-sequence polymorphisms (LSPs) in addition to SNPs [53]. LSPs are thought to mainly occur as a result of genomic deletions and rearrangements rather than through recombination following horizontal transfer. However this seems to occur rarely in *M. tuberculosis.* In the absence of horizontal gene transfer, deletions are irreversible and often unique events and therefore have been proposed for genotyping. It was found that up to 4.2% of the entire genome can be deleted in clinical isolates [55].

9.16
The Clustering Question?

The clustering of strains is based on the calculated similarities, using bootstrapping as a means for assigning a strain to a cluster. However, such clusters should always be interpreted with care since accidental clusters may exist, without any relation one to another. The use of multiple techniques, looking at different loci over the chromosome showing a different molecular clock may help to solve this problem. This however quite increases the workload. Some problems and pitfalls of clustering are exemplified here.

A large population-based molecular epidemiologic study in San Francisco (Calif., USA) and Baltimore (Md., USA) [56] revealed that an epidemiologic link between patients correlated with the designated molecular cluster in only 10–25% of the cases. Similarly, Mathema *et al.* [57] reported that only 30% of cases could be linked with clinical, demographic, and contact-tracing information.

Molecular clustering of strains depends on the specific molecular method used and on the tolerance for differences. The method and tolerance should be chosen with the appropriate study question and population in mind. Strains originating from one region and clustering with another region does not necessarily mean that both patient groups have been in contact. In these cases phylogeographic and historical influences should also be considered. In regions where the genetic diversity of the strain population is limited one can expect a gross overestimation of true clustering using spoligotyping since the discriminatory power is limited. This is exemplified in China with the Beijing strains have in general identical ST patterns [58]. Similarly, clustering based on identical IS*6110*-based RFLP patterns may yield numerous hybridization profiles with various degrees of similarity and provide an underestimate of true clustering rates, since some strains that have similar IS*6110* profiles may in fact be related by a recent common progenitor.

9.17 Conclusions

Diagnosis and subsequent molecular epidemiological studies remain problematic for *Mycobacteria*. While serological diagnosis has improved tremendously, the isolation necessary for subsequent molecular epidemiological studies remains a pitfall; the latter being extremely important in the control and elimination of tuberculosis.

The main questions to be solved by molecular epidemiology are the determination of origins and the distribution of strains. In general, by increasing the discriminatory power one may increase the utility for short-term local epidemiological investigations. Less discriminatory methods may allow strains to be grouped according to a certain characteristic. In any case, the molecular clock of the sequence should be taken into account. Molecular typing methods have been successfully applied for the determination of outbreaks and the study of *M. tuberculosis* transmission dynamics in these. The most significant practical result from performing a molecular typing of MTB is the evaluation of TB control programs [59, 61–63].

The public databases help us to determine the geographic spread of strains worldwide. They can assist in monitoring the spread of drug-resistant strains. The further improvement of global databases will improve investigations of the global evolution of *M. tuberculosis* worldwide.

References

1. World HealthWorld Health Organization (2009) WHO Report 2009: Global Tuberculosis Control 2009 – Epidemiology, Strategy, Financing. WHO/HTM/TB/2009.411.
2. Korenromp, E.L., Bierrenbach, A.L., Williams, B.G., and Dye, C. (2009) The measurement and estimation of tuberculosis mortality. *Int. J. Tuberc. Lung. Dis.*, **13**, 283–303.
3. Ryan, K.J. and Ray, C.G. (eds) (2004) *Sherris Medical Microbiology*, 4th edn, McGraw-Hill, ISBN: 0-8385-8529-9.
4. Lalvani, A., Pathan, A.A., McShane, H. *et al.* (2001) Rapid detection of Mycobacterium tuberculosis infection by enumeration of antigen-specific T cells. *Am. J. Respir. Crit. Care Med.*, **163**, 824–828.
5. Achtman, M. (2008) Evolution, population structure, and phylogeography of genetically monomorphic bacterial pathogens. *Annu. Rev. Microbiol.*, **62**, 53–70.
6. Al-Muhsen, S. and Casanova, J.L. (2008) The genetic heterogeneity of mendelian susceptibility to mycobacterial diseases. *J. Allergy Clin. Immunol.*, **122**, 1043–1051.
7. Dufour, G. (1993) Mycobacteriology. *Semin. Pediatr. Infect. Dis.*, **4**, 205–213.
8. Roberts, G.D., Koneman, E.W., and Kim, Y.K. (1991) Mycobacterium, in *Manual of Clinical Microbiology*, 7th edn (eds P.R. Murray, E.J. Baron, M.A. Pfaller, F.C. Tenover, and R.H. Yolken), American Society for Microbiology, Washington, DC, pp. 74–77.
9. Kent, P.T. and Kubica, G.P. (1985) *Public Health Mycobacteriology: A Guide for the Level III Laboratory*, Centers for Disease Control, Atlanta, GA.
10. Murray, P. (1992) Laboratory diagnosis of mycobacteriosis, *Clinical Microbiology Updates*, Hoechst-Roussel Pharmaceuticals, Somerville, NJ.
11. Hobby, G.L., Holman, A.P., Iseman, M.D., and Jones, J. (1973) Enumeration of tubercle bacilli in sputum of

patients with pulmonary tuberculosis. *Antimicrob. Agents Chemother.*, **4**, 94–104.

12. Yeager, H.J. Jr., Lacy, J., Smith, L., and LeMaistre, C. (1967) Quantitative studies of mycobacterial populations in sputum and saliva. *Am. Rev. Respir. Dis.*, **95**, 998–1004.
13. Lipsky, G.J., Gates, J., Tenover, F.C., and Plorde, J. (1984) Factors affecting the clinical value for acid-fast bacilli. *Rev. Infect. Dis.*, **6**, 214–222.
14. Morgan, M.A., Horstmeier, C.D., DeYoung, D.R., and Robers, G.D. (1983) Comparison of a radiometric method (BACTEC) and conventional culture media for recovery of mycobacteria from smear-negative specimens. *J. Clin. Microbiol.*, **18**, 384–388.
15. Ichiyama, S., Shimokata, K., and Takeuchi, J., The AMR Group (1993) Comparative study of a biphasic culture system (Roche MB check system) with a conventional egg medium for recovery of mycobacteria. *Tuberc. Lung Dis.*, **74**, 338–341.
16. Starke, J.R. (1988) Modern approach to the diagnosis and management of tuberculosis in children. *Pediatr. Clin. North Am.*, **35**, 464.
17. Braun, M. (1993) Pediatric tuberculosis, bacille Calmette-Guerin immunization, and acquired immunodeficiency syndrome. *Semin. Infect. Dis.*, **4**, 261–268.
18. American Thoracic Society (1997) Diagnosis and treatment of disease caused by nontuberculous mycobacteria. *Am. J. Respir. Crit. Care Med.*, **156**, S1–S25.
19. Runyon, E.H. (1970) Identification of mycobacterial pathogens utilizing colony characteristics. *Am. J. Clin. Pathol.*, **54**, 578–586.
20. Shinnick, T. and Good, R. (1995) Diagnostic mycobacteriology laboratory practices. *Clin. Infect. Dis.*, **21**, 291–299.
21. Butler, W.R. and Kilburn, J.O. (1988) Identification of major slowly growing pathogenic mycobacteria and *Mycobacterium gordonae* by high performance liquid chromatography of their mycolic acids. *J. Clin. Microbiol.*, **26**, 50–53.
22. American Thoracic Society (2000) Diagnostic Standards and Classification of Tuberculosis in Adults and Children. *Am. J. Respir. Crit. Care Med.*, **161**, 1376–1395.
23. Richeldi, L., Barnini, S., and Saltini, C. (1995) Molecular diagnosis of tuberculosis. *Eur. Respir. J. Suppl.*, **20**, 689–700.
24. Centers for Disease Control and Prevention (2004) *CDC M. Tuberculosis Nucleic Acid Amplification Testing Performance Evaluation Program*, CDC, Atlanta, GA. *http://www.phppo.cdc.gov/mpep/pdf/mtb/0401naa.pdf* (accessed July 21 2011).
25. U.S. Food and Drug Administration. Center for Devices and radiological Health Gen-Probe® Amplified™ *Mycobacterium Tuberculosis* Direct (MTD) Test – P940034/S008, Part 2, pp. 15–25. *http://www.fda.gov/cdrh/pdf/P940034S008b.pdf* (accessed July 21 2011).
26. Gomez, J.E. and McKinney, J.D. (2004) M. tuberculosis persistence, latency, and drug tolerance. *Tuberculosis (Edinb.)*, **84**, 29–44.
27. Huebner, R.E., Schein, M.F., and Bass, J.B. Jr. (1993) The tuberculin skin test. *Clin. Infect. Dis.*, **17**, 968–975.
28. Sorensen, A.L., Nagai, S., Houen, G., Andersen, P., and Andersen, A.B. (1995) Purification and characterization of a low-molecular-mass T-cell antigen secreted by Mycobacterium tuberculosis. *Infect. Immun.*, **631**, 710–717.
29. Mahairas, G.G., Sabo, P.J., Hickey, M.J., Singh, D.C., and Stover, C.K. (1996) Molecular analysis of genetic differences between Mycobacterium bovis BCG and virulent M. bovis. *J. Bacteriol.*, **178**, 1274–1282.
30. Behr, M.A., Wilson, M.A., Gill, W.P. *et al.* (1999) Comparative genomics of BCG vaccines by whole genome DNA microarray. *Science*, **284**, 1520–1523.
31. Van Pinxteren, L.A., Ravn, P., Agger, E.M., Pollock, J., and Andersen, P. (2000) Diagnosis of tuberculosis based on the two specific antigens ESAT-6 and CFP10. *Clin. Diagn. Lab. Immunol.*, **7**, 155–160.
32. Andersen, P., Munk, M.E., Pollock, J.M., and Doherty, T.M. (2000) Specific immune-based diagnosis of tuberculosis. *Lancet*, **356**, 1099–1104.

33. Ravn, P., Demissie, A., Eguale, T., Wondwosson, H., Lein, D., Amoudy, H.A., Mustafa, A.S., Jensen, A.K., Holm, A., Rosenkrands, I., Oftung, F., Olobo, J., von Reyn, F., and Andersen, P. (1999) Human T cell responses to the ESAT-6 antigen from Mycobacterium tuberculosis. *J. Infect. Dis.*, **179**, 637–645.
34. Arend, S.M., Andersen, P., van Meijgaarden, K.E., Skjot, R.L., Subronto, Y.W., van Dissel, J.T., and Ottenhoff, T.H. (2000) Detection of active tuberculosis infection by T cell responses to early-secreted antigenic target 6-kDa protein and culture filtrate protein 10. *J. Infect. Dis.*, **181**, 1850–1854.
35. Brock, I., Munk, M.E., Kok-Jensen, A., and Andersen, P. (2001) Performance of whole blood IFN-gamma test for tuberculosis diagnosis based on PPD or the specific antigens ESAT-6 and CFP-10. *Int. J. Tuberc. Lung. Dis.*, **5**, 462–467.
36. Aagaard, C., Govaerts, M., Okkels, L.M., Andersen, P., and Pollock, J. (2003) Genomic approach to the identification of Mycobacterium bovis diagnostic antigens in cattle. *J. Clin. Microbiol.*, **41**, 3719–3728.
37. Tufariello, J.M., Chan, J., and Flynn, J.L. (2003) Latent tuberculosis, mechanisms of host and bacillus that contribute to persistent infection. *Lancet Infect. Dis.*, **3**, 578–590.
38. Wood, P.R., Corner, L.A., and Plackett, P. (1990) Development of a simple, rapid in vitro cellular assay for bovine tuberculosis based on the production of γ interferon. *Res. Vet. Sci.*, **49**, 46–49.
39. Abadia, E., Zhang, J., dos Vultos, T., Ritacco, V., Kremer, K., Aktas, E., Matsumoto, T., Refregier, G., van Soolingen, D., Gicquel, B., and Sola, C. (2010) Resolving lineage assignation on Mycobacterium tuberculosis clinical isolates classified by spoligotyping with a new high-throughput 3R SNPs based method. *Infect. Genet. Evol.*, **10** (7), 1066–1074.
40. Filliol, I., Motiwala, A.S., Cavatore, M., Qi, W., Hazbon, M.H., Bobadilla del Valle, M.B., Fyfe, J. *et al.* (2006) Global phylogeny of Mycobacterium tuberculosis based on single nucleotide polymorphism (SNP) analysis: insights into tuberculosis evolution, phylogenetic accuracy of other DNA fingerprinting systems, and recommendations for a minimal standard SNP set. *J. Bacteriol.*, **188** (2), 759–772.
41. Hunter, P.R. and Gaston, M.A. (1988) Numerical index of the discriminatory ability of typing systems: an application of Simpson's index of diversity. *J. Clin. Microbiol.*, **26**, 2465–2466.
42. Hunter, P.R. (1990) Reproducibility and indices of discriminatory power of microbial typing methods. *J. Clin. Microbiol.*, **28**, 1903–1905.
43. Hunter, P.R. (1991) A critical review of typing methods for Candida albicans and their applications. *Crit. Rev. Microbiol.*, **17**, 417–434.
44. van Belkum, A., Tassios, P., Dijkshoorn, L., Haeggman, S., Cookson, B., Fry, N.K., Fussing, V., Green, J., Feil, E., Gerner-Smidt, P., Brisse, S., and Struelens M., for the European Society of Clinical Microbiology and Infectious Diseases (ESCMID) Study Group on Epidemiological Markers (ESGEM) (2007) Guidelines for the validation and application of typing methods for use in bacterial epidemiology. *Clin. Microbiol. Rev.*, **13** (Suppl. 3), 1–46.
45. Benjamin, W.H. Jr., Lok, K.H., Harris, R., Brook, N., Bond, L., Mulcahy, D., Robinson, N., Pruitt, V., Kirkpatrick, D.P., Kimerling, M.E., and Dunlap, N.E. (2001) Identification of a contaminating Mycobacterium tuberculosis strain with a transposition of an IS6110 insertion element resulting in an altered spoligotype. *J. Clin. Microbiol.*, **39** (3), 1092–1096.
46. Hanekom, M., van der Spuy, G.D., Gey van Pittius, N.C., McEvoy, C.R., Hoek, K.G., Ndabambi, S.L., Jordaan, A.M., Victor, T.C., van Helden, P.D., and Warren, R.M. (2008) Discordance between mycobacterial interspersed repetitive-unit-variable-number tandem-repeat typing and IS6110 restriction fragment length polymorphism genotyping for analysis of Mycobacterium tuberculosis Beijing strains in a setting of high incidence of tuberculosis. *J. Clin. Microbiol.*, **46** (10), 3338–3345.

47. Warren, R.M., Sampson, S.L., Richardson, M., Van Der Spuy, G.D., Lombard, C.J., Victor, T.C., and van Helden, P.D. (2000) Mapping of IS6110 flanking regions in clinical isolates of Mycobacterium tuberculosis demonstrates genome plasticity. *Mol. Microbiol.*, **37** (6), 1405–1416.
48. Frothingham, R. and Meeker-O'Connell, W.A. (1998) Genetic diversity in the Mycobacterium tuberculosis complex based on variable numbers of tandem DNA repeats. *Microbiology*, **144**, 1189–1196.
49. Cardoso Oelemann, M., Gomes, H.M., Willery, E., Possuelo, L., Batista Lima, K.V., Allix-Béguec, C., Locht, C., Goguet de la Salmonière, Y.O., Gutierrez, M.C., Suffys, P., and Supply, P. (2011) The forest behind the tree: phylogenetic exploration of a dominant Mycobacterium tuberculosis strain lineage from a high tuberculosis burden country. *PLoS ONE*, **6** (3), 18256.
50. Reyes, J.F. and Tanaka, M.M. (2010) Mutation rates of spoligotypes and variable numbers of tandem repeat loci in Mycobacterium tuberculosis. *Infect. Genet. Evol.*, **10** (7), 1046–1051.
51. Frothingham, R., Hills, H.G., and Wilson, K.H. (1994) Extensive DNA sequence conservation throughout the *Mycobacterium tuberculosis* complex. *J. Clin. Microbiol.*, **32**, 1639–1643.
52. Comas, I., Homolka, S., Niemann, S., and Gagneux, S. (2009) Genotyping of genetically monomorphic bacteria: DNA sequencing in Mycobacterium tuberculosis highlights the limitations of current methodologies. *PLoS One*, **12**, **4** (11), e7815.
53. Fleischmann, R.D., Alland, D., Eisen, J.A., Carpenter, L., White, O., Peterson, J., DeBoy, R., Dodson, R., Gwinn, M., Haft, D., Hickey, E., Kolonay, J.F., Nelson, W.C., Umayam, L.A., Ermolaeva, M., Salzberg, S.L., Delcher, A., Utterback, T., Weidman, J., Khouri, H., Gill, J., Mikula, A., Bishai, W., Jacobs, W.R., Venter, J.C., and Fraser, C.M. Jr. (2002) Whole-genome comparison of *Mycobacterium tuberculosis* clinical and laboratory strains. *J. Bacteriol.*, **184**, 5479–5490.
54. Maus, C.E., Plikaytis, B.B., and Shinnick, T.M. (2005) Mutation of *tlyA* confers capreomycin resistance in *Mycobacterium tuberculosis. Antimicrob. Agents Chemother.*, **49**, 571–577.
55. Tsolaki, A.G., Hirsh, A.E., DeRiemer, K., Enciso, J.A., Wong, M.Z., Hannan, M., Goguet de la Salmoniere, Y.O., Aman, K., Kato-Maeda, M., and Small, P.M. (2004) Functional and evolutionary genomics of Mycobacterium tuberculosis: insights from genomic deletions in 100 strains. *Proc. Natl. Acad. Sci. U.S.A.*, **101**, 4865–4870.
56. Bishai, W.R., Graham, N.M., Harrington, S., Pope, D.S., Hooper, N., Astemborski, J., Sheely, L., Vlahov, D., Glass, G.E., and Chaisson, R.E. (1998) Molecular and geographic patterns of tuberculosis transmission after 15 years of directly observed therapy. *J. Am. Med. A*, **280**, 1679–1684.
57. Mathema, B., Bifani, P.J., Driscoll, J., Steinlein, L., Kurepina, N., Moghazeh, S.L., Shashkina, E., Marras, S.A., Campbell, S., Mangura, B., Shilkret, K., Crawford, J.T., Frothingham, R., and Kreiswirth, B.N. (2002) Identification and evolution of an IS*6110* low-copy-number *Mycobacterium tuberculosis* cluster. *J. Infect. Dis.*, **185**, 641–649.
58. Kremer, K., Glynn, J.R., Lillebaek, T., Niemann, S., Kurepina, N.E., Kreiswirth, B.N., Bifani, P.J., and van Soolingen, D. (2004) Definition of the Beijing/W lineage of *Mycobacterium tuberculosis* on the basis of genetic markers. *J. Clin. Microbiol.*, **42**, 4040–4049.
59. Kato-Maeda, M., Metcalfe, J.Z., and Flores, L. (2011) Genotyping of Mycobacterium tuberculosis: application in epidemiologic studies. *Future Microbiol.*, **6** (2), 203–216.
60. Supply, P., Allix, C., Lesjean, S., Cardoso-Oelemann, M., Rusch-Gerdes, S., Willery, E., Savine, E., de Haas, P., van Deutekom, H., Roring, S., Bifani, P., Kurepina, N., Kreiswirth, B., Sola, C., Rastogi, N., Vatin, V., Gutierrez, M.C., Fauville, M., Niemann, S., Skuce, R., Kremer, K., Locht, C., and van Soolingen, D. (2006) Proposal for standardization of

optimized mycobacterial interspersed repetitive unit-variable-number tandem repeat typing of *Mycobacterium tuberculosis*. *J. Clin. Microbiol.*, **44**, 4498–4510.

61. Arend, S.M., van Soolingen, D., and Ottenhoff, T.H. (2009) Diagnosis and treatment of lung infection with nontuberculous mycobacteria. *Curr. Opin. Pulm. Med.*, **15** (3), 201–208.

62. Griffith, D.E., Aksamit, T., Brown-Elliott, B.A., Catanzaro, A., Daley, C., Gordin, F., Holland, S.M., Horsburgh, R., Huitt, G., Iademarco, M.F., Iseman, M., Olivier, K., Ruoss, S., von Reyn, C.F., Wallace, R.J. Jr., and Winthrop, K., ATS Mycobacterial Diseases Subcommittee; American Thoracic Society; Infectious Disease Society of America (2007) An official ATS/IDSA statement: diagnosis, treatment, and prevention of nontuberculous mycobacterial diseases. *Am. J. Respir. Crit. Care Med.*, **175** (4), 367–416.

63. Association of State and Territorial Public Health Laboratory Directors and the Centers for Disease Control and Prevention (1995) *Mycobacterium Tuberculosis*, Assessing Your Laboratory, U.S. Government Printing Office, Washington, DC.

Part II
Viruses

BSL3 and BSL4 Agents: Epidemiology, Microbiology, and Practical Guidelines, First Edition.
Edited by Mandy C. Elschner, Sally J. Cutler, Manfred Weidmann, and Patrick Butaye.

10
Influenza Virus: Highly Pathogenic Avian Influenza

Chantal J. Snoeck, Nancy A. Gerloff, Radu I. Tanasa, F. Xavier Abad, and Claude P. Muller

10.1
Introduction

Influenza A viruses have been known for decades to cause disease in avian and mammalian species. The natural reservoirs are wild birds, which host all subtypes. Stable lineages are also established in domestic poultry, humans, pigs, or horses [1]. In humans, new strains with an avian genetic background were successively introduced in 1918 (H1N1), 1957 (H2N2), and 1968 (H3N2), leading to three pandemics separated by years of seasonal activity. Each subtype superseded the previous strain, but in 1977 the H1N1 virus was reintroduced in the human population and both H3N2 and H1N1 viruses have co-circulated since [2]. Despite effective vaccines, about 250 000–500 000 people die each year of seasonal influenza [3]. In domestic poultry, certain subtypes become highly pathogenic (HP) and can be lethal for humans, but so far no highly pathogenic avian influenza (HPAI) virus has acquired the ability of sustained human to human transmission. For instance during the 2003 outbreak of HPAI H7N7 viruses in the Netherlands, 30 million birds were killed or culled [4]. Several infected individuals developed a mild disease but one veterinarian died [5, 6]. While this virus has successfully been controlled, the HPAI H5N1 virus, which emerged in Asia, spread worldwide and infected hundreds of people with a 60% mortality rate. This chapter will focus on HPAI viruses, in particular H5N1, as it was, until recently, expected to cause the next pandemic.

10.2
Characteristics of the Agent

10.2.1
Nomenclature

Influenza viruses belong to the family of *Orthomyxoviridae*, which includes the genera of *Influenzavirus A, B, C, Thogotovirus*, and *Isavirus*. Influenza A viruses are

BSL3 and BSL4 Agents: Epidemiology, Microbiology, and Practical Guidelines, First Edition.
Edited by Mandy C. Elschner, Sally J. Cutler, Manfred Weidmann, and Patrick Butaye.

further classified into subtypes based on the antigenicity of their hemagglutinin (HA) and neuraminidase (NA) proteins. There are 16 HA subtypes (H1–H16) and nine NA (N1–N9) subtypes known. The strain nomenclature describes the type of virus, the host (except for humans), the geographic origin, the strain number, and the year of detection, followed by the HA and NA subtypes [e.g., A/chicken/Nigeria/OG2/2007 (H5N1)] [7].

10.2.2
Genome and Protein Structure

Influenza A viruses are enveloped viruses containing eight negative-sense, single-stranded, segmented RNA molecules, corresponding to eight genes. The matrix (M) and nonstructural (NS) genes encode for two proteins [M1 and M2; NS1 and nuclear export protein (NEP)/NS2, respectively]. The lipid membrane of the virus is derived from the host cell and HA, NA, and M2 proteins are inserted into it, whereas the M1 protein can be found in a layer beneath this envelope. The core of the virus particle consists of ribonucleoprotein (RNP) complexes composed of viral RNA (vRNA) segments coated with nucleoprotein (NP) and is associated with the polymerase complex (PB1 – polymerase basic 1; PB2 – polymerase basic 2; and PA – polymerase acid). The NS1 is a multifunctional protein acting as an antagonist of the host cell antiviral response [8, 9]. The NEP/NS2 is also associated with the viral RNPs (vRNPs) and M1 protein [10]. A small protein of 87 amino acids (PB1-F2) is encoded by an alternative open reading frame in the PB1 gene [11].

10.2.3
Viral Replication

Host cell infection starts with the virus binding to neuraminic acids on the cell surface followed by endocytosis of the virus. Endosomal acidification induces the fusion of the viral envelope with the endosomal membrane and the release of vRNPs into the cytoplasm. The vRNPs migrate to the nucleus, where they act as a template for transcription. In the nucleus the negative-sense vRNA is transcribed into mRNA, which triggers the synthesis of viral proteins. The vRNA is replicated through a positive-sense intermediate, the complementary RNA (cRNA), which in turn is used to produce more vRNA. Viral RNPs are transported from the nucleus to the assembly site at the apical membrane of polarized cells (i.e., lung epithelial cells). Budding of complete viral particles is an active process, mediated by the enzymatic activity of NA, that removes sialic acids from the surface of the host cell [7, 12].

10.2.4
Antigenic Drift and Antigenic Shift

Due to the lack of proof-reading activity of RNA polymerases, mutations occur more often in RNA than in DNA viruses [13]. Mutations introduced in HA and NA proteins tend to modify antibody binding sites, leading to *antigenic drift*.

When a cell is co-infected with two different influenza A strains, viruses can exchange or reassort gene segments. The exchange of HA and NA segments can result in major antigenic changes and immune escape by a process referred to as *antigenic shift*. Viruses of different species may also be involved in the co-infection [14].

10.3 Pathogenesis

10.3.1 Reservoir

Wild birds, and in particular water birds, are the natural reservoir of low pathogenic avian influenza (LPAI) viruses and all HA and NA subtypes found so far occur in wild birds [1]. Migration and aquatic environments are important ecological factors for avian influenza [1, 15–17]. The aquatic environment supports efficient short-range virus transmission by the fecal–oral route [1]. Poultry (and humans) are at risk whenever wild and domestic birds intermingle [18, 19]. Moreover, during the dry season wild waterfowls are attracted by irrigated wetlands, thus increasing the chances of viral transmission to and from local poultry [20]. Influenza viruses are disseminated over long distances by migratory birds. High densities of mixed bird species at stopover and migration sites promote intra- and interspecies virus transmission [17, 21]. Thus, avian species are of major importance as a primary genetic reservoir of influenza viruses of mammalian species, including humans [21].

10.3.2 Low and Highly Pathogenic Influenza Viruses

Avian influenza viruses (AIVs) can be separated into low pathogenic (LP) or HP strains. The latter causes severe illness and mortality, which may be as high as 100% in domestic poultry. In contrast, most LPAI viruses are asymptomatic in many species of wild birds, or cause mild respiratory symptoms [22].

The factors contributing to emergence of HPAI viruses are not clearly understood. However, the current belief is that HPAI viruses emerge only during extensive circulation and adaptation of LPAI viruses in poultry [23]. This concept is based on observations of: (i) phylogenetic sublineages of avian influenza that include both HPAI and LPAI viruses [24, 25], (ii) *in vitro* selection of a HP virus from a LP virus [26], and (iii) LP precursor viruses in wild birds that become HP after introduction and circulation in poultry, as was observed during the Italian H7N1 outbreak in 1999–2000 [27, 28] and other outbreaks [29–32]. Although the main reservoirs of AIV infections in poultry are infected wild birds, transmission is not a one-way street: wild birds may also become infected by poultry, for example, by feeding on infected carcasses [33, 34].

10.3.3
Molecular Determinants of Pathogenicity

Pathogenesis and tissue tropism is largely, but not exclusively, determined by the nature of the HA protein and its cleavage site. In the host cell, the HA protein is synthesized as a precursor HA0 which is proteolytically cleaved into HA1 and HA2 subunits. The HA0 precursor of LPAI viruses is cleaved by trypsin-like proteases present in the respiratory or intestinal tract [35–38]. In ducks LPAI viruses replicate mainly in the intestinal tract, but also in the lower and upper respiratory tract. Typically the infection of ducks with most strains is asymptomatic [39–41]. In contrast, HA proteins of HPAI viruses possess multiple basic amino acids (aa) at the carboxyl terminus in the subunit HA1 (Table 10.1), a characteristic of H5 and H7 viruses only. This site can be cleaved by ubiquitous cellular proteases (e.g., furin-like proteases) [42, 43]. Thus the virus can cause systemic infections, including central nervous system involvement and death [44, 45].

The host species restriction of HPAI (H5N1) viruses is determined by multiple determinants including the receptor-binding specificity of the HA protein. Whereas most avian viruses have higher binding affinity for sialic acid α2,3 linked galactose (SAα2,3Gal), human influenza viruses bind preferentially to SAα2,6Gal [54]. In humans SAα2,6Gal oligosaccharides are more frequent on non-ciliated epithelial cells of the upper respiratory tract, which are preferentially infected by human viruses. SA2,3αGal oligosaccharides are present on ciliated cells of the lower respiratory tract [55, 56]. Receptor distribution in the respiratory tract may explain how humans can become easily infected by human strains but infections with avian viruses probably require a higher dose and are therefore

Table 10.1 Examples of avian influenza viruses and the cleavage site in their HA protein which determines pathogenicity [7,46–53].

Virus isolate	Subtype	Pathogenicity	Sequence at the HA cleavage site (*)
A/chicken/Pennsylvania/1/1983	H5N2	LPAI	PQKKKR*G
A/chicken/Mexico/31381-7/94	H5N2	LPAI	PQRERRRKKR*G
A/chicken/Queretaro/14588-19/95	H5N2	HPAI	PQRKRKTR*G
A/turkey/Italy/99 (consensus)	H7N1	LPAI	PEIPKGR*G
A/turkey/Italy/99 (consensus)	H7N1	HPAI	PEIPKGSRVRR*G
A/chicken/Chile/176822/02	H7N3	LPAI	PEKPKTR*G
A/chicken/Chile/4957/02	H7N3	HPAI	PEKPKTCSPLSRCRKTR*G
A/goose/Guangdong/1996	H5N1	HPAI	PQRERRRKKR*G
A/Viet Nam/DN-33/2004	H5N1	HPAI	PQRERRRKKR*G
A/bar headed goose/Qinghai/65/2005	H5N1	HPAI	PQGERRRKKR*G
A/chicken/Nigeria/OG2/2007	H5N1	HPAI	PQGERRRKKR*G

relatively rare. In the avian host, for example in ducks, SAα2,3Gal are found on epithelial cells of the intestine, where AIV replicates preferentially [57]. Receptor specificity is determined by the aa that form the receptor binding pocket. Glutamine in position 226 (Q226) dictates a preferential binding to sialic acid SAα2,3Gal oligosaccharides present on avian epithelial cells [58]. Mutations in this position (Q226L) of HA that allow binding to SAα2,6Gal receptor types of mammalian cells enhance viral replication in the upper respiratory tract and facilitate transmission to humans. In addition, the number and location of glycosylation sites play a role in virus–host interactions [59]. The NA protein is also involved in host range restriction and pathogenicity [60, 61]. NA activity of some avian viruses is more resistant to the low pH of the upper digestive tract than that of human- or swine-derived NA contributing to the host range restriction [62]. NA promotes viral spread within the respiratory tract by cleavage of sialic acids in the mucus [63].

Normally, avian viruses have glutamic acid in position 627 (E627) of PB2. However, a lysine 627 correlates with replication in mammalian cells, reduced host defense, and higher mortality in mice [64–66]. This mutation was not only found increasingly in H5N1 viruses that infected humans since 2001, but also in all subclade 2.2 strains (see Section 10.6), and in all human influenza strains [66–68]. However, other aa changes in PB2, PB1, and PA also interact with mammalian adaptation and virulence of HPAI viruses [69–71].

NS1 protein is also involved in viral pathogenicity by limiting host cell responses on multiple levels [8, 72–74]. Notably it targets both IFN-α/β production and antiviral effects of IFN-induced proteins [73, 75, 76]. Moreover, NS1 inhibits polyadenylation of cellular mRNA, preventing thereby its nuclear export. In parallel, translation is enhanced by NS1 in the cytoplasm, which leads to a high load of viral protein [77], whereas cellular proteins (whose cytoplasmic concentration is kept low by NS1) are poorly translated. This mechanism contributes to the limitation of the host antiviral response by NS1. Remarkably NS1 proteins of the first HPAI (H5N1) viruses conferred resistance to antiviral effects of IFN while inducing high levels of proinflammatory cytokines, such as tumor necrosis factor-α and IFN-β [78–80]. A specific alteration in the protein PB1-F2 (N66S) of HPAI (H5N1) showed increased pathogenicity in mice [81].

10.4 Clinical and Pathological Findings

10.4.1 HPAI (H5N1) Infection in Animals

In poultry, the disease induced by HPAI virus strains has historically been called *"fowl plague"*. After an incubation period of one to seven days, these strains often cause sudden death without prodromal symptoms. Morbidity and mortality often reach 90–100% within a few days. Birds that survive for 48 h develop respiratory distress, lacrimation, edema of the head and neck, sinusitis, subcutaneous hemorrhage with cyanosis of the comb, wattles and feet, and a comatose state. Extensive necrotic hemorrhagic lesions, interstitial pneumonia with edema and nephritis were reported at necropsy [82–84]. During the Qinghai Lake outbreak in 2005, neurological (tremor and opisthotonus) and gastrointestinal (diarrhea) symptoms were reported in many wild bird species. In addition, extensive pneumonia, myocardial degeneration, focal hepatitis, and pancreatic necrosis were also found in domestic geese and ducks [26]. Different mammalian species, and especially carnivores, are also susceptible to HPAI infections. Carnivores (cats, tigers, leopards, dogs, mustelids, civets) naturally infected with H5N1 subtype suffered from respiratory distress, convulsions, and death, with multiple organ hemorrhages, necrosis, and inflammation [85–88]. Infection with influenza virus H7N7 has also been associated with high mortality in harbor seals as a result of acute hemorrhagic pneumonia. The virus was directly transmitted from birds without reassortments [89].

10.4.2
HPAI (H5N1) Infection in Humans

The determination of the source of the virus is a critical component of a case investigation and it should inquire about: (i) exposure to sick, dead, or suspicious poultry or other animals from endemic areas, (ii) exposure to animal products (professional exposure), (iii) contact with suspected or confirmed human cases, and (iv) exposure to laboratory specimens [90–92].

The incubation time of H5N1 infection in humans is thought to be 2–10 days, but it is not clear when virus excretion starts [74, 93]. In humans, most infections lead to severe influenza syndrome, with fever, cough, shortness of breath, and pneumonia, but gastrointestinal symptoms are also often present [92, 94]. In severe cases the illness developed rapidly into bilateral pneumonia with acute respiratory distress symptoms requiring mechanical ventilation. Other complications included multi-organ failure and encephalitis [95]. Patients showed an early lymphopenia with an inverted CD4/CD8 ratio, thrombocytopenia, and elevated transaminase levels [74, 91, 93]. Immune pathology is thought to play an important role in H5N1 pathogenesis since high plasma levels of some proinflamatory cytokines (IL-6, IL-8, IL-10, IFN-γ, TNF-α), macrophage- and neutrophil-attractant chemokines (MCP-1, MIG, IP-10, RANTES) have been observed, predominantly in patients with fatal H5N1 subtype infection [72, 90, 96–99].

Influenza virus has been detected in brain neurons, enterocytes, and mononuclear cells of the intestinal mucosa, placenta, lymphnodes, and Kupffer cells in the liver, suggesting that extra-pulmonary dissemination is the result of viremia or infected immune cells transporting the virus to other organs [100, 101]. It is unknown whether H5N1 reaches the central nervous system through the blood–brain barrier or by spreading from peripheral nerve endings, as suggested from studies in felines and mice [86, 102].

10.5 Diagnosis

10.5.1 Direct Diagnosis

Virus isolation is the gold standard of influenza diagnosis and follows recommended protocols. Specimens collected into viral transport medium are inoculated into cell cultures of, for example, Madin Darby canine kidney (MDCK) or into the amnio-allantoic cavity of 9–11 day old embryonated chicken eggs. Inoculated eggs are incubated at 35–37 °C for 48–72 h. The allantoic fluid is tested for virus by hemagglutination and hemagglutination inhibition test (HAI). The HAI with reference antisera or monoclonal antibodies directed against one of the 16 different HAs and nine NAs subtypes is still the method of choice for the identification of influenza isolates [103, 104].

Some methods are based on antigen detection in clinical specimens by immunofluorescence, enzyme immunoassays, or immunochromatography using monoclonal antibodies against conserved epitopes of NP or M protein. Some of these methods are available also as rapid point of care assays that provide results within 15 min. Such assays are widely used for the rapid diagnosis of human influenza infection [105]. Although these assays are fast and simple to use, they lack sensitivity and normally do not differentiate between subtypes of avian and human influenza viruses [106]. More recently, such assays for the specific detection of H5N1 became available [107–109].

Molecular methods based on polymerase chain reaction (PCR) offer a rapid, sensitive, and subtype specific diagnosis of HPAI. Specific primers target regions that are highly conserved in all influenza A viruses, typically within the M (or NP) gene. Subtype specific primers identify HA and/or NA subtypes, for example, H5 and H7. Because of the continuing evolution of HPAI (H5N1) into distinct genetic clades, primer sequences should be updated periodically to maintain assay sensitivity and specificity [82, 106, 110, 111]. Viral RNA can be detected by several PCR approaches, including one-step PCRs combining a reverse transcription step followed by a PCR in a single reaction [103, 110, 112, 113], real time PCRs or multiplex PCRs which allow parallel detection of several gene segments, subtypes, or viruses in a single run [63, 90, 91, 114]. However, since nucleic acid methods cannot distinguish between live and inactivated viruses, PCR may not be appropriate to test the environment, for example, to certify permises free of replicating virus [111].

10.5.2 Indirect Diagnosis

Conventional techniques such as agar gel immunodiffusion (AGID) and HAI assays continue to be used in many countries [111]. While AGID detects antibodies against NP and M proteins, HAI assays identify and characterize serum antibodies to HA. Neuraminidase inhibition (NI) assays characterize the serological response to NA. NI assays can differentiate infected from vaccinated animals (DIVA strategy) [115, 116]. ELISA is used to detect antibodies against avian influenza of different subtypes including H5 or H7. Although these tests can easily be automated for large numbers of samples, they are of limited use in the case of acute infections and HPAI outbreaks [111, 117, 118]. Nevertheless, serology is useful for epidemiological studies and retrospective confirmation of LPAI infections [111, 114, 117–119]. A fourfold or greater increase of the antibody titer in the convalescent serum compared to acute serum (two to three weeks apart) or the presence of virus-specific IgM response in the acute phase indicates a recent infection [90, 91, 105]. Serological diagnoses of influenza in humans include neutralization or microneutralization (MN) assays using reference HPAI viruses [120]. The MN assay is highly sensitive but requires live virus and BSL-3 laboratory facilities [63]. Similar assays can be carried out in BSL-2 facilities using viral pseudotypes (retroviral virions carrying influenza H5 HA) [121–123].

10.5.3
Pathotyping

Pathotyping is essential to contain the spread of HPAI. The conventional method requires experimental inoculation of infective allantoic fluid intravenously into six-week old specific pathogen free (SPF) chickens. A virus is then classified as HP if it has an intravenous pathogenicity index (IVPI) greater than 1.2 or alternatively by at least 75% mortality in four to eight week old chicken. Viruses that do not fulfill these criteria should be sequenced to exclude the presence of multiple basic amino acids motifs at the proteolytic cleavage site. If the motif is similar to those of other HPAI isolates, the strain is considered as HP [103]. Depending on pathotype classification, the OIE recommends different interventions [82, 103].

10.6
Evolution and Geographic Spread of HPAI (H5N1) Viruses

10.6.1
Chronology of H5N1 Virus

10.6.1.1 First Wave

HPAI H5N1 was first isolated in Hong Kong in 1997 from a child with fatal respiratory illness. This was the first of 18 patients who became infected from poultry at live bird markets; six of them died [74]. Surveillance and epidemiological studies established that several AIV subtypes, including H5N1, co-circulated in chickens, ducks, and geese in live bird markets in Hong Kong [124]. The initial outbreaks were associated with a relatively low mortality in chickens, but the mortality in humans was as high as 30%. Intensive poultry culling started in December 1997 in markets and farms, contained the outbreaks, and no new cases were found until 2000 [125]. HPAI H5N1 probably resulted from a reassortment between a H5N1-like virus (HA gene, A/goose/Guangdong/1/96), a H9N2-like virus (internal genes, A/quail/Hong Kong/G1/97) [126] and/or a H6N1-like virus (NA gene and/or internal genes, A/teal/Hong Kong/W312/97) [127].

10.6.1.2 Second Wave

Poultry culling did not interrupt the continuous circulation of Gs/Gd/96-like viruses [128, 129], some of which reassorted with unknown viruses from an aquatic bird reservoir. Several genotypes with distinct internal genes emerged in 2001 and 2002, causing a number of outbreaks in China [130, 131]. One of these genotypes (Z) became dominant in southern China and eventually differentiated into the distinct H5N1 clades that continue to circulate today. In February 2003, the first human case was reported in Hong Kong since 1997. This virus (Z^+) was similar to genotype Z but lacked the NA stalk deletion characteristic of genotype Z strains [132]. During the same year, HPAI H5N1 started to spread to other Asian countries: by the end of 2004, the Republic of Korea, Thailand, Vietnam, Japan, Cambodia, Lao PDR, Indonesia, and Malaysia had experienced H5N1 outbreaks in poultry and on rare occasions the virus was also detected in dead wild birds [133, 134].

10.6.1.3 Third Wave

In 2005, a large outbreak affected thousands of waterfowl at Qinghai Lake, an important breeding site for migratory birds in western China. At least four genotypes were detected but one genotype became dominant [63, 135]. By the end of 2005, this virus was reported from Russia, Kazakhstan, Mongolia, Turkey, Romania, Croatia, Ukraine, and Kuwait. But this was only the beginning and this third wave eventually spread to more countries in Europe, the Middle East, and Africa [134]. Wild birds, especially ducks, may have contributed to this long distance spread since some species were relatively resistant to H5N1 morbidity [18, 136], and some European countries reported HPAI H5N1 cases in wild birds, without outbreaks in poultry [17].

During 2005 and later, outbreaks continued in eastern, southeastern and southern Asia, and phylogenetic analyzes revealed the co-circulation of several genotypes, named clades. Although various gene constellations resulting from reassortments were observed, the HA gene was still derived from an A/goose/Guangdong/1/96 H5N1-like virus evolving by genetic drift. This genetic diversity led to the adoption of an international standard nomenclature which is regularly adapted [137]. Starting with the initial H5N1 strain A/goose/Guangdong/1/96 (clade 0), HPAI H5N1 has now evolved into 10 major clades (0–9) and additional subclades (Figure 10.1). Clade 1 strains were found in Southeastern and Eastern Asia while clades 3–9 viruses are mainly restricted to Vietnam, Hong Kong and China. Clade 2 strains spread further in Asia, the Middle East, Europe, and Africa and diversified into second- and third-order groups (Figure 10.1). All clade 2.2 viruses were derived from early strains (A/bar-headed goose/Qinghai/A1/2005-like strain), transmitted through aquatic birds at Qinghai Lake [137]. The emergence of multiple

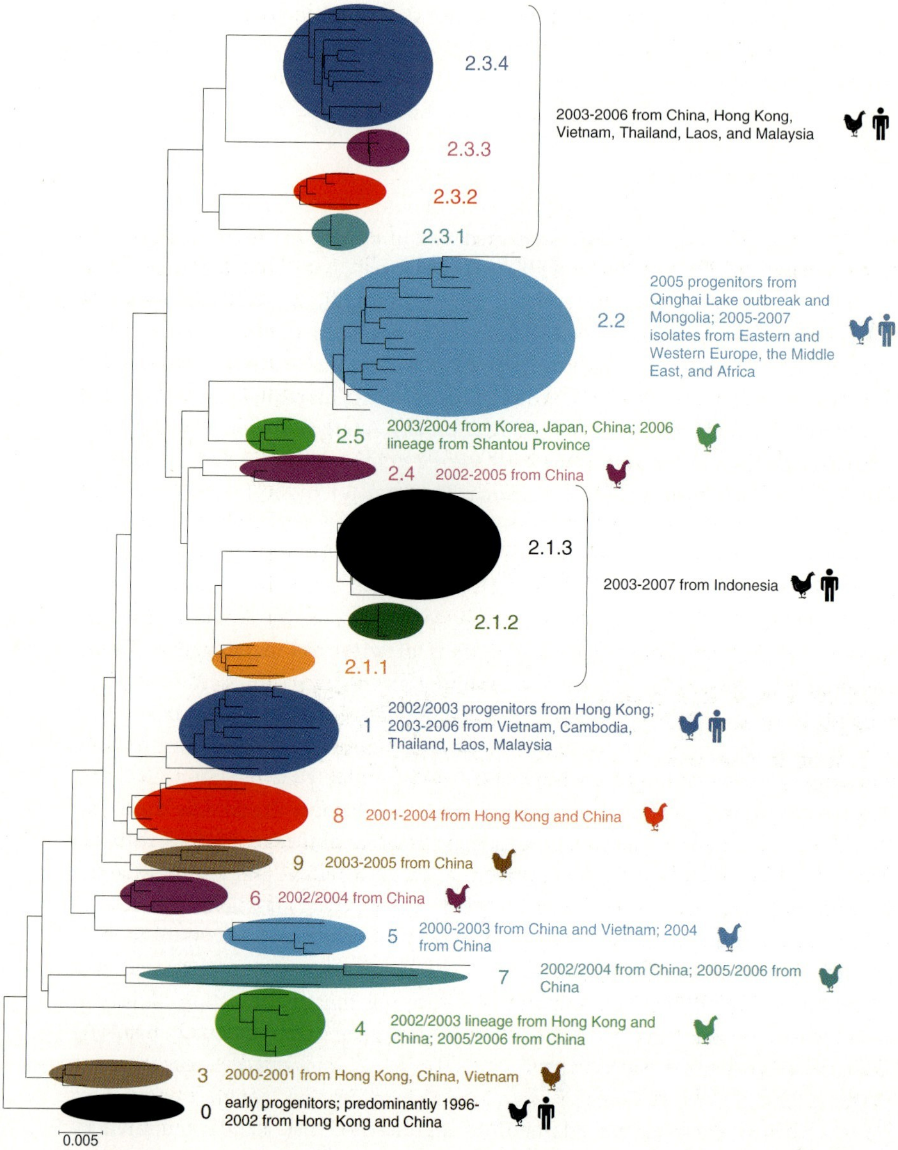

Figure 10.1 Genetic diversity of HPAI H5N1 virus since its emergence in 1997 (adapted from Ref. [137].)

clades and subclades in Asia reflects the uninterrupted circulation of H5N1 despite culling and vaccination measures [138]. In 2009, 11 Asian countries detected HPAI H5N1 in wild birds, farms, and live bird markets [139]. The movement of birds between live bird markets and backyard farms, free ranging ducks as an interface between wild and domestic birds, a large waterfowl population, legal or illegal bird movements, and poor biosafety measures all contribute to a favorable breeding ground for influenza viruses in Asia and beyond [17, 138].

10.6.2 Focus on Africa

A seroprevalence study in sub-Saharan Africa conducted in commercial poultry in Nigeria between 1999 and 2004 did not detect antibodies to influenza viruses. Since Nigeria has the largest and the most active poultry industry in that region, this may suggest that at least LPAI viruses did not enzootically circulate in sub-Saharan poultry populations [140]. However, when HPAI H5N1 swept from Asia across Russia to Europe, it also reached Africa. The first officially reported case occurred in commercial poultry farms in northern Nigeria in February 2006 [141], although one report based on a single strain suggests that HPAI H5N1 was already introduced in 2005 [142]. Soon afterward, HPAI H5N1 infections were reported throughout most Nigerian Federal States (*http://empres-i.fao.org*). The genetic diversity, the timeline, the observed substitution rates, and the phylogenetic relationship suggested that three sublineages (A, B, C) of clade 2.2 were independently introduced into the country [143, 144]. HA gene sequences clustered with strains found in Europe, Russia, and western China, but were distinct from strains in the rest of China and Southeast Asia [34, 143, 144]. The route of H5N1 introduction into Africa is difficult to establish. While illegal animal trade is not unusual, the presence of several bird sanctuaries along migratory pathways in West Africa are compatible with introductions by migratory birds [145]. Within three months, outbreaks were reported in Egypt (sublineage B), Niger (sublineage A), Cameroon, Burkina Faso (sublineage C), Sudan (sublineage C), Ivory Coast (sublineage C), and Djibouti (sublineage B), and all strains were most closely related to viruses found earlier in Nigeria [143, 144, 146, 147] (Figure 10.2). In Nigeria, the co-circulation of several sublineages led to multiple reassortment events between sublineage A and C viruses. One reassortant AC $_{PB1/HA/NP/NS}$ virus, found only once in 2006, had four genes (PB1/HA/NP/NS) from sublineage C and four other genes from sublineage A [148]. In 2007, one AC_{NS} (all genes except the NS gene derived from sublineage A) [51] and 15 $AC_{HA/NS}$ reassortants emerging from several independent reassortment events were found in several Nigerian states (Figure 10.3) [51, 149]. Interestingly, NS of all reassorted viruses originated from sublineage C viruses, which may reflect an improved fitness and adaptation to the African ecology [51]. In 2007, H5N1 spread also to three other countries, Ghana (sublineage C), Togo (sublineage A), and Benin ($AC_{HA/NS}$ reassortant) [150], probably by bird trade. Reassortant strains from Benin were closely related to Nigerian reassortant from 2007 [150]. In 2008, only four African countries (Nigeria, Togo, Egypt, Benin) reported H5N1 outbreaks [134, 151]. In Nigeria a H5N1 virus, phylogenetically most closely related to European strains, was found [147] (Figure 10.3). This constitutes the first evidence of a new virus introduction since the first HPAI H5N1 outbreaks in Nigeria and suggests that the introduction of HPAI H5N1 in Africa is a rare event. In 2009 only Egypt

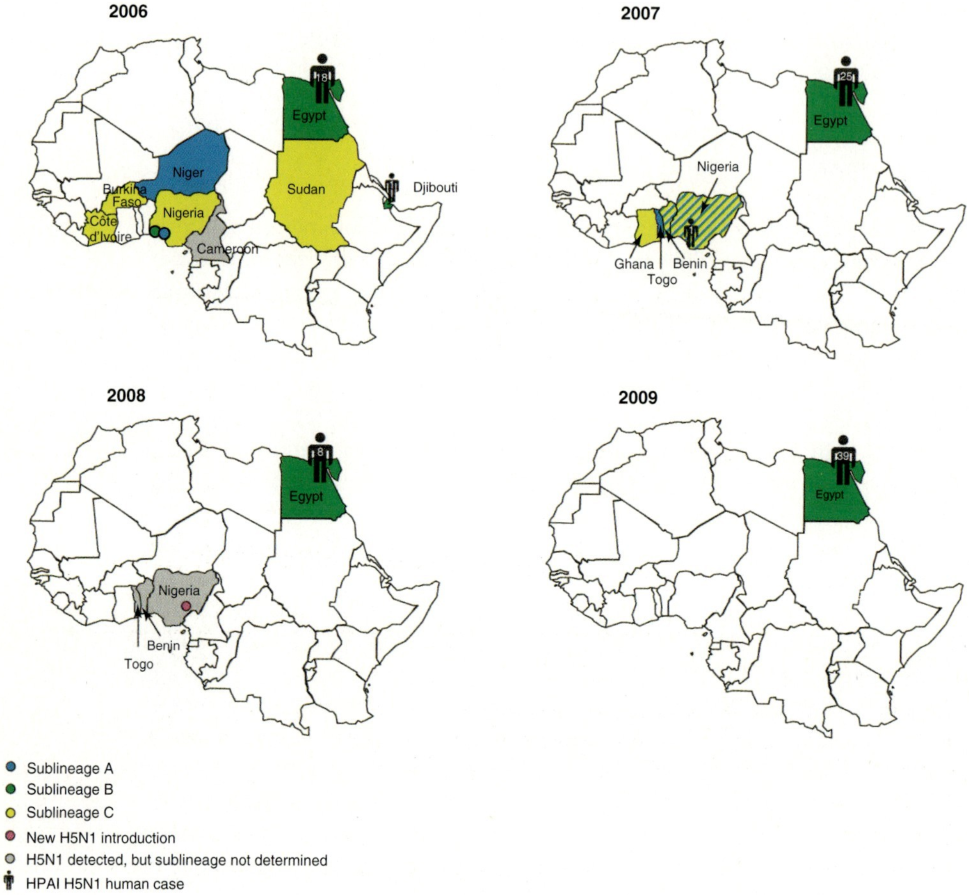

Figure 10.2 Epidemiology of HPAI H5N1 viruses in Africa since their first introduction in 2006 (based on data collected from *http://www.oie.int*, *http://www.who.int*, *http://empres-i.fao.org*, and EpiFlu Database *http://epiflu.vital-it.ch*.)

continued to struggle with outbreaks in domestic poultry [134]. Because of its high diversity, sublineage B (primarily found in Egypt) has been defined as a third-order clade 2.2.1 [152].

In Egypt, human cases of HPAI H5N1 (sublineage B) were reported every year since 2006 (Figure 10.2). In 2009, 39 of worldwide 52 human cases were reported from Egypt but the mortality in Egypt dropped from 45% (2006–2008) to 10% (2009) [153], possibly due to viral and/or public health factors. Some Egyptian viruses may also have evolved toward the receptor usage (SAα2,6Gal) of human viruses, increasing their replication efficiency in the upper respiratory tract [154]. In addition to Egypt, human cases of HPAI H5N1 occurred also in Djibouti in 2006

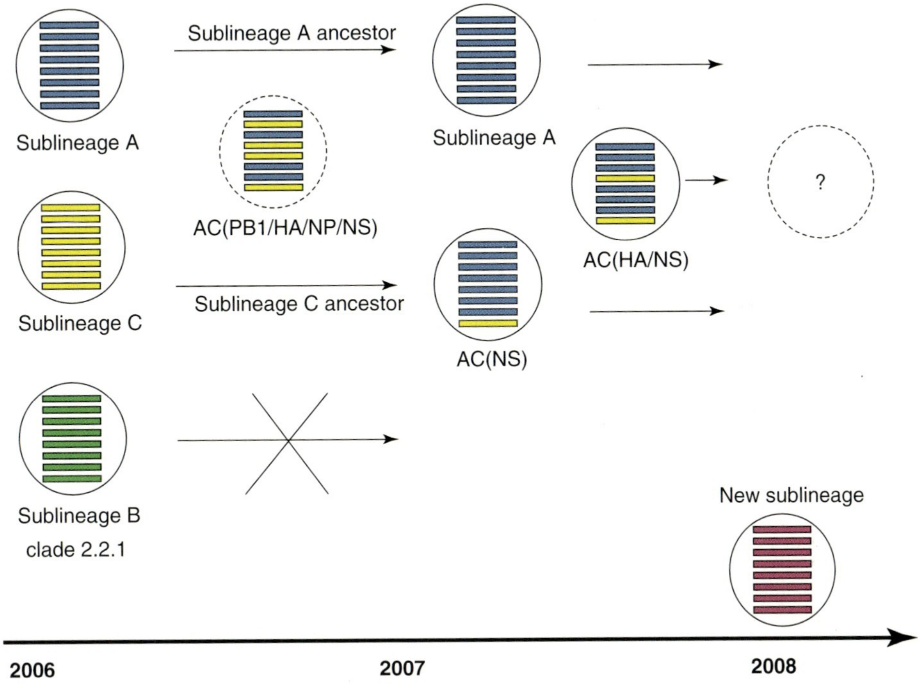

Figure 10.3 Circulation of HPAI (H5N1) sublineages in Nigeria since their introduction in 2006 and their derived reassortants [51, 143, 144, 147–149]. The same color code as used in Figure 10.2 was applied.

($n = 1$; sublineage B) [144] and in Nigeria in 2007 ($n = 1$; $AC_{HA/NS}$ reassortant; unpublished data).

10.7 Epidemiology of Other Influenza Subtypes

10.7.1 HPAI Virus Outbreaks

Besides the Asian HPAI H5N1, 29 other HPAI outbreaks caused by several H5 or H7 genotypes have occurred since the first confirmed HPAI outbreak in 1959 in Scotland [155, 156]. Supposedly all H5 and H7 viruses have the possibility to become HP in poultry; however, the majority of the HPAI outbreaks were caused by H5N1, H5N2, H7N3, and H7N7 viruses and one by each of the subtypes H5N3 (1963, South Africa), H5N8 (1963, Ireland), H5N9 (1966, USA), H7N1 (1999–2000, Italy), and H7N4 (1997, Australia) [156]. Several of these viruses were responsible for the death or culling of millions of birds, like H5N2 in the

United States in 1983–1985 [157], H5N2 in Mexico in 1994–1995 [158], three H7N3 outbreaks in Pakistan in 1994–1995 [159], 2000, and 2003–2004 [160], H7N1 in Italy in 1999–2000 [27], H7N3 in Chile in 2002 [161], H7N7 in the Netherlands in 2003 [4], and H7N3 in Canada in 2004 [162]. The frequency of HPAI outbreaks seems to increase with time although this could be biased by improved disease reporting.

10.7.2
LPAI Virus Outbreaks

LPAI viruses are mostly resident in wild birds but some also infect domestic birds. LPAI H5 and H7 subtypes generate most concern because of their potential to mutate into HPAI strains and therefore are notifiable influenza viruses (LPNAI). All outbreaks in poultry have economical implications, because of the loss of productivity and the need for large vaccination campaigns or culling. Recently, LPAI H5N2 (Japan, 2005 [163]; Italy, 2005 [115]), H7N1 (Italy, 1999–2001 [164]), H7N2 (USA, 2002 [165]; UK, 2007), and H7N3 (Italy, 2002–2003 [115]; UK, 2006 [166]) viruses were responsible for several outbreaks. But also other non-H5 or H7 subtypes caused large outbreaks such as H6N1 in Asia [167], H6N2 in South Africa [168], or in the United States [169].

In the mid 1990s, H9N2 viruses became established in poultry and occurred widely in Asia. They also caused several outbreaks in Europe, South Africa, the United States, and the Middle East [23] and are now considered enzootic in the Middle East and Asia. Although H9N2 viruses do not have a polybasic cleavage site in the HA, they have the ability to cause severe respiratory distress, high morbidity and mortality, and a drop in egg productivity (depending on co-infections), representing another threat to the poultry industry [170]. Several cases of poultry to human H9N2 virus transmission have also been reported [171]. Surprisingly, a large percentage of H9N2 viruses circulating in poultry have a human virus-like receptor specificity (Q226L substitution in HA protein) [172–174]. The co-circulation of H9N2 with other influenza subtypes, paired with their affinity for SAα2,6Gal type receptors, represents a dangerous breeding ground for yet another panzootic or pandemic virus [172, 174, 175].

10.8
Conclusion

Although influenza viruses have been extensively studied over the last century, the recent epidemics and pandemics show that we still have many lessons to learn about their pathogenicity and host range restriction. Although pigs were considered as mixing vessels allowing the introduction of new AIVs into humans, the Asian HPAI H5N1 virus has shown that a "fully" avian virus can be directly transmitted to humans without a passage through the pig host. The recent emergence of the swine-originating A/H1N1 and its pandemic spread has also opened new opportunities for the emergence of mixed viruses. This A/H1N1 virus infects not only humans but also pigs and poultry. The co-circulation and overlapping host ranges of a pandemic virus, highly contagious in humans, and a HPAI virus causing high mortality in humans paves the way for new and dangerous scenarios in the ongoing influenza saga. Although major pathogenicity and host species determinants are both located on the HA protein, reassortments of A/H1N1 and HPAI H5N1 may result in viruses whose properties are difficult to predict. So far H5N1 has not become pandemic but preparedness plans elaborated for this eventuality have provided the blueprint for rapid public health interventions in response to A/H1N1. Different vaccine strategies and vaccine prototypes are available thanks to the efforts that have been made to fight HPAI H5N1. Thus, extensive studies of HPAI H5N1 have prepared the world to face a pandemic, but it has also been partially misleading because of the low pathogenicity of the A/H1N1 pandemic strain.

References

1. Webster, R.G., Bean, W.J., Gorman, O.T., Chambers, T.M., and Kawaoka, Y. (1992) Evolution and ecology of influenza A viruses. *Microbiol. Rev.*, **56**, 152–179.
2. Neumann, G., Noda, T., and Kawaoka, Y. (2009) Emergence and pandemic potential of swine-origin H1N1 influenza virus. *Nature*, **459**, 931–939.
3. WHO (2009) Influenza (Seasonal) Fact Sheet N°211, *http://www.who.int/mediacentre/factsheets/fs211/en/* (accessed 23 November 2009).
4. Elbers, A.R., Fabri, T.H., de Vries, T.S., de Wit, J.J., Pijpers, A., and Koch, G. (2004) The highly pathogenic avian influenza A (H7N7) virus epidemic in The Netherlands in 2003--lessons learned from the first five outbreaks. *Avian. Dis.*, **48**, 691–705.
5. Fouchier, R.A., Schneeberger, P.M., Rozendaal, F.W., Broekman, J.M., Kemink, S.A., Munster, V., Kuiken, T., Rimmelzwaan, G.F., Schutten, M., Van Doornum, G.J., Koch, G., Bosman, A., Koopmans, M., and Osterhaus, A.D. (2004) Avian influenza A virus (H7N7) associated with human conjunctivitis and a fatal case of acute respiratory distress syndrome. *Proc. Natl. Acad. Sci. U.S.A.*, **101**, 1356–1361.
6. Koopmans, M., Wilbrink, B., Conyn, M., Natrop, G., van der Nat, H., Vennema, H., Meijer, A., van Steenbergen, J., Fouchier, R., Osterhaus, A., and Bosman, A. (2004) Transmission of H7N7 avian influenza A virus to human beings during a large outbreak in commercial poultry farms in the Netherlands. *Lancet*, **363**, 587–593.
7. Wright, P.F., Neumann, G., and Kawaoka, Y. (2007) in *Fields' Virology*, 5th edn, vol. **2** (eds D.M. Knipe and P.M. Howley), Lippincott Williams & Wilkins, a Wolters Kluwer Business, Philadelphia, PA, pp. 1691–1740.
8. Garcia-Sastre, A. (2001) Inhibition of interferon-mediated antiviral responses by influenza A viruses and other negative-strand RNA viruses. *Virology*, **279**, 375–384.
9. Katze, M.G., He, Y., and Gale, M. Jr. (2002) Viruses and interferon: a fight for supremacy. *Nat. Rev. Immunol.*, **2**, 675–687.
10. Richardson, J.C. and Akkina, R.K. (1991) NS2 protein of influenza virus is found in purified virus and phosphorylated in infected cells. *Arch. Virol.*, **116**, 69–80.
11. Chen, W., Calvo, P.A., Malide, D., Gibbs, J., Schubert, U., Bacik, I., Basta, S., O'Neill, R., Schickli, J., Palese, P., Henklein, P., Bennink, J.R.,

and Yewdell, J.W. (2001) A novel influenza A virus mitochondrial protein that induces cell death. *Nat. Med.*, **7**, 1306–1312.

12. Swayne, D.E. and Halvorson, D.A. (2003) in *Diseases of Poultry*, 11th edn (ed. Y.M. Saif), Iowa State Press, Ames, Iowa, pp. 135–160.
13. Domingo, E. and Holland, J.J. (1997) RNA virus mutations and fitness for survival. *Annu. Rev. Microbiol.*, **51**, 151–178.
14. Bouvier, N.M. and Palese, P. (2008) The biology of influenza viruses. *Vaccine*, **26** (Suppl. 4), D49–D53.
15. Garamszegi, L.Z. and Moller, A.P. (2007) Prevalence of avian influenza and host ecology. *Proc. Biol. Sci.*, **274**, 2003–2012.
16. Ito, T., Okazaki, K., Kawaoka, Y., Takada, A., Webster, R.G., and Kida, H. (1995) Perpetuation of influenza A viruses in Alaskan waterfowl reservoirs. *Arch. Virol.*, **140**, 1163–1172.
17. Olsen, B., Munster, V.J., Wallensten, A., Waldenstrom, J., Osterhaus, A.D., and Fouchier, R.A. (2006) Global patterns of influenza a virus in wild birds. *Science*, **312**, 384–388.
18. Hulse-Post, D.J., Sturm-Ramirez, K.M., Humberd, J., Seiler, P., Govorkova, E.A., Krauss, S., Scholtissek, C., Puthavathana, P., Buranathai, C., Nguyen, T.D., Long, H.T., Naipospos, T.S., Chen, H., Ellis, T.M., Guan, Y., Peiris, J.S., and Webster, R.G. (2005) Role of domestic ducks in the propagation and biological evolution of highly pathogenic H5N1 influenza viruses in Asia. *Proc. Natl. Acad. Sci. U.S.A.*, **102**, 10682–10687.
19. Sims, L.D., Domenech, J., Benigno, C., Kahn, S., Kamata, A., Lubroth, J., Martin, V., and Roeder, P. (2005) Origin and evolution of highly pathogenic H5N1 avian influenza in Asia. *Vet. Rec.*, **157**, 159–164.
20. Gilbert, M., Xiao, X., Domenech, J., Lubroth, J., Martin, V., and Slingenbergh, J. (2006) Anatidae migration in the western Palearctic and spread of highly pathogenic avian influenza H5NI virus. *Emerg. Infect. Dis.*, **12**, 1650–1656.
21. Baigent, S.J. and McCauley, J.W. (2003) Influenza type A in humans, mammals and birds: determinants of virus virulence, host-range and interspecies transmission. *Bioessays*, **25**, 657–671.
22. Alexander, D.J. (2000) A review of avian influenza in different bird species. *Vet. Microbiol.*, **74**, 3–13.
23. Alexander, D.J. (2007) An overview of the epidemiology of avian influenza. *Vaccine*, **25**, 5637–5644.
24. Banks, J., Speidel, E.C., McCauley, J.W., and Alexander, D.J. (2000) Phylogenetic analysis of H7 haemagglutinin subtype influenza A viruses. *Arch. Virol.*, **145**, 1047–1058.
25. Rohm, C., Horimoto, T., Kawaoka, Y., Suss, J., and Webster, R.G. (1995) Do hemagglutinin genes of highly pathogenic avian influenza viruses constitute unique phylogenetic lineages? *Virology*, **209**, 664–670.
26. Liu, J., Xiao, H., Lei, F., Zhu, Q., Qin, K., Zhang, X.W., Zhang, X.L., Zhao, D., Wang, G., Feng, Y., Ma, J., Liu, W., Wang, J., and Gao, G.F. (2005) Highly pathogenic H5N1 influenza virus infection in migratory birds. *Science*, **309**, 1206.
27. Capua, I., Marangon, S., and Cancellotti, F.M. (2003) The 1999–2000 avian influenza (H7N1) epidemic in Italy. *Vet. Res. Commun.*, **27**, 123–127.
28. Capua, I., Mutinelli, F., Terregino, C., Cattoli, G., Manvell, R.J., and Burlini, F. (2000) Highly pathogenic avian influenza (H7N1) in ostriches farmed in Italy. *Vet. Rec.*, **146**, 356.
29. Campitelli, L., Di Martino, A., Spagnolo, D., Smith, G.J., Di Trani, L., Facchini, M., De Marco, M.A., Foni, E., Chiapponi, C., Martin, A.M., Chen, H., Guan, Y., Delogu, M., and Donatelli, I. (2008) Molecular analysis of avian H7 influenza viruses circulating in Eurasia in 1999–2005: detection of multiple reassortant virus genotypes. *J. Gen. Virol.*, **89**, 48–59.
30. Krauss, S., Walker, D., Pryor, S.P., Niles, L., Chenghong, L., Hinshaw, V.S., and Webster, R.G. (2004) Influenza A viruses of migrating wild aquatic birds in North America. *Vector Borne Zoonotic Dis.*, **4**, 177–189.

31. Lupiani, B. and Reddy, S.M. (2009) The history of avian influenza. *Comp. Immunol. Microbiol. Infect. Dis.*, **32**, 311–323.
32. Munster, V.J., Wallensten, A., Baas, C., Rimmelzwaan, G.F., Schutten, M., Olsen, B., Osterhaus, A.D., and Fouchier, R.A. (2005) Mallards and highly pathogenic avian influenza ancestral viruses, northern Europe. *Emerg. Infect. Dis.*, **11**, 1545–1551.
33. Desvaux, S., Marx, N., Ong, S., Gaidet, N., Hunt, M., Manuguerra, J.C., Sorn, S., Peiris, M., Van der Werf, S., and Reynes, J.M. (2009) Highly pathogenic avian influenza virus (H5N1) outbreak in captive wild birds and cats, Cambodia. *Emerg. Infect. Dis.*, **15**, 475–478.
34. Ducatez, M.F., Tarnagda, Z., Tahita, M.C., Sow, A., De Landtsheer, S., Londt, B.Z., Brown, I.H., Osterhaus, A.D.M.E., Fouchier, R.A.M., Ouedraogo, J.B., and Muller, C.P. (2007) Genetic characterization of HA1 of HPAI H5N1 viruses from poultry and wild vultures in Burkina Faso. *Emerg. Infect. Dis.*, **13**, 611–613.
35. Horimoto, T. and Kawaoka, Y. (1994) Reverse genetics provides direct evidence for a correlation of hemagglutinin cleavability and virulence of an avian influenza A virus. *J. Virol.*, **68**, 3120–3128.
36. Horimoto, T. and Kawaoka, Y. (2001) Pandemic threat posed by avian influenza A viruses. *Clin. Microbiol. Rev.*, **14**, 129–149.
37. Klenk, H.D. and Rott, R. (1988) The molecular biology of influenza virus pathogenicity. *Adv. Virus Res.*, **34**, 247–281.
38. Klenk, H.D., Rott, R., Orlich, M., and Blodorn, J. (1975) Activation of influenza A viruses by trypsin treatment. *Virology*, **68**, 426–439.
39. Sturm-Ramirez, K.M., Ellis, T., Bousfield, B., Bissett, L., Dyrting, K., Rehg, J.E., Poon, L., Guan, Y., Peiris, M., and Webster, R.G. (2004) Reemerging H5N1 influenza viruses in Hong Kong in 2002 are highly pathogenic to ducks. *J. Virol.*, **78**, 4892–4901.
40. Webster, R.G., Hinshaw, V.S., Bean, W.J., Turner, B., and Shortridge, K.F. Jr. (1977) Influenza viruses from avian and porcine sources and their possible role in the origin of human pandemic strains. *Dev. Biol. Stand.*, **39**, 461–468.
41. Webster, R.G., Yakhno, M., Hinshaw, V.S., Bean, W.J., and Murti, K.G. (1978) Intestinal influenza: replication and characterization of influenza viruses in ducks. *Virology*, **84**, 268–278.
42. Senne, D.A., Panigrahy, B., Kawaoka, Y., Pearson, J.E., Suss, J., Lipkind, M., Kida, H., and Webster, R.G. (1996) Survey of the hemagglutinin (HA) cleavage site sequence of H5 and H7 avian influenza viruses: amino acid sequence at the HA cleavage site as a marker of pathogenicity potential. *Avian Dis.*, **40**, 425–437.
43. Wood, G.W., McCauley, J.W., Bashiruddin, J.B., and Alexander, D.J. (1993) Deduced amino acid sequences at the haemagglutinin cleavage site of avian influenza A viruses of H5 and H7 subtypes. *Arch. Virol.*, **130**, 209–217.
44. Londt, B.Z., Nunez, A., Banks, J., Nili, H., Johnson, L.K., and Alexander, D.J. (2008) Pathogenesis of highly pathogenic avian influenza A/turkey/Turkey/1/2005 H5N1 in Pekin ducks (Anas platyrhynchos) infected experimentally. *Avian Pathol.*, **37**, 619–627.
45. Swayne, D.E. (2007) Understanding the complex pathobiology of high pathogenicity avian influenza viruses in birds. *Avian Dis.*, **51**, 242–249.
46. Banks, J., Speidel, E.S., Moore, E., Plowright, L., Piccirillo, A., Capua, I., Cordioli, P., Fioretti, A., and Alexander, D.J. (2001) Changes in the haemagglutinin and the neuraminidase genes prior to the emergence of highly pathogenic H7N1 avian influenza viruses in Italy. *Arch. Virol.*, **146**, 963–973.
47. Chen, H., Smith, G.J., Zhang, S.Y., Qin, K., Wang, J., Li, K.S., Webster, R.G., Peiris, J.S., and Guan, Y. (2005) Avian flu: H5N1 virus outbreak in

migratory waterfowl. *Nature*, **436**, 191–192.

48. Garcia, M., Crawford, J.M., Latimer, J.W., Rivera-Cruz, E., and Perdue, M.L. (1996) Heterogeneity in the haemagglutinin gene and emergence of the highly pathogenic phenotype among recent H5N2 avian influenza viruses from Mexico. *J. Gen. Virol.*, **77**, 1493–1504.
49. Horimoto, T., Rivera, E., Pearson, J., Senne, D., Krauss, S., Kawaoka, Y., and Webster, R.G. (1995) Origin and molecular changes associated with emergence of a highly pathogenic H5N2 influenza virus in Mexico. *Virology*, **213**, 223–230.
50. Kawaoka, Y., Naeve, C.W., and Webster, R.G. (1984) Is virulence of H5N2 influenza viruses in chickens associated with loss of carbohydrate from the hemagglutinin? *Virology*, **139**, 303–316.
51. Owoade, A.A., Gerloff, N.A., Ducatez, M.F., Taiwo, J.O., Kremer, J.R., and Muller, C.P. (2008) Replacement of sublineages of avian influenza (H5N1) by reassortments, sub-Saharan Africa. *Emerg. Infect. Dis.*, **14**, 1731–1735.
52. Suarez, D.L., Senne, D.A., Banks, J., Brown, I.H., Essen, S.C., Lee, C.W., Manvell, R.J., Mathieu-Benson, C., Moreno, V., Pedersen, J.C., Panigrahy, B., Rojas, H., Spackman, E., and Alexander, D.J. (2004) Recombination resulting in virulence shift in avian influenza outbreak, Chile. *Emerg. Infect. Dis.*, **10**, 693–699.
53. Xu, X., Subbarao P., Cox, N.J., and Guo, Y. (1999) Genetic characterization of the pathogenic influenza A/Goose/Guangdong/1/96 (H5N1) virus: similarity of its hemagglutinin gene to those of H5N1 viruses from the 1997 outbreaks in Hong Kong. *Virology*, **261**, 15–19.
54. Rogers, G.N., Paulson, J.C., Daniels, R.S., Skehel, J.J., Wilson, I.A., and Wiley, D.C. (1983) Single amino acid substitutions in influenza haemagglutinin change receptor binding specificity. *Nature*, **304**, 76–78.
55. Matrosovich, M.N., Matrosovich, T.Y., Gray, T., Roberts, N.A., and Klenk, H.D. (2004) Human and avian influenza viruses target different cell types in cultures of human airway epithelium. *Proc. Natl. Acad. Sci. U.S.A.*, **101**, 4620–4624.
56. Shinya, K., Ebina, M., Yamada, S., Ono, M., Kasai, N., and Kawaoka, Y. (2006) Avian flu: influenza virus receptors in the human airway. *Nature*, **440**, 435–436.
57. Ito, T., Suzuki, Y., Suzuki, T., Takada, A., Horimoto, T., Wells, K., Kida, H., Otsuki, K., Kiso, M., Ishida, H., and Kawaoka, Y. (2000) Recognition of N-glycolylneuraminic acid linked to galactose by the alpha2,3 linkage is associated with intestinal replication of influenza A virus in ducks. *J. Virol.*, **74**, 9300–9305.
58. Connor, R.J., Kawaoka, Y., Webster, R.G., and Paulson, J.C. (1994) Receptor specificity in human, avian, and equine H2 and H3 influenza virus isolates. *Virology*, **205**, 17–23.
59. Claas, E.C., Osterhaus, A.D., van Beek, R., De Jong, J.C., Rimmelzwaan, G.F., Senne, D.A., Krauss, S., Shortridge, K.F., and Webster, R.G. (1998) Human influenza A H5N1 virus related to a highly pathogenic avian influenza virus. *Lancet*, **351**, 472–477.
60. Goto, H. and Kawaoka, Y. (1998) A novel mechanism for the acquisition of virulence by a human influenza A virus. *Proc. Natl. Acad. Sci. U.S.A.*, **95**, 10224–10228.
61. Palese, P., Tobita, K., Ueda, M., and Compans, R.W. (1974) Characterization of temperature sensitive influenza virus mutants defective in neuraminidase. *Virology*, **61**, 397–410.
62. Takahashi, T., Suzuki, Y., Nishinaka, D., Kawase, N., Kobayashi, Y., Hidari, K.I., Miyamoto, D., Guo, C.T., Shortridge, K.F., and Suzuki, T. (2001) Duck and human pandemic influenza A viruses retain sialidase activity under low pH conditions. *J. Biochem.*, **130**, 279–283.
63. Peiris, J.S., de Jong, M.D., and Guan, Y. (2007) Avian influenza virus (H5N1): a threat to human health. *Clin. Microbiol. Rev.*, **20**, 243–267.

64. Crescenzo-Chaigne, B., van der Werf, S., and Naffakh, N. (2002) Differential effect of nucleotide substitutions in the 3' arm of the influenza A virus vRNA promoter on transcription/replication by avian and human polymerase complexes is related to the nature of PB2 amino acid 627. *Virology*, **303**, 240–252.
65. Hatta, M., Gao, P., Halfmann, P., and Kawaoka, Y. (2001) Molecular basis for high virulence of Hong Kong H5N1 influenza A viruses. *Science*, **293**, 1840–1842.
66. Shinya, K., Hamm, S., Hatta, M., Ito, H., Ito, T., and Kawaoka, Y. (2004) PB2 amino acid at position 627 affects replicative efficiency, but not cell tropism, of Hong Kong H5N1 influenza A viruses in mice. *Virology*, **320**, 258–266.
67. Subbarao, E.K., Kawaoka, Y., and Murphy, B.R. (1993) Rescue of an influenza A virus wild-type PB2 gene and a mutant derivative bearing a site-specific temperature-sensitive and attenuating mutation. *J. Virol.*, **67**, 7223–7228.
68. Subbarao, E.K., London, W., and Murphy, B.R. (1993) A single amino acid in the PB2 gene of influenza A virus is a determinant of host range. *J. Virol.*, **67**, 1761–1764.
69. Gabriel, G., Dauber, B., Wolff, T., Planz, O., Klenk, H.D., and Stech, J. (2005) The viral polymerase mediates adaptation of an avian influenza virus to a mammalian host. *Proc. Natl. Acad. Sci. U.S.A.*, **102**, 18590–18595.
70. Li, Z., Chen, H., Jiao, P., Deng, G., Tian, G., Li, Y., Hoffmann, E., Webster, R.G., Matsuoka, Y., and Yu, K. (2005) Molecular basis of replication of duck H5N1 influenza viruses in a mammalian mouse model. *J. Virol.*, **79**, 12058–12064.
71. Salomon, R., Franks, J., Govorkova, E.A., Ilyushina, N.A., Yen, H.L., Hulse-Post, D.J., Humberd, J., Trichet, M., Rehg, J.E., Webby, R.J., Webster, R.G., and Hoffmann, E. (2006) The polymerase complex genes contribute to the high virulence of the human H5N1 influenza virus isolate A/Vietnam/1203/04. *J. Exp. Med.*, **203**, 689–697.
72. To, K.F., Chan, P.K., Chan, K.F., Lee, W.K., Lam, W.Y., Wong, K.F., Tang, N.L., Tsang, D.N., Sung, R.Y., Buckley, T.A., Tam, J.S., and Cheng, A.F. (2001) Pathology of fatal human infection associated with avian influenza A H5N1 virus. *J. Med. Virol.*, **63**, 242–246.
73. Wang, X., Li, M., Zheng, H., Muster, T., Palese, P., Beg, A.A., and Garcia-Sastre, A. (2000) Influenza A virus NS1 protein prevents activation of NF-kappaB and induction of alpha/beta interferon. *J. Virol.*, **74**, 11566–11573.
74. Yuen, K.Y., Chan, P.K., Peiris, M., Tsang, D.N., Que, T.L., Shortridge, K.F., Cheung, P.T., To, W.K., Ho, E.T., Sung, R., and Cheng, A.F. (1998) Clinical features and rapid viral diagnosis of human disease associated with avian influenza A H5N1 virus. *Lancet*, **351**, 467–471.
75. Ludwig, S., Wang, X., Ehrhardt, C., Zheng, H., Donelan, N., Planz, O., Pleschka, S., Garcia-Sastre, A., Heins, G., and Wolff, T. (2002) The influenza A virus NS1 protein inhibits activation of Jun N-terminal kinase and AP-1 transcription factors. *J. Virol.*, **76**, 11166–11171.
76. Talon, J., Horvath, C.M., Polley, R., Basler, C.F., Muster, T., Palese, P., and Garcia-Sastre, A. (2000) Activation of interferon regulatory factor 3 is inhibited by the influenza A virus NS1 protein. *J. Virol.*, **74**, 7989–7996.
77. Hale, B.G., Randall, R.E., Ortin, J., and Jackson, D. (2008) The multifunctional NS1 protein of influenza A viruses. *J. Gen. Virol.*, **89**, 2359–2376.
78. Cheung, C.Y., Poon, L.L., Lau, A.S., Luk, W., Lau, Y.L., Shortridge, K.F., Gordon, S., Guan, Y., and Peiris, J.S. (2002) Induction of proinflammatory cytokines in human macrophages by influenza A (H5N1) viruses: a mechanism for the unusual severity of human disease? *Lancet*, **360**, 1831–1837.
79. Seo, S.H., Hoffmann, E., and Webster, R.G. (2002) Lethal H5N1 influenza viruses escape host anti-viral cytokine responses. *Nat. Med.*, **8**, 950–954.

80. Seo, S.H., Hoffmann, E., and Webster, R.G. (2004) The NS1 gene of H5N1 influenza viruses circumvents the host anti-viral cytokine responses. *Virus Res.*, **103**, 107–113.
81. Conenello, G.M., Zamarin, D., Perrone, L.A., Tumpey, T., and Palese, P. (2007) A single mutation in the PB1-F2 of H5N1 (HK/97) and 1918 influenza A viruses contributes to increased virulence. *PLoS Pathog.*, **3**, 1414–1421.
82. Alexander, D.J. (2008) Avian influenza – diagnosis. *Zoonoses Public Health*, **55**, 16–23.
83. Chen, H., Deng, G., Li, Z., Tian, G., Li, Y., Jiao, P., Zhang, L., Liu, Z., Webster, R.G., and Yu, K. (2004) The evolution of H5N1 influenza viruses in ducks in southern China. *Proc. Natl. Acad. Sci. U.S.A.*, **101**, 10452–10457.
84. Swayne, D.E. and Pantin-Jackwood, M. (2008) in *Avian Influenza* (ed. D.E. Swayne), Blackwell Publishing Ltd, Ames, Iowa, pp. 87–122.
85. Keawcharoen, J., Oraveerakul, K., Kuiken, T., Fouchier, R.A., Amonsin, A., Payungporn, S., Noppornpanth, S., Wattanodorn, S., Theambooniers, A., Tantilertcharoen, R., Pattanarangsan, R., Arya, N., Ratanakorn, P., Osterhaus, D.M., and Poovorawan, Y. (2004) Avian influenza H5N1 in tigers and leopards. *Emerg. Infect. Dis.*, **10**, 2189–2191.
86. Rimmelzwaan, G.F., van Riel, D., Baars, M., Bestebroer, T.M., van Amerongen, G., Fouchier, R.A., Osterhaus, A.D., and Kuiken, T. (2006) Influenza A virus (H5N1) infection in cats causes systemic disease with potential novel routes of virus spread within and between hosts. *Am. J. Pathol.*, **168**, 176–183; quiz 364.
87. Songserm, T., Amonsin, A., Jam-on, R., Sae-Heng, N., Meemak, N., Pariyothorn, N., Payungporn, S., Theamboonlers, A., and Poovorawan, Y. (2006) Avian influenza H5N1 in naturally infected domestic cat. *Emerg. Infect. Dis.*, **12**, 681–683.
88. Songserm, T., Amonsin, A., Jam-on, R., Sae-Heng, N., Pariyothorn, N., Payungporn, S., Theamboonlers, A., Chutinimitkul, S., Thanawongnuwech, R., and Poovorawan, Y. (2006) Fatal avian influenza A H5N1 in a dog. *Emerg. Infect. Dis.*, **12**, 1744–1747.
89. Lang, G., Gagnon, A., and Geraci, J.R. (1981) Isolation of an influenza A virus from seals. *Arch. Virol.*, **68**, 189–195.
90. Abdel-Ghafar, A.N., Chotpitayasunondh, T., Gao, Z., Hayden, F.G., Nguyen, D.H., de Jong, M.D., Naghdaliyev, A., Peiris, J.S., Shindo, N., Soeroso, S., and Uyeki, T.M. (2008) Update on avian influenza A (H5N1) virus infection in humans. *N. Engl. J. Med.*, **358**, 261–273.
91. Chotpitayasunondh, T., Thisyakorn, U., Pancharoen, C., Pepin, S., and Nougarede, N. (2008) Safety, humoral and cell mediated immune responses to two formulations of an inactivated, split-virion influenza A/H5N1 vaccine in children. *PLoS One*, **3**, e4028.
92. Sandrock, C. and Kelly, T. (2007) Clinical review: update of avian influenza A infections in humans. *Crit. Care*, **11**, 209.
93. Tran, T.H., Nguyen, T.L., Nguyen, T.D., Luong, T.S., Pham, P.M., Nguyen, V.C., Pham, T.S., Vo, C.D., Le, T.Q., Ngo, T.T., Dao, B.K., Le, P.P., Nguyen, T.T., Hoang, T.L., Cao, V.T., Le, T.G., Nguyen, D.T., Le, H.N., Nguyen, K.T., Le, H.S., Le, V.T., Christiane, D., Tran, T.T., Menno de, J., Schultsz, C., Cheng, P., Lim, W., Horby, P., and Farrar, J. (2004) Avian influenza A (H5N1) in 10 patients in Vietnam. *N. Engl. J. Med.*, **350**, 1179–1188.
94. Hui, D.S. (2008) Review of clinical symptoms and spectrum in humans with influenza A/H5N1 infection. *Respirology*, **13** (Suppl. 1), S10–S13.
95. de Jong, M.D. and Hien, T.T. (2006) Avian influenza A (H5N1). *J. Clin. Virol.*, **35**, 2–13.
96. de Jong, M.D., Simmons, C.P., Thanh, T.T., Hien, V.M., Smith, G.J., Chau, T.N., Hoang, D.M., Chau, N.V., Khanh, T.H., Dong, V.C., Qui, P.T., Cam, B.V., Ha do, Q., Guan, Y., Peiris, J.S., Chinh, N.T., Hien, T.T., and Farrar, J. (2006) Fatal outcome of human influenza A (H5N1) is associated with high viral load and hypercytokinemia. *Nat. Med.*, **12**, 1203–1207.

97. Hui, K.P., Lee, S.M., Cheung, C.Y., Ng, I.H., Poon, L.L., Guan, Y., Ip, N.Y., Lau, A.S., and Peiris, J.S. (2009) Induction of proinflammatory cytokines in primary human macrophages by influenza A virus (H5N1) is selectively regulated by IFN regulatory factor 3 and p38 MAPK. *J. Immunol.*, **182**, 1088–1098.
98. La Gruta, N.L., Kedzierska, K., Stambas, J., and Doherty, P.C. (2007) A question of self-preservation: immunopathology in influenza virus infection. *Immunol. Cell Biol.*, **85**, 85–92.
99. Peiris, J.S., Yu, W.C., Leung, C.W., Cheung, C.Y., Ng, W.F., Nicholls, J.M., Ng, T.K., Chan, K.H., Lai, S.T., Lim, W.L., Yuen, K.Y., and Guan, Y. (2004) Re-emergence of fatal human influenza A subtype H5N1 disease. *Lancet*, **363**, 617–619.
100. de Jong, M.D., Bach, V.C., Phan, T.Q., Vo, M.H., Tran, T.T., Nguyen, B.H., Beld, M., Le, T.P., Truong, H.K., Nguyen, V.V., Tran, T.H., Do, Q.H., and Farrar, J. (2005) Fatal avian influenza A (H5N1) in a child presenting with diarrhea followed by coma. *N. Engl. J. Med.*, **352**, 686–691.
101. Korteweg, C. and Gu, J. (2008) Pathology, molecular biology, and pathogenesis of avian influenza A (H5N1) infection in humans. *Am. J. Pathol.*, **172**, 1155–1170.
102. Tanaka, H., Park, C.H., Ninomiya, A., Ozaki, H., Takada, A., Umemura, T., and Kida, H. (2003) Neurotropism of the 1997 Hong Kong H5N1 influenza virus in mice. *Vet. Microbiol.*, **95**, 1–13.
103. OIE (2005) Chapter 2.7.12. Avian Influenza, Manual for Diagnostic Tests and Vaccines for Terrestrial Animals, 5th edn World Organisation for Animal Health, *http://www.oie.int/eng/normes/MMANUAL/A_00037.htm* (accessed 21 July 2009).
104. WHO (2005) WHO Manual on Animal Influenza Diagnosis and Surveillance, *http://www.who.int/vaccine_research/diseases/influenza/WHO_manual_on_animal-diagnosis_and_surveillance_2002_5.pdf* (accessed 22 November 2009).
105. Zambon, M. and Potter, C.W. (2009) in *Principles and Practice of Clinical Virology*, 6th edn (eds A.J. Zuckerman, J.E. Banatvala, B.D. Schoub, P.D. Griffiths, and P. Mortimer), John Wiley & Sons, Ltd, Chichester, pp. 373–408.
106. Spackman, E., Ip, H.S., Suarez, D.L., Slemons, R.D., and Stallknecht, D.E. (2008) Analytical validation of a real-time reverse transcription polymerase chain reaction test for Pan-American lineage H7 subtype Avian influenza viruses. *J. Vet. Diagn. Invest.*, **20**, 612–616.
107. He, Q., Velumani, S., Du, Q., Lim, C.W., Ng, F.K., Donis, R., and Kwang, J. (2007) Detection of H5 avian influenza viruses by antigen-capture enzyme-linked immunosorbent assay using H5-specific monoclonal antibody. *Clin. Vaccine Immunol.*, **14**, 617–623.
108. Ho, H.T., Qian, H.L., He, F., Meng, T., Szyporta, M., Prabhu, N., Prabakaran, M., Chan, K.P., and Kwang, J. (2009) Rapid detection of H5N1 subtype influenza viruses by antigen capture enzyme-linked immunosorbent assay using H5- and N1-specific monoclonal antibodies. *Clin. Vaccine Immunol.*, **16**, 726–732.
109. Tanasa, R.I., Lupulescu, E., Stavaru, C., Baetel, A.E., Necula, G., Vuta, V., Barboi, G., Popovici, A., Banica, L., Besliu, A., and Marineata, S. (2009) COST ACTION B28 "Array technologies for BSL3 and BSL4 pathogens". 7th Management Committee and WG1, WG2, WG3, WG4 and WG5 Meetings, Belgrade, Serbia, April 22–24.
110. Charlton, B., Crossley, B., and Hietala, S. (2009) Conventional and future diagnostics for avian influenza. *Comp. Immunol. Microbiol. Infect. Dis.*, **32**, 341–350.
111. Spackman, E., Suarez, D.L., and Senne, D.A. (2008) in *Avian Influenza* (ed. D.E. Swayne), Blackwell Publishing Ltd, Ames, IA, pp. 299–307.
112. Pasick, J. (2008) Advances in the molecular based techniques for the diagnosis and characterization of avian influenza virus infections. *Transbound. Emerg. Dis.*, **55**, 329–338.

113. Spackman, E., Senne, D.A., Myers, T.J., Bulaga, L.L., Garber, L.P., Perdue, M.L., Lohman, K., Daum, L.T., and Suarez, D.L. (2002) Development of a real-time reverse transcriptase PCR assay for type A influenza virus and the avian H5 and H7 hemagglutinin subtypes. *J. Clin. Microbiol.*, **40**, 3256–3260.

114. Gambotto, A., Barratt-Boyes, S.M., de Jong, M.D., Neumann, G., and Kawaoka, Y. (2008) Human infection with highly pathogenic H5N1 influenza virus. *Lancet*, **371**, 1464–1475.

115. Capua, I., Cattoli, G., Terregina, C., and Marangon, S. (2008) in *Avian Influenza* (eds H.-D. Klenk, M.N. Matrosovich, and J. Stech), Karger, Basel, pp. 59–70.

116. Capua, I., Cattoli, G., and Marangon, S. (2004) DIVA--a vaccination strategy enabling the detection of field exposure to avian influenza. *Dev. Biol. (Basel)*, **119**, 229–233.

117. Brown, J.D., Stallknecht, D.E., Berghaus, R.D., Luttrell, M.P., Velek, K., Kistler, W., Costa, T., Yabsley, M.J., and Swayne, D. (2009) Evaluation of a commercial blocking enzyme-linked immunosorbent assay to detect avian influenza virus antibodies in multiple experimentally infected avian species. *Clin. Vaccine Immunol.*, **16**, 824–829.

118. Prabakaran, M., Ho, H.T., Prabhu, N., Velumani, S., Szyporta, M., He, F., Chan, K.P., Chen, L.M., Matsuoka, Y., Donis, R.O., and Kwang, J. (2009) Development of epitope-blocking ELISA for universal detection of antibodies to human H5N1 influenza viruses. *PLoS ONE*, **4**, e4566.

119. Cheng, X., Wu, C., He, J., Lv, X., Zhang, S., Zhou, L., Wang, J., Deng, R., Long, Q., Wang, X., and Cheng, J. (2008) Serologic and genetic characterization analyses of a highly pathogenic influenza virus (H5N1) isolated from an infected man in Shenzhen. *J. Med. Virol.*, **80**, 1058–1064.

120. Rowe, T., Abernathy, R.A., Hu-Primmer, J., Thompson, W.W., Lu, X., Lim, W., Fukuda, K., Cox, N.J., and Katz, J.M. (1999) Detection of antibody to avian influenza A (H5N1) virus in human serum by using a combination of serologic assays. *J. Clin. Microbiol.*, **37**, 937–943.

121. Nefkens, I., Garcia, J.M., Ling, C.S., Lagarde, N., Nicholls, J., Tang, D.J., Peiris, M., Buchy, P., and Altmeyer, R. (2007) Hemagglutinin pseudotyped lentiviral particles: characterization of a new method for avian H5N1 influenza sero-diagnosis. *J. Clin. Virol.*, **39**, 27–33.

122. Temperton, N.J., Hoschler, K., Major, D., Nicolson, C., Manvell, R., Hien, V.M., Ha do, Q., de Jong, M., Zambon, M., Takeuchi, Y., and Weiss, R.A. (2007) A sensitive retroviral pseudotype assay for influenza H5N1-neutralizing antibodies. *Influenza Other Respir. Viruses*, **1**, 105–112.

123. Wang, W., Butler, E.N., Veguilla, V., Vassell, R., Thomas, J.T., Moos, M., Ye, Z., Hancock, K., and Weiss, C.D. Jr. (2008) Establishment of retroviral pseudotypes with influenza hemagglutinins from H1, H3, and H5 subtypes for sensitive and specific detection of neutralizing antibodies. *J. Virol. Methods*, **153**, 111–119.

124. Shortridge, K.F., Zhou, N.N., Guan, Y., Gao, P., Ito, T., Kawaoka, Y., Kodihalli, S., Krauss, S., Markwell, D., Murti, K.G., Norwood, M., Senne, D., Sims, L., Takada, A., and Webster, R.G. (1998) Characterization of avian H5N1 influenza viruses from poultry in Hong Kong. *Virology*, **252**, 331–342.

125. Shortridge, K.F., Gao, P., Guan, Y., Ito, T., Kawaoka, Y., Markwell, D., Takada, A., and Webster, R.G. (2000) Interspecies transmission of influenza viruses: H5N1 virus and a Hong Kong SAR perspective. *Vet. Microbiol.*, **74**, 141–147.

126. Guan, Y., Shortridge, K.F., Krauss, S., and Webster, R.G. (1999) Molecular characterization of H9N2 influenza viruses: were they the donors of the "internal" genes of H5N1 viruses in Hong Kong? *Proc. Natl. Acad. Sci. U.S.A.*, **96**, 9363–9367.

127. Hoffmann, E., Stech, J., Leneva, I., Krauss, S., Scholtissek, C., Chin, P.S., Peiris, M., Shortridge, K.F., and Webster, R.G. (2000) Characterization of the influenza A virus gene pool in avian species in southern China: was H6N1 a derivative or a precursor of H5N1? *J. Virol.*, **74**, 6309–6315.

128. Guan, Y., Peiris, M., Kong, K.F., Dyrting, K.C., Ellis, T.M., Sit, T., Zhang, L.J., and Shortridge, K.F. (2002) H5N1 influenza viruses isolated from geese in Southeastern China: evidence for genetic reassortment and inter-species transmission to ducks. *Virology*, **292**, 16–23.

129. Webster, R.G., Guan, Y., Peiris, M., Walker, D., Krauss, S., Zhou, N.N., Govorkova, E.A., Ellis, T.M., Dyrting, K.C., Sit, T., Perez, D.R., and Shortridge, K.F. (2002) Characterization of H5N1 influenza viruses that continue to circulate in geese in southeastern China. *J. Virol.*, **76**, 118–126.

130. Guan, Y., Peiris, J.S., Lipatov, A.S., Ellis, T.M., Dyrting, K.C., Krauss, S., Zhang, L.J., Webster, R.G., and Shortridge, K.F. (2002) Emergence of multiple genotypes of H5N1 avian influenza viruses in Hong Kong SAR. *Proc. Natl. Acad. Sci. U.S.A.*, **99**, 8950–8955.

131. Li, K.S., Guan, Y., Wang, J., Smith, G.J., Xu, K.M., Duan, L., Rahardjo, A.P., Puthavathana, P., Buranathai, C., Nguyen, T.D., Estoepangestie, A.T., Chaisingh, A., Auewarakul, P., Long, H.T., Hanh, N.T., Webby, R.J., Poon, L.L., Chen, H., Shortridge, K.F., Yuen, K.Y., Webster, R.G., and Peiris, J.S. (2004) Genesis of a highly pathogenic and potentially pandemic H5N1 influenza virus in eastern Asia. *Nature*, **430**, 209–213.

132. Guan, Y., Poon, L.L., Cheung, C.Y., Ellis, T.M., Lim, W., Lipatov, A.S., Chan, K.H., Sturm-Ramirez, K.M., Cheung, C.L., Leung, Y.H., Yuen, K.Y., Webster, R.G., and Peiris, J.S. (2004) H5N1 influenza: a protean pandemic threat. *Proc. Natl. Acad. Sci. U.S.A.*, **101**, 8156–8161.

133. FAO (2004) Avian Influenza Disease: Emergency Bulletin, Issue 4, *ftp://ftp.fao.org/docrep/fao/011/aj046e/aj046e00.pdf* (accessed 13 August 2009).

134. WHO (2009) H5N1 Avian Influenza: Timeline of Major Events, 27 July 2009, *http://www.who.int/csr/disease/avian_influenza/Timeline090727.pdf* (accessed 13 November 2009).

135. Chen, H., Li, Y., Li, Z., Shi, J., Shinya, K., Deng, G., Qi, Q., Tian, G., Fan, S., Zhao, H., Sun, Y., and Kawaoka, Y. (2006) Properties and dissemination of H5N1 viruses isolated during an influenza outbreak in migratory waterfowl in western China. *J. Virol.*, **80**, 5976–5983.

136. Sturm-Ramirez, K.M., Hulse-Post, D.J., Govorkova, E.A., Humberd, J., Seiler, P., Puthavathana, P., Buranathai, C., Nguyen, T.D., Chaisingh, A., Long, H.T., Naipospos, T.S., Chen, H., Ellis, T.M., Guan, Y., Peiris, J.S., and Webster, R.G. (2005) Are ducks contributing to the endemicity of highly pathogenic H5N1 influenza virus in Asia? *J. Virol.*, **79**, 11269–11279.

137. WHO/OIE/FAO H5N1 Evolution Working Group (2008) Toward a unified nomenclature system for highly pathogenic avian influenza virus (H5N1). *Emerg. Infect. Dis.*, **14**, e1 *http://wwwnc.cdc.gov/eid/article/14/7/07-1681_article.html* (accessed 10 November 2009).

138. Chen, H., Bu, Z., and Wang, J. (2008) in *Avian Influenza* (eds H.-D. Klenk, M.N. Matrosovich, and J. Stech), Karger, Basel, pp. 27–40.

139. OIE (2009) Point sur la situation de l'Influenza Aviaire hautement pathogène chez les animaux (Type H5 et H7), *http://www.oie.int/downld/AVIAN%20-INFLUENZA/F2009_AI.php* (accessed 06 January 2010).

140. Owoade, A.A., Ducatez, M.F., and Muller, C.P. (2006) Seroprevalence of avian influenza virus, infectious bronchitis virus, reovirus, avian pneumovirus, infectious laryngotracheitis virus, and avian leukosis virus in Nigerian poultry. *Avian Dis.*, **50**, 222–227.

141. ProMED (2006) Avian Influenza – Worldwide, OIE, Nigeria, archive number 20060208.0409, *http://www.promedmail.org* (accessed 22 November 2009).

142. Aiki-Raji, C.O., Aguilar, P.V., Kwon, Y.K., Goetz, S., Suarez, D.L., Jethra, A.I., Nash, O., Adeyefa, C.A., Adu, F.D., Swayne, D., and Basler, C.F. (2008) Phylogenetics and pathogenesis of early avian influenza viruses (H5N1), Nigeria. *Emerg. Infect. Dis.*, **14**, 1753–1755.
143. Ducatez, M.F., Olinger, C.M., Owoade, A.A., De Landtsheer, S., Ammerlaan, W., Niesters, H.G., Osterhaus, A.D., Fouchier, R.A., and Muller, C.P. (2006) Avian flu: multiple introductions of H5N1 in Nigeria. *Nature*, **442**, 37.
144. Ducatez, M.F., Olinger, C.M., Owoade, A.A., Tarnagda, Z., Tahita, M.C., Sow, A., De Landtsheer, S., Ammerlaan, W., Ouedraogo, J.B., Osterhaus, A.D., Fouchier, R.A., and Muller, C.P. (2007) Molecular and antigenic evolution and geographical spread of H5N1 highly pathogenic avian influenza viruses in western Africa. *J. Gen. Virol.*, **88**, 2297–2306.
145. Fishpool, L.D.C. and Evans, M.I. (2001) *Important Bird Areas in Africa and Associated Islands: Priority Sites for Conservation*, Pisces Publications and BirdLife International, Newbury and Cambridge.
146. De Benedictis, P., Joannis, T.M., Lombin, L.H., Shittu, I., Beato, M.S., Rebonato, V., Cattoli, G., and Capua, I. (2007) Field and laboratory findings of the first incursion of the Asian H5N1 highly pathogenic avian influenza virus in Africa. *Avian Pathol.*, **36**, 115–117.
147. Fusaro, A., Joannis, T., Monne, I., Salviato, A., Yakubu, B., Meseko, C., Oladokun, T., Fassina, S., Capua, I., and Cattoli, G. (2009) Introduction into Nigeria of a distinct genotype of avian influenza virus (H5N1). *Emerg. Infect. Dis.*, **15**, 445–447.
148. Salzberg, S.L., Kingsford, C., Cattoli, G., Spiro, D.J., Janies, D.A., Aly, M.M., Brown, I.H., Couacy-Hymann, E., De Mia, G.M., Dung do, H., Guercio, A., Joannis, T., Maken Ali, A.S., Osmani, A., Padalino, I., Saad, M.D., Savic, V., Sengamalay, N.A., Yingst, S., Zaborsky, J., Zorman-Rojs, O., Ghedin, E., and Capua, I. (2007) Genome analysis linking recent European and African influenza (H5N1) viruses. *Emerg. Infect. Dis.*, **13**, 713–718.
149. Monne, I., Joannis, T.M., Fusaro, A., De Benedictis, P., Lombin, L.H., Ularamu, H., Egbuji, A., Solomon, P., Obi, T.U., Cattoli, G., and Capua, I. (2008) Reassortant avian influenza virus (H5N1) in poultry, Nigeria, 2007. *Emerg. Infect. Dis.*, **14**, 637–640.
150. Cattoli, G., Monne, I., Fusaro, A., Joannis, T.M., Lombin, L.H., Aly, M.M., Arafa, A.S., Sturm-Ramirez, K.M., Couacy-Hymann, E., Awuni, J.A., Batawui, K.B., Awoume, K.A., Aplogan, G.L., Sow, A., Ngangnou, A.C., El Nasri Hamza, I.M., Gamatie, D., Dauphin, G., Domenech, J.M., and Capua, I. (2009) Highly pathogenic avian influenza virus subtype H5N1 in Africa: a comprehensive phylogenetic analysis and molecular characterization of isolates. *PLoS ONE*, **4**, e4842.
151. OIE (2008) Review Report: Six-Monthly Report on the Notification of the Absence or Presence of OIE-Listed Diseases, July-December 2008, Benin, *http://www.oie.int/wahis/public.php?page=home* (OIE ref: 52409, accessed 18 December 2009).
152. WHO/OIE/FAO H5N1 Evolution Working Group (2009) Continuing progress towards a unified nomenclature for the highly pathogenic H5N1 avian influenza viruses: divergence of clade 2.2 viruses. *Influenza Other Respir. Viruses*, **3**, 59–62.
153. WHO (2009) Cumulative Number of Confirmed Human Cases of Avian Influenza A/(H5N1) Reported to WHO, 21 December 2009, *http://www.who.int/csr/disease/avian_influenza/country/cases_table_2009_09_24/en/index.html* (accessed 04 January 2010).
154. Veljkovic, V., Veljkovic, N., Muller, C.P., Muller, S., Glisic, S., Perovic, V., and Kohler, H. (2009) Characterization of conserved properties of hemagglutinin of H5N1 and human influenza viruses: possible consequences for therapy and infection control. *BMC Struct. Biol.*, **9**, 21.
155. Berhane, Y., Hisanaga, T., Kehler, H., Neufeld, J., Manning, L., Argue,

C., Handel, K., Hooper-McGrevy, K., Jonas, M., Robinson, J., Webster, R.G., and Pasick, J. (2009) Highly pathogenic avian influenza virus A (H7N3) in domestic poultry, Saskatchewan, Canada, 2007. *Emerg. Infect. Dis.*, **15**, 1492–1495.

156. Perdue, M.L. and Swayne, D.E. (2005) Public health risk from avian influenza viruses. *Avian Dis.*, **49**, 317–327.

157. Bean, W.J., Kawaoka, Y., Wood, J.M., Pearson, J.E., and Webster, R.G. (1985) Characterization of virulent and avirulent A/chicken/Pennsylvania/83 influenza A viruses: potential role of defective interfering RNAs in nature. *J. Virol.*, **54**, 151–160.

158. Villarreal-Chavez, C. and Rivera-Cruz, E. (2003) An update on avian influenza in Mexico. *Avian Dis.*, **47**, 1002–1005.

159. Naeem, K. and Hussain, M. (1995) An outbreak of avian influenza in poultry in Pakistan. *Vet. Rec.*, **137**, 439.

160. Naeem, K., Siddique, N., Ayaz, M., and Jalalee, M.A. (2007) Avian influenza in Pakistan: outbreaks of low- and high-pathogenicity avian influenza in Pakistan during 2003–2006. *Avian Dis.*, **51**, 189–193.

161. Rojas, H., Moreira, R., Avalos, P., Capua, I., and Marangon, S. (2002) Avian influenza in poultry in Chile. *Vet. Rec.*, **151**, 188.

162. Bowes, V.A., Ritchie, S.J., Byrne, S., Sojonky, K., Bidulka, J.J., and Robinson, J.H. (2004) Virus characterization, clinical presentation, and pathology associated with H7N3 avian influenza in British Columbia broiler breeder chickens in 2004. *Avian Dis.*, **48**, 928–934.

163. Okamatsu, M., Saito, T., Yamamoto, Y., Mase, M., Tsuduku, S., Nakamura, K., Tsukamoto, K., and Yamaguchi, S. (2007) Low pathogenicity H5N2 avian influenza outbreak in Japan during the 2005–2006. *Vet. Microbiol.*, **124**, 35–46.

164. Marangon, S., Bortolotti, L., Capua, I., Bettio, M., and Dalla Pozza, M. (2003) Low-pathogenicity avian influenza (LPAI) in Italy (2000–2001): epidemiology and control. *Avian Dis.*, **47**, 1006–1009.

165. Senne, D.A. (2007) Avian influenza in North and South America, 2002–2005. *Avian Dis.*, **51**, 167–173.

166. Nguyen-Van-Tam, J.S., Nair, P., Acheson, P., Baker, A., Barker, M., Bracebridge, S., Croft, J., Ellis, J., Gelletlie, R., Gent, N., Ibbotson, S., Joseph, C., Mahgoub, H., Monk, P., Reghitt, T.W., Sundkvist, T., Sellwood, C., Simpson, J., Smith, J., Watson, J.M., Zambon, M., and Lightfoot, N. (2006) Outbreak of low pathogenicity H7N3 avian influenza in UK, including associated case of human conjunctivitis. *Euro Surveill.*, **11**, E060504.2.

167. Chin, P.S., Hoffmann, E., Webby, R., Webster, R.G., Guan, Y., Peiris, M., and Shortridge, K.F. (2002) Molecular evolution of H6 influenza viruses from poultry in Southeastern China: prevalence of H6N1 influenza viruses possessing seven A/Hong Kong/156/97 (H5N1)-like genes in poultry. *J. Virol.*, **76**, 507–516.

168. Abolnik, C. (2007) Molecular characterization of H5N2 avian influenza viruses isolated from South African ostriches in 2006. *Avian Dis.*, **51**, 873–879.

169. Woolcock, P.R., Suarez, D.L., and Kuney, D. (2003) Low-pathogenicity avian influenza virus (H6N2) in chickens in California, 2000–2002. *Avian Dis.*, **47**, 872–881.

170. Bano, S., Naeem, K., and Malik, S.A. (2003) Evaluation of pathogenic potential of avian influenza virus serotype H9N2 in chickens. *Avian Dis.*, **47**, 817–822.

171. Lin, Y.P., Shaw, M., Gregory, V., Cameron, K., Lim, W., Klimov, A., Subbarao, K., Guan, Y., Krauss, S., Shortridge, K., Webster, R., Cox, N., and Hay, A. (2000) Avian-to-human transmission of H9N2 subtype influenza A viruses: relationship between H9N2 and H5N1 human isolates. *Proc. Natl. Acad. Sci. U.S.A.*, **97**, 9654–9658.

172. Iqbal, M., Yaqub, T., Reddy, K., and McCauley, J.W. (2009) Novel genotypes of H9N2 influenza A viruses isolated from poultry in Pakistan containing NS genes similar to highly pathogenic

H7N3 and H5N1 viruses. *PLoS ONE*, **4**, e5788.

173. Matrosovich, M.N., Krauss, S., and Webster, R.G. (2001) H9N2 influenza A viruses from poultry in Asia have human virus-like receptor specificity. *Virology*, **281**, 156–162.

174. Zhang, P., Tang, Y., Liu, X., Peng, D., Liu, W., Liu, H., and Lu, S. (2008) Characterization of H9N2 influenza viruses isolated from vaccinated flocks in an integrated broiler chicken operation in eastern China during a 5 year period (1998–2002). *J. Gen. Virol.*, **89**, 3102–3112.

175. Alexander, D.J. (2007) Summary of avian influenza activity in Europe, Asia, Africa, and Australasia, 2002–2006. *Avian Dis.*, **51**, 161–166.

11
Variola: Smallpox

Andreas Nitsche and Hermann Meyer

11.1
Introduction

The genus *Orthopoxvirus* contains four species that infect humans: variola virus (VARV), vaccinia virus (VACV), monkeypox virus, and cowpox virus. The most notorious member is VARV, causing smallpox, which was declared eradicated by a Global Commission of the World Health Organization (WHO) in December 1979. VACV, once used as vaccine to eradicate smallpox, led to vaccine-related severe complications, and global vaccination was finally halted after the eradication was accomplished. Human monkeypox, an emerging zoonotic smallpox-like disease caused by monkeypox virus, with recurrent and likely endemic disease in the Congo Basin countries of Africa, is now regarded as the most severe human poxvirus infection occurring naturally. Cowpox virus causes pustular lesions in various animals and humans that can become fatal in immunosuppressed patients. Cowpox infections are restricted to the Old World with wild rodents as the natural reservoir. Taking into account the serious consequences of the diagnosis "smallpox" or even the implication of a misdiagnosis, there is a need to unambiguously, rapidly, and reliably identify smallpox and to differentiate it from other similar clinical entities. This chapter briefly describes the four human pathogenic orthopoxviruses and focuses on the evaluation of published real-time polymerase chain reaction (PCR)-based laboratory procedures for identifying smallpox.

11.2
Variola Virus

VARV has a strict human host range and no animal reservoir [1]. The virus was most often transmitted between humans by large-droplet respiratory particles inhaled by susceptible persons who had close, face to face contact. Variola major strains produced a severe prodrome, fever, prostration, and rash with case-fatality rates of up to 30%, with secondary attack rates among unvaccinated contacts of 30–80%. Variola minor strains (alastrim, amass, or kaffir viruses) produced infections with a less severe course and case-fatality rates of <1%. After an asymptomatic incubation

BSL3 and BSL4 Agents: Epidemiology, Microbiology, and Practical Guidelines, First Edition.
Edited by Mandy C. Elschner, Sally J. Cutler, Manfred Weidmann, and Patrick Butaye.

period of 10–14 days (range 7–17 days) the most common clinical presentation for ordinary smallpox was fever, sometimes with dermal petechia. Associated symptoms included backache, headache, vomiting, and prostration. Within a day or two after incubation, a systemic rash appeared that was characteristically distributed centrifugally. Initially, the rash lesions appeared as a macule, then papule, enlarging and progressing to a vesicle by day 4–5 and a pustule by day 7; lesions were encrusted and scabby by day 14 and sloughed off. Skin lesions were deep-seated and in the same stage of development in any one area of the body.

Prior to its eradication, smallpox as a clinical entity was relatively easy to recognize, but other exanthematous illnesses could be mistaken for this disease [2]. The rash of severe chickenpox, caused by varicella-zoster virus, was often misdiagnosed. Other diseases confused with vesicular-stage smallpox included monkeypox, generalized vaccinia, disseminated herpes virus infections, drug eruptions, erythema multiforme, enteroviral infections, cutaneous anthrax, insect bites, impetigo, and molluscum contagiosum. The Centers for Disease Control and Prevention (CDC), in collaboration with numerous professional organizations, developed an algorithm for evaluating patients (*www.cdc.gov/nip/smallpox and www.bt.cdc.gov/EmContact/index.asp*).

11.3 Human Monkeypox

Monkeypox was first recognized by Von Magnus in Copenhagen in 1958 as an exanthemic disease of primates in captivity. Later, the disease was seen in other captive animals, including primates in zoos and animal import centers. Human monkeypox was first reported in 1970 in the Democratic Republic of Congo (DRC). Serosurveys and virologic investigations during the 1980s in the DRC by the WHO indicated that monkeys are sporadically infected, as are humans, and that 75% of cases, mainly young children, were attributed to animal contact. The clinical appearance of human monkeypox is much like that of smallpox, with fever, a centrifugally distributed vesiculopustular rash (also on the palms and soles), respiratory distress, and, in some cases, death. Like VARV, monkeypox virus appears to enter through skin abrasions or the mucosa of the upper respiratory tract, where it produces an enanthem and cough. During primary viremia the virus migrates to regional lymph nodes, while during secondary viremia it is disseminated throughout the body and the skin rash appears. In the prodrome, lymphadenopathy (generally inguinal) with fever and headache is common. Individual skin lesions develop through the stages of macule, papule, vesicle, and pustule. Sequelae involve secondary infections, permanent scarring and pitting at the sites of the lesions, and sometimes corneal opacities. The emergence of monkeypox in the United States in 2003 demonstrated the ability of this zoonotic disease to exploit new ecologic niches: the prairie dog, diseased after exposure to imported infected African rodent(s), subsequently infected and caused illness in United States human populations which in general had a mild course.

11.4
Vaccinia Virus

VACV is the prototype orthopoxvirus and the poxvirus most thoroughly studied. The origin of VACV is uncertain. The smallpox vaccine, also called *vaccinia vaccine*, is a live virus preparation that mediates cross-protection to other orthopoxvirus infections. The most recent recommendations of the Advisory Committee on Immunization Practices (ACIP) on vaccinia vaccination are available at *www.cdc.gov/mmwr//preview/mmwrhtml/rr5010a1.htm*. The ACIP recommends vaccination as a safeguard for those laboratory and health care workers who are at high risk of orthopoxvirus infection. Vaccination is done by using a multiple puncture technique that causes a local lesion which develops and recedes in a distinctive manner in primary vaccines during a three-week period. At the site of percutaneous vaccination, a papule forms within 2–5 days, and the lesion reaches maximum size (about 1 cm diameter) by 8–10 days post vaccination after evolving through vesicle and pustule stages. The pustule dries to a scab which usually separates by 14–21 days after vaccination. Anecdotally, there have been reports of laboratory exposure of laboratory personnel to VACV, some with significant clinical pathology. VACV infections are not generally regarded as occurring naturally. Sporadic outbreaks of infection caused by the VACV subspecies buffalopox virus that involve transmission between milking buffalo and people have been reported, mainly in India. Vaccinia-like lesions have been observed on the animals' teats and the milkers' hands. Biological data and limited DNA analyses of isolates from an outbreak in India in 1985 suggest that buffalopox virus may be derived from VACV strains transmitted from humans to livestock during the smallpox vaccination era. Quite interestingly, multiple distinct VACVs, possibly related to the vaccine strain used during smallpox eradication in Brazil, were recently found in cattle and farm workers in rural Rio de Janeiro. Inadvertent exposure to vaccinia-vectored recombinant rabies virus vaccine dispersed to control rabies in wildlife has resulted in at least one instance of human infection.

11.5
Cowpox Virus

Cowpox, sometimes a rare occupational infection of humans, can be acquired by contact with infected cows; more often, other animals, for example, infected rats, pet cats, and zoo/circus elephants, have been sources of the disease. Cowpox virus is a rather diverse species and has been isolated from humans and a variety of animals in Europe and Western Asia. A serosurvey of wild animals in the United Kingdom found orthopoxvirus antibodies in bank and field voles and wood mice, which is consistent with small rodents being reservoir hosts for cowpox virus. Lesions in humans occur mainly on the fingers and in the face, with reddening and swelling, and systemic severe infections have been reported.

11.6
Collection of Specimens

Suitable specimens for laboratory testing are at least two to four scabs and/or material from vesicular lesions. Scabs can be separated from the underlying intact skin with a scalpel or a needle, and each specimen should be stored in a separate container to avoid cross-contamination. Coexistent infectious rash illnesses, including simultaneous chickenpox and monkeypox infections, have been noted. Lesions should be sampled so that both the vesicle fluid and overlying skin are collected. Once the overlying skin is lifted off and placed in a specimen container, the base of the vesicle should be vigorously swabbed by using a swab. Sample storage in transport medium is discouraged, largely because this complicates nucleic acid extraction. Biopsy of lesions may also provide material suitable for evaluation. Blood and throat swabs obtained during the prodromal febrile phase and early in the rash phase are also a potential source of virus. Virus-containing specimens can be shipped at standard refrigerator temperatures.

11.7
Real-Time Polymerase Chain Reaction

PCR exploits natural DNA replication mechanism and results in the in vitro production of large quantities of a desired sequence of DNA from a complex mixture of heterogeneous sequences. PCR does not differentiate between viable and nonviable organisms or incomplete pieces of genomic DNA, and this may complicate the interpretation of results. It is also important to check for potentially "negative" results caused by the inclusion of a control template known to produce a PCR product. Application of these precautions allows the PCR to become a promising realistic option for the diagnostician due to an unmatched sensitivity. Assays described in the past to identify and differentiate species of the genus Orthopoxvirus (OPV) [3] are based on sequences of the hemagglutinin [4], the cytokine response modifier B [5], and the A-type inclusion protein gene [6]. These conventional PCR methods are now being replaced by real-time PCR assays which combine amplification and detection of target DNA in one reaction vessel. By this, time-consuming post-PCR procedures can be avoided, thereby decreasing the risk of cross-contamination. In addition, real-time PCR provides quantitative information. The numerous advantages of the real-time PCR technique led to its introduction into the field of poxvirus diagnostics in 2002. Recent developments of portable real-time PCR machines and lyophilized reagents [7] raise the exciting prospect of these techniques being used for rapid diagnosis of disease outbreaks in the field.

However, for any PCR-based diagnostic test, the knowledge of at least part of the poxvirus genome is necessary to design valid and reliable assays. With regard to VARV a total of 45 epidemiologically varied VARV isolates from the past 30 years of the smallpox era were sequenced [8]. The genome is a linear DNA of about

186 kbp with covalently closed ends. Low sequence diversity suggests that there is probably very little difference in the isolates' functional gene content. This increases the likelihood that sequence-based detection methods will efficiently identify a re-emerging VARV. Poxvirus genome sequences are accessible at *www.poxvirus.org*, representing all eight genera of the subfamily *Chordopoxvirinae*, including 46 VARV strains. It will be important, as additional sequence information becomes available from related viruses, that the various PCR primers and probes are periodically reviewed *in silico*, if not via practical laboratory testing, for their actual specificity and sensitivity. For that purpose, screening of large OPV strain collections is essential to demonstrate the utility and establish the performance characteristics of the assays developed.

11.8
Evaluation of Real-Time PCR Assays

In the following we present data on *in silico* and laboratory tests of several real-time PCR assays published recently (Table 11.1). A theoretical evaluation of the assay

Table 11.1 Summary of in silico and laboratory evaluation of real-time PCR assays for the identification and differentiation of variola virus (VARV).

VARV identification method[a]	Target[b]	Number of VARV strains tested	Reference	False positives[c]	Reliable for VARV identification
FMCA	HA	Artificial construct	[9]	Yes (BLAST)	No
	HA	Artificial construct	[10]	Yes (24/26)	No
	Rpo18	12	[11]	No (0/110)	Yes
	VETF	12	[11]	No (0/110)	Yes
	A13L	12	[11]	No (0/110)	Yes
	14 kda	46	[12]	No (0/190)	Yes
	HA	Artificial construct	[13]	No (BLAST)	Yes[d]
	HA	46	[14, 15]	No (BLAST)	Yes
	B9R	8	[15]	Yes (BLAST)	No
Detection	B10R	8	[15]	Yes (5/110)	No
	crmB	Artificial construct	[16]	No (BLAST)	Yes
	14 kDa	12	[17]	No (0/85)	Yes[d]

[a]FMCA: Fluorescent melting curve analysis.
[b]HA: Hemagglutinin gene; crmB: cytokine response modifier B gene; VETF: viral early transcription factor; rpo18: reverse polymerase subunit 18 gene; 14 kDa: fusion protein gene; B9R, B10R, and A13L: designations of genes according to the nomenclature of vaccinia virus strain Copenhagen.
[c]Generation of false-positive reactions was either deduced from BLAST search or by amplifying a poxvirus test panel (Table 11.2). The number of positive amplifications versus the number of total strains tested is given in parentheses.
[d]See text for comment.

Table 11.2 Poxvirus test panel to evaluate real-time PCR assays.

Genus	Species	Number
Orthopoxvirus	Vaccinia virus	21
	Cowpox virus	70
	Mousepox virus	4
	Monkeypox virus	12
	Raccoon poxvirus	1
Parapoxvirus	Orf virus	1
Avipoxvirus	Fowlpox virus	1
Total		110

design was performed by using a BLAST search to check the specificity of primers and detection probes. A practical evaluation of those assays that met theoretical demands was done by testing up to 110 nonvariola poxviruses (Table 11.2).

11.9
Real-Time PCR Assays with Hybridization Probes

Espy and colleagues [9] were the first to introduce real-time LightCycler PCR as a diagnostic tool for OPV differentiation. A generic OPV-specific PCR followed by fluorescence melting curve analysis (FMCA) allows discrimination of smallpox virus due to its different melting temperature as compared to nonVARVs. However, a blast search using the sequences of the probes revealed not only identity to sequences in the hemagglutinin (HA) gene of VARV, but also to the HA gene of 12 camelpox, 2 taterapox, and 17 cowpox virus strains. This would result in false-positive results and as a consequence lead to considerable confusion.

Panning *et al.* [10] described a generic real-time LightCycler PCR targeting the HA gene. Subsequent FMCA should differentiate smallpox virus from other OPVs. A blast search using the target sequences of the probes revealed identical sequences in the HA gene of several cowpox virus strains. And indeed, this was proven in our experiments with 24 out of 26 cowpox viruses showing the same melting temperature as VARV.

In 2004 Nitsche *et al.* [11] published three LightCycler-based real-time PCR assays. These assays used FMCA for discrimination of VARV from other OPV, with two assays being based on generic amplification [rpo18 and viral early transcription factor (VETF)]. The generic assays were subjected to practical validation with 12 VARV isolates and the above-mentioned 110 samples of the test panel. Both assays revealed VARV-specific melting temperatures that allowed discrimination from all other OPV tested so far. The third assay (A13L) is VARV-specific by using a VARV-specific forward primer combined with an LC Red640 internally labeled reverse primer and a fluorescein-labeled hybridization probe. Correct amplification observed with the A13L assay can be confirmed by FMCA, thus introducing a second level of specificity.

Real-time LightCycler PCR identification of VARV DNA was developed and compiled in a kit system under GMP conditions with standardized reagents [12]. A single nucleotide mismatch resulting in a unique amino acid substitution in a total of 64 VARV strains was used to design a hybridization probe pair with a specific sensor probe that allows differentiation of VARV from other OPVs via FMCA. The applicability of this method was demonstrated by successful amplification of 180 strains belonging to the OPV species VARV, monkeypox, cowpox, vaccinia, camelpox, and mousepox viruses. The melting temperature determined for VARV (57 °C) differed significantly from those obtained for the other OPV strains (>62 °C).

In 2009 Putkuri *et al.* [13] published a LightCycler assay using generic OPV primers to amplify part of the HA gene. Identification of VARV is achieved by FMCA. The presented melting curves and the respective T_m show that discrimination between VARV and nonVARV OPVs is based on a difference of 2 °C only. A blast search demonstrated that VARV strains as compared to 30 other OPVs (including camelpox and cowpox viruses) differed within the probe-binding region by only one mismatch, which can be sufficient when highly discriminating mismatches are produced. Considering the broadened melting peaks due to the long hybridization probes applied, this small difference in T_m of 2 °C may lead to difficulties in the correct genotyping of VARV. Due to this publication being so recent [13], we have not yet performed experimental analysis of this assay with the poxvirus test panel.

11.10
Real-Time PCR Assays with 5′ Nuclease Probes

In 2003 a VARV-specific assay targeting the HA gene was described and evaluated by Ibrahim *et al.* [14]. Later this assay was modified by Kulesh *et al.* [15] who used identical primers but a slightly modified, shortened probe. This assay had been successfully evaluated with 322 coded samples that included genomic DNAs from 48 different isolates of VARV and 25 different strains other than VARV. We analyzed 110 OPV strains and isolates (Table 11.2) with this assay and we did not see any amplification. However, when using the primers only (probe excluded), there was amplification in 100 specimens. In our view the assay is well suited for specific VARV detection, although based on the design of the probe, some VARVs may not be detected equally sensitively.

Two further assays described in the same paper [15], supposed to be VARV-specific, target short stretches of the B9R and B10R genes (87 and 88 bp, respectively). This part of the genome is regarded to be unique for VARV. We used the B10R assay to analyze the OPV panel mentioned above and did see specific amplification with DNA of five cowpox virus strains. The sequence of the primers showed VARV-specific bases in the middle of the primers, not suitable for discrimination of other OPV. The sequence of MGB probe of the B9R assay was identical to the sequences of two cowpox and one taterapox virus strain as proven by blast search, therefore no further experimental testing of the B9R assay was performed.

In 2007 Scaramozzino *et al.* [17] published a 5′ nuclease-based assay using a combination of two probes differing in one nucleotide of the 14 kDA gene only. One probe matches perfectly with nonvariola OPVs, the other probe matches perfectly with VARVs. Since the VARV-specific probe reacts to some extent with monkeypox virus DNA, results have to be interpreted by analysis of the height of the fluorescence signal. Only the perfectly matching probe will give a higher fluorescence signal. Although the differences in signal intensity are shown to be significant for the presented viruses, in a screening run of unknown specimens every run should contain appropriate controls that allow a clear classification of VARV and nonVARV OPV.

Fedele and colleagues [16] proposed an assay that targets the cytokine response modifier B (crmB) gene with OPV generic primers and two 5′ nuclease probes modified by MGB. One MGB probe is designed to recognize VARV only, whereas the other detects OPVs generically. Although there is always an assumed risk targeting heterogeneous genomic stretches like *crmB*, a theoretical evaluation by blast search showed that the VARV-specific probe will detect all VARV strains known so far and miss all nonVARV OPV. In contrast, the OPV probe will certainly miss individual OPVs like some CPXV strains.

11.11
Other Real-Time PCR Formats

One of the great benefits of real-time PCR is the application of additional hybridizing oligonucleotides, called *probes*, which contribute to a second level of specificity. Since poxviruses display comparably stably conserved genomic sequences, the use of probes is possible, as documented by several publications. Therefore, for diagnostic screening purposes, probe-based assays facilitate more reliable results than assays that exclusively rely on specific primers. In the following, two publications are briefly described that do not use hybridizing probes and are therefore not considered to be state of the art.

Carletti *et al.* [18] presented an assay that detects all OPVs and discriminates herpesviruses based on staining amplicons with the dye SybrGreen followed by melting curve analysis. Since analysis of different OPV species leads to hardly distinguishable melting temperatures, OPV typing must be performed by enzymatic digestion of amplicons. Therefore, this approach lacks the general benefits of real-time PCR.

Aitichou *et al.* [7] used an LUX primer-based system that needs no probe to specifically detect VARV DNA in the HA gene region. A blast search of the F primer revealed that this primer will not cross-bind to nonVARV OPV, but will, on the other hand, fail to bind perfectly to all known VARV strains. The reverse primer will bind to all VARV sequences known. However, this approach lacks confirmation of the specificity of the amplicon which is the benefit of hybridization- or nuclease-based real-time PCR assays.

A most recent paper [19] used a rather new format, called *Eclipse probes,* to detect VARV major and minor. Eclipse probe are dual-labeled probes that contain a reporter at the 3′ end and a quencher in addition to an MGB modification at the 5′ end that mediates tight binding. These probes are not degraded by the polymerase and produce fluorescence signals by binding in a linear manner to the target. Therefore, these probes can be used for FMCA subsequently to PCR. The authors emphasize that this assay and a second one also described in their paper have not been developed for screening purposes but are to be used to differentiate VARV major from VARV minor strains when a positive VARV diagnosis has already been obtained from other assays. We therefore excluded these assays from our analysis.

11.12 Conclusions

Nevertheless, it needs to be stressed that a positive VARV PCR result must be confirmed by amplifying other parts of the genome. The use of multiple assays which target various portions of the genome, in addition to non-nucleic acid detection/diagnostic assays, will increase the confidence in a laboratory-based diagnosis of smallpox, especially in the context of the lack of naturally occurring disease. In our view a "state of the art" real-time PCR strategy is a combination of: (i) a generic OPV assay which allows VARV identification by melting analysis, and (ii) a VARV-specific assay. The parallel application of different detection formats, like 5′ nuclease and hybridization probes, can be beneficial by overcoming format-related shortcomings. Of course, an internal control to monitor inhibition has to be included and a modified VARV DNA as positive control would be of great advantage. By following this strategy and by selecting well-evaluated real-time PCR assays, VARV can be reliably detected even in mixtures of variola and nonvariola OPVs. Only a correct recognition will provide healthcare officials with critical insight into the management of a smallpox outbreak and allow the implementation of extensive emergency control measures, if necessary.

References

1. Damon, I.K. and Esposito, J.J. (2003) in *Manual of Clinical Microbiology*, 8th edn (eds P.R. Murray, E.J. Baron, J.H. Jorgensen, M.A. Pfaller, and M.H. Yolker), ASM Press, Washington, DC, pp. 1583–1591.
2. Henderson, D.A., Inglesby, T.V., Bartlett, J.G., Ascher, M.S., Eitzen, E., Jahrling, P.B., Hauer, J., Layton, M., McDade, J., Osterholm, T.M. *et al.* (1999) Smallpox as a biological weapon: medical and public health management. Working Group on Civilian Biodefense. *J. Am. Med. Assoc.*, **281**, 2127–2137.
3. Li, Y., Ropp, S.L., Zhao, H., Damon, I.K., and Esposito, J.J. (2007) Orthopoxvirus pan-genomic DNA assay. *J. Virol. Methods*, **141**(2), 154–165.
4. Ropp, S.L., Jin, Q., Knight, J.C., Massung, R.F., and Esposito, J.J. (1995)

PCR strategy for identification and differentiation of small pox and other orthopoxviruses. *J. Clin. Microbiol.*, **33**, 2069–2076.

5. Loparev, V.N., Massung, R.F., Esposito, J.J., and Meyer, H. (2001) Detection and differentiation of old world orthopoxviruses: restriction fragment length polymorphism of the crmB gene region. *J. Clin. Microbiol.*, **39**, 94–100.
6. Meyer, H., Ropp, S.L., and Esposito, J.J. (1997) Gene for A-type inclusion body protein is useful for a polymerase chain reaction assay to differentiate orthopoxviruses. *J. Virol. Methods*, **64**, 217–221.
7. Aitichou, M., Saleh, S., Kyusung, P., Huggins, J., O'Guinn, M., Jahrling, P., and Ibrahim, S. (2008) Dual-probe real time PCR assay for detection of variola or other orthopoxviruses with dried reagents. *J. Virol. Methods*, **153**, 190–195.
8. Esposito, J.J., Sammons, S.A., Frace, A.M., Osborne, J.D., Olsen-Rasmussen, M., Zhang, M., Govil, D., Damon, I.K., Kline, R., Laker, M., Li, Y., Smith, G.L., Meyer, H., LeDuc, J.W., and Wohlhueter, R.M. (2006) Genome sequence diversity and clues to the evolution of variola virus. *Science*, **313**, 807–812.
9. Espy, M.J., Cockerill, I.F., Meyer, R.F., Bowen, M.D., Poland, G.A., Hadfield, T.L., and Smith, T.F. (2002) Detection of smallpox virus DNA by LightCycler PCR. *J. Clin. Microbiol.*, **40**, 1985–1988.
10. Panning, M., Asper, M., Kramme, S., Schmitz, H., and Drosten, C. (2004) Rapid detection and differentiation of human pathogenic orthopox viruses by a fluorescence resonance energy transfer real-time PCR assay. *Clin. Chem.*, **50**, 702–708.
11. Nitsche, A., Ellerbrok, H., and Pauli, G. (2004) Detection of orthopoxvirus DNA by real-time PCR and identification of variola virus DNA by melting analysis. *J. Clin. Microbiol.*, **42**, 1207–1213.
12. Olson, V.A., Laue, T., Laker, M.T., Babkin, I.V., Drosten, C., Shchelkunov, S.N., Niedrig, M., Damon, I.K., and Meyer, H. (2004) Real-time PCR system for detection of orthopoxviruses and simultaneous identification of smallpox virus. *J. Clin. Microbiol.*, **42**, 1940–1946.
13. Putkuri, N., Piiparinen, H., Vaheri, A., and Vapalahti, O. (2009) Detection of human orthopoxvirus infections and differentiation of smallpox virus with real time PCR. *J. Med. Virol.*, **81**, 146–152.
14. Ibrahim, M.S., Kulesh, D.A., Saleh, S.S., Damon, I.K., Esposito, J.J., Schmaljohn, A.L., and Jahrling, P.B. (2003) Real-time PCR assay to detect smallpox virus. *J. Clin. Microbiol.*, **41**, 3835–3839.
15. Kulesh, D.A., Baker, R.O., Loveless, B.M., Norwood, D., Zwiers, S.H., Mucker, E., Hartmann, C., Herrara, R., Miller, D., Christensen, D. *et al.* (2004) Smallpox and pan-orthopox virus detection by real-time 3′-minor groove binder TaqMan assays on the Roche LightCycler and the Cepheid smart Cycler platforms. *J. Clin. Microbiol.*, **42**, 601–609.
16. Fedele, C.G., Negredo, A., Molero, F., Sánchez-Seco, M.P., and Tenorio, A. (2006) Use of internally controlled real-time genome amplification for detection of variola virus and other Orthopoxviruses infecting humans. *J. Clin. Microbiol.*, **44**, 4464–4470.
17. Scaramozzino, N., Ferrier-Rembert, A., Favier, A., Rothlisberger, C., Richard, S., Crance, J.M., Meyer, H., and Garin, D. (2007) Real-time PCR to identify variola virus or other human pathogenic orthopox viruses. *Clin. Chem.*, **53** (4), 606–613.
18. Carletti, F., Di Caro, A., Calcaterra, S., Grolla, A., Czub, M., Ippolito, G., Capobianchi, M.R., and Horejsh, D. (2005) Rapid, differential diagnosis of orthopox- and herpesviruses based upon real-time PCR product melting temperature and restriction enzyme analysis of amplicons. *J. Virol. Methods*, **129**, 97–100.
19. Loveless, B.M., Mucker, E.M., Hartmann, C., Craw, P.D., Huggins, J., and Kulesh, D.A. (2009) Differentiation of Variola major and Variola minor variants by MGB-Eclipse probe melt curves and genotyping analysis. *Mol. Cell. Probes*, **4**, 64–70.

12
Arenaviruses: Hemorrhagic Fevers

Amy C. Shurtleff, Steven B. Bradfute, Sheli R. Radoshitzky, Peter B. Jahrling, Jens H. Kuhn, and Sina Bavari

12.1
Characteristics

The family *Arenaviridae* contains a single genus, *Arenavirus*, which currently contains 23 recognized viruses and several viruses that have yet to be assigned [1–6]. Most of these viruses are not pathogenic for humans. Based on their antigenic properties, arenaviruses have been divided into two distinct groups. The Old World arenaviruses (Lassa–lymphocytic choriomeningitis serocomplex) include Lassa virus (LASV), which causes a viral hemorrhagic fever (Lassa fever) in Africa and the ubiquitous lymphocytic choriomeningitis virus (LCMV). The New World arenaviruses (Tacaribe serocomplex) include viruses indigenous to the Americas, such as the viral hemorrhagic fever-causing Junín (JUNV), Chapare (CHAPV), Guanarito (GTOV), Machupo (MACV), and Sabiá (SABV) viruses [2, 7–9]. Recently, a novel pathogenic arenavirus, Lujo virus (LUJV), was isolated from patients suffering from viral hemorrhagic fever. Phylogenetically, LUJV branches off the root of the Old World arenaviruses, which suggests it may represent a third arenavirus lineage [10].

Arenaviruses are enveloped viruses that are pleomorphic, ranging from 50 to 300 nm in diameter (reviewed in [1, 4, 11, 12]). The name "arenavirus" is derived from the particles' "sandy" appearance in electron microscopy sections, historically thought to be due to the incorporation of cellular ribosomes (*arenosus* in Latin means sand). The arenavirus genome consists of two single-stranded RNA molecules, designated L (large) and S (small). Each segment encodes two different proteins in two non-overlapping reading frames of opposite polarities. The L segment (≈7200 bp) encodes a viral RNA-dependent RNA polymerase (L) and a zinc-binding matrix protein (Z). The S segment (≈3500 bp) encodes the structural viral proteins: the nucleoprotein (NP) and the envelope glycoprotein precursor (GPC), which is cleaved into the envelope glycoproteins GP1 and GP2. The two open reading frames on each of the two genomic segments are separated by an intergenic noncoding region (IGR) with the potential to form one or more energetically stable stem-loop (hairpin) structures [13, 14]. The IGR serves

BSL3 and BSL4 Agents: Epidemiology, Microbiology, and Practical Guidelines, First Edition.
Edited by Mandy C. Elschner, Sally J. Cutler, Manfred Weidmann, and Patrick Butaye.

individual functions in structure-dependent transcription termination for enhanced gene expression [15–17] and in virion assembly and/or budding [18]. The 5′ and 3′ ends of the segments are noncoding untranslated regions (UTRs) and contain conserved reverse complementary sequences of 19 or 20 nucleotides [19, 20], which are thought to base-pair and form panhandle-like structures [21–23]. Genomic RNA extracted from arenavirions is not infectious and therefore arenaviruses are considered by some taxonomists as negative-sense RNA viruses.

Arenaviruses enter cells by first attaching to their cellular receptors, followed by endocytosis into clathrin-coated (JUNV, LASV) [24, 25] or smooth-walled (LCMV) [26] vesicles. Acidification of the endosome facilitates the pH-dependent fusion of the virions with cellular membranes resulting in delivery of viral nucleocapsids into the cell cytoplasm, where arenavirus replication exclusively occurs [12]. The L protein associates with the viral nucleocapsids and initiates transcription from the genome promoter located at the 3′ end. Primary transcription results in NP and L expression from the S and L segments, respectively [15, 27, 28]. Replication of the viral genome follows, and L "reads through" the IGR to generate full-length antigenomic RNA [29]. This RNA serves as a template for GPC and Z mRNA synthesis, as well as amplification of genomic RNA. Consequently, NP mRNA and NP protein accumulate earlier than GPC mRNA and the GP1 and GP2 glycoproteins [12, 30]. The newly synthesized full-length antigenomic and genomic RNA species are encapsidated by the NP to generate the RNP complexes for further mRNA transcription and for the production of virus progeny [31]. Following the association of electron-dense structures with the plasma membrane, budding occurs from the cellular membrane [32].

12.2 Epidemiology

12.2.1 Old World Arenaviruses

Lassa fever was first described in 1969 in a patient who was hospitalized in Jos, Nigeria. The caretaker of this patient also became severely ill and died [33]. It is a severe disease common in areas of western sub-Saharan Africa (Nigeria, Liberia, Guinea, Sierra Leone). Its etiologic agent is LASV, which is harbored in endemic areas by its rodent reservoir host, the Natal praomys (*Praomys natalensis*) [34], which is persistently and asymptomatically infected. Humans become infected by contact with infected rodent excreta or saliva (rodents are also part of the local diet) [35]. Person to person transmission is less frequent but possible by direct contact with the blood, urine, feces, or other bodily secretions and excretions of infected patients. Lassa fever cases tend to be more common in February to April compared to the rest of the year. The case-fatality rate is about 1–2% in the endemic areas, with an estimated 300 000–500 000 infections annually. The disease is especially severe late in pregnancy [34]. Antibodies against LASV were detected in 8–52% of tested humans in Sierra Leone despite the absence of reported Lassa fever.

LUJV was discovered during a recent nosocomial viral hemorrhagic fever outbreak in September and October of 2008. The index case became infected in *Lu*saka, Zambia, and was transferred to *Jo*hannesburg, South Africa, for medical management (hence the virus name: Lujo). There were three secondary and one tertiary infection. Four of the five patients died, but the origin of infection of the index patient remains unclear [10, 36].

12.2.2
New World Arenaviruses

Argentinian hemorrhagic fever (AHF) cases were first described in the humid Pampas of Argentina in 1955 [37], but AHF epidemics may have occurred as early as 1943. Its etiologic agent, JUNV, was later isolated from humans [38, 39]. Since the 1950s, JUNV is estimated to have caused around 30 000 cases of AHF. The case-fatality rate in the absence of any treatment is approximately 20%. The AHF endemic region has expanded progressively into north-central Argentina to the extent that five million people are estimated to be at risk of infection today [40]. AHF is typically a seasonal disease, with a peak of frequency occurring during the corn-harvesting season (March–June), when rodent populations are active. During this period, 75% of infected cases are male agricultural workers harvesting corn. Infection occurs by the inhalation of aerosols produced from rodent excreta or from rodents caught in mechanical harvesters. The principal host of AHF is the drylands laucha (*Calomys musculinus*) [41].

In 1964, Mackenzie and coworkers [42] described a new hemorrhagic fever in the Beni region in northeastern Bolivia. The disease, named Bolivian hemorrhagic fever (BHF), was first recognized in 1959 on the island of Orobayaya. A total of 470 cases were reported in the years up to 1962. MACV, named after a river close to the outbreak area, was isolated in 1963 from the spleen of a fatal human case studied in the town of San Joaquín [43]. The virus, transmitted by the big laucha (*Calomys callosus*) is responsible for a series of localized outbreaks from 1962 to 1964 involving more than 1000 patients, of whom 180 died. Similarly to AHF, the outbreak frequency peaked during the annual harvest (April–July). The case-fatality rate of BHF is approximately 5–30%. After 20 years of no reported cases, mainly as a result of rodent control measures [44], an outbreak of 19 cases was reported in 1994. Eight more cases were described in 1999, and 18 cases in 2000. A larger outbreak, with suspected 200 cases, occurred in 2008 [45].

CHAPV was discovered in 2003–2004. It was recovered from a single fatal case of viral hemorrhagic fever in the Chapare River region in rural Bolivia, outside of the area in which MACV is endemic. Additional cases were reported from this outbreak, but details and laboratory confirmation are lacking [46].

GTOV emerged in 1989 as the cause of "Venezuelan hemorrhagic fever" when settlers moved into a cleared forest area in the Guanarito Municipality on the central plains of Venezuela. Between 1990 and 1991, a total of 104 cases were reported, with a 25% case-fatality rate [47]. The virus was later isolated from the spleen of a male farm worker during autopsy [48]. The disease has appeared in subsequent years, mainly in the harvest season, between November and January, occurring predominantly among male agricultural workers who come in contact with GTOV's rodent reservoir, the short-tailed zygodont (*Zygodontomys brevicauda*). To date, approximately 200 cases of this disease have been reported [49]. After a spontaneous drop in human cases between 1989 and 1992, a new outbreak occurred in 2002 with 18 reported cases [1].

SABV, the cause of "Brazilian hemorrhagic fever," was isolated in 1990 from a single fatal infection in São Paulo, Brazil [50]. Subsequently, two laboratory infections were reported. One of the infected was successfully treated with ribavirin [51].

12.3 Clinical Signs

12.3.1 Old World Arenaviral Hemorrhagic Fevers

The symptoms of Lassa fever vary, but some of the best predicting symptoms are fever, pharyngitis, retrosternal pain, and proteinuria [52]. The incubation period ranges from 1 to 24 days with an average of 7–18 days. The disease commonly presents with fever, malaise, abdominal pain, pharyngitis, sore throat, cough, vomiting, diarrhea, high aspartate aminotransferase to alanine aminotransferase (AST to ALT) concentration ratios, and proteinuria. Clinical complications include pleural and pericardial effusions, neurological manifestations, bleeding from mucosal surfaces, and shock. A few patients with severe disease will experience hemorrhages, adult respiratory distress syndrome (ARDS), facial and neck edema, convulsions, and coma. Viremia has been noted by day 3 of disease, and high risk of death is closely associated with high viremia. As a general rule, survival is lowest in patients with both high viremia and high AST concentrations. Neurological complications can be frequent in critically ill patients, although it has never been specifically proven that LASV has any tropism for neurological tissues. Sensorineural deafness is a unique hallmark of disease, occurring uni- or bilaterally in about 30% of convalescent patients [52, 53].

Patients infected with LUJV presented with severe headache, malaise, diarrhea, vomiting, fever, retrosternal pain, severe sore throat, macropapular rash, and facial edema. Terminal features were ARDS, neurologic signs, and circulatory collapse. There was no overt hemorrhage besides gingival bleeding, petechial rash, and some oozing from injection sites [10, 36].

12.3.2 New World Arenaviral Hemorrhagic Fevers

Argentinian, Bolivian, "Venezuelan," and "Brazilian" hemorrhagic fevers are clinically similar [37, 49, 54–57]. Disease begins insidiously after an incubation period of one to two weeks with fever and malaise, headache, myalgia, epigastric pain, and anorexia. After three to four days the symptoms become increasingly severe with multisystem involvement: prostration, abdominal pain, nausea and vomiting, constipation, or mild diarrhea. Dizziness, photophobia, retro-orbital pain, and disorientation may also appear in some cases, as well as the earliest signs of vascular damage, such as conjunctival injection, flushing over the head and upper torso, skin petechiae, and mild (postural) hypotension. The more severe cases ($\approx$30% of patients) begin to display hemorrhagic or neurologic manifestations and secondary bacterial infections during the second week of illness. Neurological signs may be prominent and include tremor of the hands and tongue, coma, and convulsions. Hemorrhages from mucous membranes (gums, nose, vagina/uterus, gastrointestinal tract) and ecchymoses at needle puncture sites develop, and shock follows. Capillary leakiness is a hallmark of disease, with minor blood loss overall, and elevated hematocrit during the peak of capillary leak syndrome [1]. Death usually occurs 7–12 days after disease onset. Patients who survive begin to improve during the second week, as the appearance of neutralizing antibodies signals the onset of an immune response. Convalescence often lasts several weeks, with fatigue, dizziness, Beau lines in nails, and hair loss. "Venezuelan hemorrhagic fever" symptoms also include sore throat and pharyngitis. Diarrhea is also commonly seen in this illness [49, 58, 59]. However, petechiae, erythema, facial edema, and hyperesthesia are less common signs of the disease.

12.4
Pathological Findings

12.4.1
Old World Arenaviral Hemorrhagic Fevers

The histopathology of fatal human Lassa fever reveals minimal necrosis of hepatocytes, necrosis of the splenic red pulp and adrenal gland necrosis, focal tubular necrosis of the kidney, and interstitial pneumonia [60–62]. Macroscopic changes observed during Lassa fever in humans are pulmonary edema, pleural effusion, ascites, and signs of gastrointestinal hemorrhage [63]. Microscopically, hepatocellular necrosis with phagocytic macrophage reaction, but with minimal lymphocyte infiltration, is a typical finding, and splenic red pulp necrosis is common. Renal tubular injury, interstitial nephritis, interstitial pneumonitis, and limited myocarditis are additional hallmarks of the disease, but central nervous system lesions are generally absent. LASV is present in many diverse tissue types, which probably can be explained by almost ubiquitous tissue expression of α-dystroglycan, the principal LASV cell surface receptor [64]. Autopsies of patients who succumbed to LUJV infection revealed thrombocytopenia, granulocytosis, raised serum ALT and AST concentrations, hepatocyte necrosis without prominent inflammatory cell infiltrates, and skin vasculitis [10, 36].

Animal models of Lassa fever include infection of domesticated guinea pigs (*Cavia porcellus*), common marmosets (*Callithrix jacchus*), and rhesus monkeys (*Macaca mulatta*), but as of yet there are no models for LUJV infection.

The guinea pig model of LASV infection relies on strain 13 inbred guinea pigs that are susceptible to lethal infection with only two plaque-forming units of LASV strain Josiah [65]. Outbred Hartley guinea pigs are not uniformly susceptible to lethal infection [65]. In the strain 13 model, there is pantropism of LASV in the extraneuronal tissues as is seen in human LASV infection. For example, the spleen and lymph nodes replicate virus to high titers (10^7 and 10^8 PFU/g tissue) by days 8 and 9 of infection. Virus is present in salivary gland, lung, liver, adrenal gland, kidney, pancreas, and heart tissue. Low viral titers are common in plasma and the brain (10^4 PFU/g tissue). The most severe lesion in the LASV-infected guinea pig is interstitial pneumonia [65].

LASV infection in common marmosets mimics the disease course seen in fatally infected humans [66]. Histological lesions are mostly limited to the organs of the reticuloendothelial system (liver, spleen, lymph nodes). Multifocal hepatic necrosis, mild to moderate interstitial pneumonitis, and edematous changes in lung vasculature are typical findings, while lymphoid depletion occurred in the spleen and lymph nodes [67].

The rhesus monkey model of LASV infection is the most similar to human disease both in terms of clinical course of severe disease progression and histopathological findings [68]. The onset of viremia in rhesus monkeys occurs typically at an average of four days past infection, and it remains detectable at various levels for the subsequent two weeks [69]. As in natural infections, fatality in this model directly correlates with the level of viremia. Cellular immune responses do not appear to halt viral replication, which proceeds to a high titer and spreads through all tissues, leading to extensive infection. Interferon (IFN) levels at days 4–6 post infection are high, at or above 400 U/ml, and this elevation does not appear to influence viremia or disease outcome [68, 70]. Despite the wide range of tissues infected, the histological lesions are minor or largely absent and are not predictive of risk for death [71].

LASV, as most other arenaviruses discussed in this chapter, is a WHO Risk Group 4 pathogen. This classification limits research with replication-competent agent restricted to maximum-containment (BSL-4) laboratories. Pichindé virus (PICV), a New World arenavirus, causes a disease similar to human Lassa fever in guinea pigs, providing an accepted animal model for LASV infection [72]. PICV, serially passaged in strain 13 guinea pigs, was found to be uniformly lethal upon infection with less than 10 PFU after the eighth passage. The animals died about 13–19 days post-inoculation, and serum viremias reached greater than 100 000 PFU/ml of serum. The animals experienced severe lymphopenia, and the virus had a pantropism for almost all organ tissues, except those of the brain and nervous system. There was moderate hepatocellular necrosis in the liver, and necrosis of splenic red pulp and adrenal gland tissues. The lung tissue demonstrated interstitial pneumonia. The similarity in pathogenesis of PICV-infected guinea pigs to Lassa fever in humans and macaques makes this an acceptable surrogate.

12.4.2 New World Arenaviral Hemorrhagic Fevers

The most common macroscopic abnormalities in severe cases of New World arenaviral hemorrhagic fevers are widespread hemorrhage, particularly in the skin and mucous membranes (gastrointestinal tract), as well as intracranial hemorrhages (Virchow–Robin space) and hemorrhages in the kidney, pericardium, spleen, adrenal glands, and lung (the last four common in AHF). Microscopic lesions include acidophilic bodies and focal necrosis in the liver, acute tubular and papillary necrosis in the kidney, reticular hyperplasia of the spleen and lymph nodes, and secondary bacterial lung infections in the case of AHF (acute bronchitis, bronchopneumonia, myocardial and lung abscesses) and interstitial pneumonia in the case of BHF [73, 74]. In AHF the sites of cellular necrosis (hepatocytes, renal tubular epithelium, macrophages, dendritic reticular cells of the spleen, lymph nodes) have been shown to correspond to sites of viral antigen accumulation, and both JUNV and MACV could be isolated from the blood, spleen, and lymph nodes of patients [41, 75–77]. However, the overall histopathological findings in human and animal model infections are relatively subtle with comparatively little necrosis and do not reflect the severity of disease [47, 73, 74, 78–80].

Patterns of clinical AHF illness are JUNV strain-specific and can be "hemorrhagic" (Espindola strain), "neurologic" (Ledesma strain), "mixed" (P-3551 strain), and "common" (Romero strain) [41]. Guinea pigs and nonhuman primates infected with JUNV Espindola strain ("hemorrhagic") demonstrate a pronounced bleeding diathesis, with disseminated cutaneous and mucous membrane hemorrhage. In contrast, Ledesma-infected animals ("neurologic") show little or no hemorrhagic manifestations, but develop overt and generally progressive signs of neurologic dysfunction: limb paresis, ataxia, tremors, and hyperactive startle reflexes. In guinea pigs, the Espindola strain replicates predominantly in spleen, lymph nodes, and bone marrow, the major sites of necrosis, while lower levels of virus are present in blood and brain. The Ledesma strain, however, is found predominantly in the brain and only low amounts of virus are recovered from the spleen and lymph nodes. In the rhesus monkey, the pathological picture is slightly different with higher serum viremias in animals infected with the Espindola strain. These animals also shed more virus from the oropharynx and respond more slowly with the production of antibodies than do those infected with the Ledesma strain [81–83].

Humans suffering from "Venezuelan hemorrhagic fever" present with high fever, weakness, headache, toxicity, diarrhea, conjunctivitis, pharyngitis, leukopenia, and thrombocytopenia [47, 49]. The severe disease and hemorrhagic manifestations are associated with marked thrombocytopenia [47]. Patient autopsies revealed pulmonary edema, congestion in the lungs, and cardiomegaly. Liver tissue was also congested, with focal hepatic necrosis and organ yellowing. Spleens were enlarged and congested, and there was the presence of blood in the gastrointestinal tract and other hollow visceral organs. In the guinea pig model, hemorrhage is minimal, but the histopathology is similar in both Hartley and strain 13 guinea pigs [84]. There are several notable lesions that appear to be due to virus infection, such as epithelial necrosis in the gastrointestinal tract, lung interstitial pneumonia and hemorrhage, and necrosis or depletion of lymphoid tissues of spleen, lymph nodes, intestines, and lungs. There is notable adrenal cortical necrosis and congestion, and splenic congestion. Viral antigen is present in all lymphoid cells of all organs. In the spleen, viral antigen is abundant in large macrophages and lymphoblastic cells of the marginal zone. Virus levels are higher in homogenates of organ tissues than in guinea pig serum, suggesting viral replication in these organ tissues. The splenic tissue has the highest viral titer, with up to 7.9 $\log_{10}$ PFU/g of tissue from strain 13 guinea pigs [84]. In general, these findings indicate that GTOV infection pathologically resembles infection with other arenaviruses, leading to lymphoid necrosis, bone marrow depletion interstitial pneumonia, and platelet thrombi. Comparatively, PICV and LASV seem to cause liver necrosis to a higher extent than GTOV infection in this guinea pig model. Infection of guinea pigs with GTOV closely resembles infection with viscerotropic strains of Junín, where there are similar virus levels and distribution of virus in the tissues [81].

12.5 Pathogenesis of Old and New World Hemorrhagic Fevers

Arenaviruses presumably enter the body by inhalation and deposit in the lung terminal respiratory bronchioles. They gain entry to the lymphoid system and spread systematically, but usually do not leave a detectable pneumonic focus [85]. There is also prominent infection of mesothelial surfaces *in vivo* (perhaps a source of some of the observed effusions) and the parenchymal cells of several organs, particularly lymphoid tissues. Macrophages are usually identified as early and prominent targets of arenavirus infection, and as their infection becomes more abundant, the epithelial cells of many organs begin to show signs of infection as well. Widespread infection of the marginal zone and necrosis of lymphoid follicles of the spleen and lymph nodes are common, with the potential to inhibit an effective curative immune response [30, 51, 86].

New and Old World arenaviruses infect endothelial cells in culture [87–89]. However, there is relatively little evidence of vascular lesions *in vivo*. Therefore, indirect effects might be responsible for the increased vascular permeability seen in patients [30], and the profuse bleeding often seen is presumably a consequence of vascular damage caused by both cytokines and virus replication. Thrombocytopenia, which is commonly found in human patients and animal models, and elevated amounts of factor VIII-related antigen (von Willebrand factor, vWF), which is synthesized and released from endothelial cells, could contribute to the observed endothelial dysfunction [90–94].

Platelet counts are significantly depressed following infection of rhesus macaques with Old and New World arenaviruses. In LASV infection, platelet aggregation is abnormal by day 6 post infection, and virtually absent by day 13 [95]. It was hypothesized that platelet and endothelial cell function may fail due to an imbalance of prostacyclin and thromboxane A2. For arenavirus diseases, thrombocytopenia is progressive, with counts reaching their lowest point at or near the time of death. Coincident with the dropping platelet count, progressive necrosis of bone marrow occurs, suggesting that the decrease in the number of platelets may also be related to impaired production [80, 93, 94, 96]. Plasma from patients with AHF has been shown to contain an inhibitor of platelet function [53]. The platelet inhibition appears to be reversible *in vitro* [97, 98]. JUNV infection of human $CD34^+$ hematopoietic stem cells decreases their ability to generate platelets *in vitro* [99]. Thus, the available evidence suggests that abnormal platelet function in patients with arenaviral infections is not due to an intrinsic platelet defect, but to an inhibition of function and production induced by a number of mechanisms. The central involvement of coagulopathy in arenaviral disease has been ruled out as there is no general correlation of specific coagulation abnormalities with the severity of disease. Furthermore, there are only four reported observations of disseminated intravascular coagulation (DIC) in AHF cases [74, 100], indicating that DIC is not an important pathogenic phenomenon in arenaviral disease, although there are several modest abnormalities of clotting factors and activation of fibrinolysis [90–93, 101–104]. Studies of JUNV infection have revealed that Factor V is uniformly elevated (starting from day 8), and fibrinogen is normal in mild cases and elevated in severe cases in the later stages of infection (after day 10). Activated partial thromboplastin time (APTT) is prolonged during the acute phase of illness. A lower ratio of factor VIII:C to factor VIII-related antigen (vWF) is noted during the illness, but returns to normal during the convalescence period. In the guinea pig model, and to a lesser extent in humans, factors IX and XI are slightly reduced [90, 92–94, 104, 105]. Levels of prothrombin fragment 1 + 2 and thrombin–antithrombin III complexes are increased. In the guinea pig model, thrombocytopenia begins before abnormalities in the coagulation system, indicating that activation of the blood coagulation system is not the cause for peripheral platelet consumption but can in theory contribute to the enhancement and maintenance of the decreased platelet count in later stages.

Plasminogen activity is below normal in AHF patients during the earlier stages of the disease (days 6–11) [106], although normal or slightly elevated levels of α_2-antiplasmin are detected [90, 105]. Tissue plasminogen activator (t-PA) and D-dimer levels are high during the early stages of the disease, while plasminogen activator inhibitor-I (PAI-1) is increased only in severe cases during the second week of illness [105]. Fibrin monomers can be detected in the AHF guinea pig model but disappear progressively toward terminal stages of illness. Fibrin degradation products (FDPs) cannot be detected in the serum and urine, indicating no intravascular fibrin deposition [93]. In MACV-infected rhesus monkeys, low levels of FDPs are observed in the serum [94]. Histopathologic studies of AHF have also not provided any evidence of intravascular deposition of fibrin [74, 101]. In human patients, neither fibrin monomers nor FDPs are observed [104]. While the presence of fibrin monomers and abnormalities in the coagulation system occur during the early phases of the disease, there are no changes in fibrinogen levels at this time. Moreover, fibrinogen levels increase toward the final stages of the disease. This is hypothesized to be a result of higher rate of fibrinogen synthesis [90, 93, 94].

Limited complement activation through the classical pathway is observed during the acute phase of AHF and BHF [90, 91, 107–110]. There is a moderate reduction of total complement activity, with a variety of activation products from both the classical and alternate complement pathways of activation found in the serum. Complement activation coincides with the appearance of coagulation abnormalities, but to date there are no conclusive studies demonstrating a cause–effect relationship between complement activation and the coagulation cascade [93].

The acute phase of arenaviral hemorrhagic fevers is associated with a significant depression of host immunity from the virus and/or cytokines. The frequency of pyogenic secondary bacterial infections in humans and animal models of JUNV and LASV infections [65, 73, 74, 82] suggests that PMN leukocyte function is compromised. In addition, there are many observations of focal necrotic lesions in the liver and kidney tissues of Lassa fever and New World arenavirus patients where inflammatory cells are absent. Leukocyte dysfunction may be a result of direct interactions of virus with PMN cells, [95, 111, 112] bone marrow necrosis, maturation arrest, and leukopenia [113, 114]. AHF is associated with a profound decrease in recall of delayed hypersensitivity, diminished responsiveness of lymphocytes to nonspecific mitogens, decreased levels of circulating B- and T-cells, lymphoid necrosis, and inversion of CD4/CD8 lymphocyte ratios [115–118]. Abnormalities reported in animal models include necrosis of macrophages, T- and B-lymphocyte depletion (bone marrow inhibition or destruction), decreased primary and secondary antibody responses, blunted Arthus reaction and anergy after established tuberculin sensitivity [41, 113, 119–122]. IFN and defective macrophage function are highly plausible causes for these observed abnormalities. Virulent JUNV strains damage macrophages extensively *in vivo* and are able to replicate in both dendritic cells and macrophages isolated from infected guinea pigs spleens, whereas attenuated strains, which are not immunosuppressive, replicate only in dendritic cells [76, 111]. During the acute period of AHF, viral antigens are also detected in lymphatic tissues, and JUNV can be regularly isolated from peripheral blood mononuclear cells (PBMCs) by co-cultivation [123]. Decrease in B and $CD4^+$ lymphocyte values are compatible with diminished or, at least, delayed B-cell function, which seems to be involved in the outcome of the disease.

There is some debate on the importance of antibody responses to control arenavirus infection. High-quality neutralizing antibodies are sometimes demonstrable in serum within a few days of resolution of fever [108, 124, 125]. However, in LASV infection, neutralizing antibodies may not be required for protection against infection, since they are not readily detected in protective vaccination; additionally, neutralizing antibodies are rarely found in unvaccinated survivors until long after resolution of disease ([126–129], reviewed in [63]). However, passive transfer of immune plasma or sera has long been used to treat infected human patients, with various results [130, 131]. Passive transfer of immune plasma or sera can protect nonhuman primates after LASV infection, and this protection correlates with neutralizing activity [69, 132]. Similarly, post-infection treatment of JUNV infection with immune plasma or sera is effective in humans and nonhuman primates [133, 134], although a low percentage of recipients develop neurological problems after disease resolution. The protection of plasma transfer is directly related to neutralizing activity of the antibodies (reviewed in [135]). MACV infection in nonhuman primates can also be treated with passive antibody transfer, although neurologic symptoms are also found in some survivors [109]. Therefore, it is likely that neutralizing antibody is required for arenavirus protection conferred by passive antibody therapy, but is not required for protection in vaccination [134, 136, 137].

The role of nonneutralizing antibody in combating arenavirus infection is not clear. Nonhuman primates that survive LASV infection have a more rapid nonneutralizing immunoglobulin (IgG) response compared to nonsurvivors [138]. However, nonneutralizing antibody titers had no correlation with survivors in vaccinated nonhuman primates [126, 127]. Because of the questions surrounding the importance of antibody, it has been speculated that T cells play the main role in protection against LASV infection. Indeed, T cells from nonfatal infections have increased antigen-specific T cell proliferation compared to cells from animals that eventually succumb to infection [138]. Conversely, mice that express human HLA-A2 in place of mouse MHC I molecules are partially susceptible to LASV infection, but the depletion of $CD8^+$ T cells reverses this pathogenesis, suggesting that $CD8^+$ T cells may contribute to pathogenesis in some cases [139]. However, preliminary studies have shown that the generation of anti-arenavirus T cell responses has no deleterious effects and protects mice from pseudotyped virus challenge [140–143]. Since neutralizing antibody is not required for protection against LASV, it is likely that T cell responses are required for successful responses against infection, but may contribute to pathogenesis in certain cases. However, additional work needs to be done before any solid conclusions can be reached.

In addition to antibody therapy, several vaccines are being developed to combat arenaviral infection. A live attenuated vaccine for JUNV infection, Candid 1, is effective and safe in nonhuman primates [144, 145]. Further studies in Argentina demonstrated that the vaccine is also safe and effective in humans [146] and can boost antibody titers in individuals who have already been exposed to the virus [147]. Vaccines against LASV are currently in the developmental stage. A vaccinia virus-based platform is efficacious in nonhuman primates [126, 127]. A vesicular stomatitis Indiana virus-based vaccine expressing the LASV glycoprotein is also effective in nonhuman primates [128]. Additionally, a live attenuated reassortant vaccine, ML29, comprised of both LASV and Mopeia virus genes, can protect nonhuman primates from LASV infection [129]. To date, these vaccines have not been tested in humans. More preliminary peptide-based vaccines are in development, with the hope for development of a cross-protective vaccine based on T cell responses to a cocktail of epitopes or cross-reactive epitopes from many arenaviruses [140, 141, 143].

Ribavirin, a non-immunogenic nucleoside analog with broad-spectrum antiviral activity, has also been tested in a nonhuman primate model of LASV infection [69].

All monkeys tested received virus and then drug treatment on either day 1 or day 5 of infection. None died, but all animals were mildly sick, with lower viremias than untreated controls. Further studies found a synergistic effect of the use of ribavirin and immune plasma in treatment of LASV infection in crab-eating macaques. Treatment with both agents was capable of protecting animals even when administered as late as day 10 of infection, where either treatment alone would not work as late in infection as day 7 post infection. Ribavirin has been tested in a few MACV- and SABV-infected humans in noncontrolled studies and may be protective [51, 148]; the effectiveness against JUNV is unclear [135]. At this time, there are a variety of other small molecule or antisense therapeutics in various stages of early discovery and preclinical testing. These therapeutics are being developed against viral targets such as genes or proteins critical for virus replication or against host cell pathways which are essential for the viruses' life cycles [149–156].

There are conflicting reports regarding the activation of pro-inflammatory cytokines during arenavirus infections. Studies with AHF patients demonstrate the activation of pro- as well as anti-inflammatory cytokine pathways in this disease. Patients and animal models of JUNV and LASV infection can have high levels of circulating IFN-α and IFN-γ in serum samples [63]. Tumor necrosis factor α (TNF-α), (interleukin) IL-6, and IL-8 levels are also elevated and correlate with fatal outcome of JUNV infection. However, levels of IL-1β remain normal [85, 103, 157–159]. Elevated IL-6 is also correlated with fatal infection in LASV infection, while higher early IFN-α levels are correlated with survival [138]. Interestingly, IFN-α and IFN-γ decreases LASV replication *in vitro* [160]. However, the Z protein of GTOV, JUNV, MACV, and SABV, but not of LASV, decreases the IFN-β response by binding to RIG-I [161].

In vitro studies with human monocyte-derived dendritic cells infected with LASV demonstrate productive infection with LASV, but no elaboration of pro-inflammatory cytokines, such as IL-8 and IFN-inducible protein 10. In addition, surviving LASV infected patients had higher levels of IL-8 and IFN inducible protein 10 compared to those who succumbed to infection [162]. Rhesus monkeys infected with the WE strain of LCMV, which causes a hemorrhagic disease similar to LASV infection, had increases in IL-6, sIL6R, sTNFR1 and -2, and IFN-γ [163]. Overall, immunosuppression due to infection with arenaviruses is a factor in disease, but there are inconsistent reports among individual arenaviruses as to which cytokines are induced or suppressed in patients, animal models or *in vitro* cultures. It appears that there could be a pathophysiological cascade of events which is triggered by high levels of arenaviral replication in many organ and endothelial tissues. Several experts hypothesized that deregulated cytokine expression, similar to that seen during sepsis or severe inflammation response syndrome, could be responsible for events that could lead to vascular leakiness and shock in these patients [63].

There have been some studies of pathophysiology during the infection of guinea pigs with adapted PICV, toward an understanding of events during severe arenaviral hemorrhagic fever. These studies were uniquely designed to measure cardiac and pulmonary function during the infection of catheterized guinea pigs [164]. Symptoms such as severe hypotension, circulatory shock, and pulmonary edema could all be attributed to cardiovascular disturbances. Cardiac output, heart rate, cardiac work, cardiac power, and stroke volume were all significantly decreased in the guinea pig model, while pulmonary functions were not particularly altered [164]. There is marked wasting of greater than 25% of the animal's body weight over the course of the infection, with cardiac decompensation, acidosis, and renal tubular defect. Along with pulmonary distress syndrome, there is also a circulatory response to the fluid load which is probably due to capillary leakiness syndrome. Despite the cardiac involvement, no viral antigen was found in cardiac tissues or cells at day 14 of infection [72, 164].

The pathogenesis of arenaviral hemorrhagic fevers is clearly complicated and may involve many factors that lead to inflammation, cytokine release, vascular leakiness, cardio-pulmonary involvement, and other similar phenomena. It is not clear whether many of these factors are released due to direct interaction with the virus, or whether viral infection triggers an event which leads to a cascade of inflammatory mediators and cytokines to be released at variable times post-infection.

12.6 Diagnostics

12.6.1 Serological Tests

Enzyme-linked immunosorbent assays (ELISAs) using recombinant protein or infected cells as antigen or IgM and IgG have been developed [165–167]. At least a fourfold rise in the titer should be observed to prove acute infection, if only specific IgG is detectable [34, 168]. This is because the prevalence of virus-specific IgGs in the population of endemic areas may be high due to previous exposure. Specific IgM might also be used for detection, but might not be present at the early stages of arenaviral infection. Indirect immunofluorescence antibody (IFA) tests with virus-infected cells are also available [52, 168]. However, the interpretation of IFA is complicated by positive staining results in both the acute and convalescent phases of infection, as well as the subjective nature of the assay. ELISAs are thought to be more sensitive and specific [166, 169, 170]. Cross-reactions can occur between arenaviruses in these tests. In contrast, virus neutralization tests are highly specific, but neutralizing antibodies may appear too late to be useful in immediate diagnosis. For example, patients with Lassa fever do not usually develop neutralizing antibodies until weeks after they became ill. Arenaviral antigens can be detected in blood or tissues, using antigen-capture ELISAs.

12.7
PCR

Diagnostic reverse transcriptase PCR and real-time PCR tests are available or have been published [171–177]. However, there is limited experience with these assays in a clinical or environmental situation for the early detection of human cases. Some PCR tests detect a wide range of arenaviruses by targeting the highly conserved termini of the S RNA segment. Others are more specific. Serum, plasma, CSF, throat washings, and urine can be used for sample preparation. Real-time PCR might provide an advantage as the risk of contamination is greatly reduced and quantification of viral RNA in serum is possible. Of note, specimens containing high concentration of viral RNA may produce false negative results due to inhibition of the enzymatic reaction. Furthermore, PCR techniques may fail to amplify arenavirus strains even with limited sequence deviations.

12.7.1
Virus Culture and Antigen Testing

Virus isolation can also be used for diagnosis if a BSL-4 laboratory is available. Arenaviruses can be recovered in cell cultures, particularly from Vero cells. Alternatively, initial passaging of the isolate in laboratory rodents such as suckling mice, guinea pigs, or newborn hamsters may be more sensitive. The presence of virus can then be confirmed by PCR or by the detection of virus antigen in cells using immunohistochemical or immunofluorescence assays. The disadvantages are the time required for virus isolation (days to weeks) as well as the need for special facilities, which are unavailable in many arenavirus-endemic areas [178]. Recent discoveries of novel arenaviruses relied on the use of IFA, PCR, as well as pyrosequencing technology [5, 10, 46, 179, 180].

12.8
Disclaimer

The content of this publication does not necessarily reflect the views or policies of the United States (US) Department of Defense, the US Department of the Army, the US Department of Health and Human Services, or the institutions and companies affiliated with the authors.

References

1. Charrel, R.N. and de Lamballerie, X. (2003) Arenaviruses other than Lassa virus. *Antiviral Res.*, **57**, 89–100.
2. Clegg, J.C. (2002) Molecular phylogeny of the arenaviruses. *Curr. Top. Microbiol. Immunol.*, **262**, 1–24.
3. Erickson, B.R., Delgado, S., Aguda, R., Vallejo, E., Vergas, J., Blair, P.J., Albarino, C., Comer, J.A., Rollin, P.E., Ksiazek, T.G., Olson, J.G., and Nichol, S.T. (2006) Presented at the Abstracts of the XIIIth International Conference

on Negative Strand Viruses, Salamanca, Spain, June 17–22.
4. Jay, M.T., Glaser, C., and Fulhorst, C.F. (2005) The arenaviruses. *J. Am. Vet. Med. Assoc.*, **227**, 904–915.
5. Lecompte, E., Ter Meulen, J., Emonet, S., Daffis, S., and Charrel, R.N. (2007) Genetic identification of Kodoko virus, a novel arenavirus of the African pigmy mouse (Mus Nannomys minutoides) in West Africa. *Virology*, **364**, 178–183.
6. Oldstone, M.B. (2002) Arenaviruses. I. The epidemiology molecular and cell biology of arenaviruses. Introduction. *Curr. Top. Microbiol. Immunol.*, **262**, V–XII.
7. Bowen, M.D., Peters, C.J., and Nichol, S.T. (1996) The phylogeny of New World (Tacaribe complex) arenaviruses. *Virology*, **219**, 285–290.
8. Cajimat, M.N. and Fulhorst, C.F. (2004) Phylogeny of the venezuelan arenaviruses. *Virus Res.*, **102**, 199–206.
9. Rowe, W.P., Pugh, W.E., Webb, P.A., and Peters, C.J. (1970) Serological relationship of the Tacaribe complex of viruses to lymphocytic choriomeningitis virus. *J. Virol.*, **5**, 289–292.
10. Briese, T., Paweska, J.T., McMullan, L.K., Hutchison, S.K., Street, C., Palacios, G., Khristova, M.L., Weyer, J., Swanepoel, R., Egholm, M., Nichol, S.T., and Lipkin, W.I. (2009) Genetic detection and characterization of Lujo virus, a new hemorrhagic fever-associated arenavirus from southern Africa. *PLoS Pathog.*, **5**, e1000455.
11. Buchmeier, M.J. (2002) Arenaviruses: protein structure and function. *Curr. Top. Microbiol. Immunol.*, **262**, 159–173.
12. Meyer, B.J., de la Torre, J.C., and Southern, P.J. (2002) Arenaviruses: genomic RNAs, transcription, and replication. *Curr. Top. Microbiol. Immunol.*, **262**, 139–157.
13. Auperin, D.D., Galinski, M., and Bishop, D.H. (1984) The sequences of the N protein gene and intergenic region of the S RNA of pichinde arenavirus. *Virology*, **134**, 208–219.
14. Wilson, S.M. and Clegg, J.C. (1991) Sequence analysis of the S RNA of the African arenavirus Mopeia: an unusual secondary structure feature in the intergenic region. *Virology*, **180**, 543–552.
15. Meyer, B.J. and Southern, P.J. (1993) Concurrent sequence analysis of 5′ and 3′ RNA termini by intramolecular circularization reveals 5′ nontemplated bases and 3′ terminal heterogeneity for lymphocytic choriomeningitis virus mRNAs. *J. Virol.*, **67**, 2621–2627.
16. Meyer, B.J. and Southern, P.J. (1994) Sequence heterogeneity in the termini of lymphocytic choriomeningitis virus genomic and antigenomic RNAs. *J. Virol.*, **68**, 7659–7664.
17. Tortorici, M.A., Albarino, C.G., Posik, D.M., Ghiringhelli, P.D., Lozano, M.E., Rivera Pomar, R., and Romanowski, V. (2001) Arenavirus nucleocapsid protein displays a transcriptional antitermination activity in vivo. *Virus Res.*, **73**, 41–55.
18. Pinschewer, D.D., Perez, M., and de la Torre, J.C. (2005) Dual role of the lymphocytic choriomeningitis virus intergenic region in transcription termination and virus propagation. *J. Virol.*, **79**, 4519–4526.
19. Auperin, D., Dimock, K., Cash, P., Rawls, W.E., Leung, W.C., and Bishop, D.H. (1982) Analyses of the genomes of prototype pichinde arenavirus and a virulent derivative of Pichinde Munchique: evidence for sequence conservation at the 3′ termini of their viral RNA species. *Virology*, **116**, 363–367.
20. Auperin, D.D., Compans, R.W., and Bishop, D.H. (1982) Nucleotide sequence conservation at the 3′ termini of the virion RNA species of New World and Old World arenaviruses. *Virology*, **121**, 200–203.
21. Harnish, D.G., Polyak, S.J., and Rawls, W.E. (1993) in *The Arenaviridae* (ed M.S. Salvato), Plenum Press, New York, pp. 157–174.
22. Salvato, M., Shimomaye, E., and Oldstone, M.B. (1989) The primary structure of the lymphocytic choriomeningitis virus L gene encodes a putative RNA polymerase. *Virology*, **169**, 377–384.
23. Young, P.R. and Howard, C.R. (1983) Fine structure analysis of Pichinde

virus nucleocapsids. *J. Gen. Virol.*, **64** (Pt 4), 833–842.

24. Martinez, M.G., Cordo, S.M., and Candurra, N.A. (2007) Characterization of Junin arenavirus cell entry. *J. Gen. Virol.*, **88**, 1776–1784.

25. Vela, E.M., Zhang, L., Colpitts, T.M., Davey, R.A., and Aronson, J.F. (2007) Arenavirus entry occurs through a cholesterol-dependent, non-caveolar, clathrin-mediated endocytic mechanism. *Virology*, **211**, 67–71.

26. Borrow, P. and Oldstone, M.B. (1994) Mechanism of lymphocytic choriomeningitis virus entry into cells. *Virology*, **198**, 1–9.

27. Singh, M.K., Fuller-Pace, F.V., Buchmeier, M.J., and Southern, P.J. (1987) Analysis of the genomic L RNA segment from lymphocytic choriomeningitis virus. *Virology*, **161**, 448–456.

28. Southern, P.J., Singh, M.K., Riviere, Y., Jacoby, D.R., Buchmeier, M.J., and Oldstone, M.B. (1987) Molecular characterization of the genomic S RNA segment from lymphocytic choriomeningitis virus. *Virology*, **157**, 145–155.

29. Leung, W.C., Ghosh, H.P., and Rawls, W.E. (1977) Strandedness of Pichinde virus RNA. *J. Virol.*, **22**, 235–237.

30. Buchmeier, M.J., de la Torre, J.C., and Peters, C.J. (2006) *Arenaviridae*: the viruses and their replication, in *Fields Virology*, 5th edn, vol. 2 (eds D.M. Knipe and P.M. Howley), Lippincott Williams & Wilkins, Philadelphia, pp. 299–307.

31. Raju, R., Raju, L., Hacker, D., Garcin, D., Compans, R., and Kolakofsky, D. (1990) Nontemplated bases at the 5′ ends of Tacaribe virus mRNAs. *Virology*, **174**, 53–59.

32. Dalton, A.J., Rowe, W.P., Smith, G.H., Wilsnack, R.E., and Pugh, W.E. (1968) Morphological and cytochemical studies on lymphocytic choriomeningitis virus. *J. Virol.*, **2**, 1465–1478.

33. Frame, J.D., Baldwin, J.M. Jr., Gocke, D.J., and Troup, J.M. (1970) Lassa fever, a new virus disease of man from West Africa. I. Clinical description and pathological findings. *Am. J. Trop. Med. Hyg.*, **19**, 670–676.

34. McCormick, J.B., Webb, P.A., Krebs, J.W., Johnson, K.M., and Smith, E.S. (1987) A prospective study of the epidemiology and ecology of Lassa fever. *J. Infect. Dis.*, **155**, 437–444.

35. Keenlyside, R.A., McCormick, J.B., Webb, P.A., Smith, E., Elliott, L., and Johnson, K.M. (1983) Case-control study of Mastomys natalensis and humans in Lassa virus-infected households in Sierra Leone. *Am. J. Trop. Med. Hyg.*, **32**, 829–837.

36. Paweska, J.T., Sewlall, N.H., Ksiazek, T.G., Blumberg, L.H., Hale, M.J., Lipkin, W.I., Weyer, J., Nichol, S.T., Rollin, P.E., McMullan, L.K., Paddock, C.D., Briese, T., Mnyaluza, J., Dinh, T.H., Mukonka, V., Ching, P., Duse, A., Richards, G., de Jong, G., Cohen, C., Ikalafeng, B., Mugero, C., Asomugha, C., Malotle, M.M., Nteo, D.M., Misiani, E., Swanepoel, R., and Zaki, S.R. (2009) Nosocomial outbreak of novel arenavirus infection, southern Africa. *Emerging Infect. Dis.*, **15**, 1598–1602.

37. Arribalzaga, R.A. (1955) New epidemic disease due to unidentified germ: nephrotoxic, leukopenic and enanthematous hyperthermia. *Dia Med.*, **27**, 1204–1210.

38. Parodi, A.S., Greenway, D.J., Rugiero, H.R., Frigerio, M., de la Barrera, J.M., Mettler, N., Garzon, F., Boxaca, M., Guerrero, L., and Nota, N. (1958) Concerning the epidemic outbreak in Junin. *Dia Med.*, **30**, 2300–2301.

39. Pirosky, I., Zuccarini, J., Molinelli, E.A., Di Pietro, A., Barrera-Oron, J.G., Martini, P., and Copello, A.R. (1959) *Virosis Hemorrágica del Noroeste Bonaerense (Endemo-Epidémica, Febril, Enantémática, y Leucopenica), Ministerio de Asistencia Social y Salud Pública,* Buenos Aires Instituto Nacional de Microbiologia.

40. Enria, D.A. and Feuillade, M.R. (1998) in *An Overview of Arbovirology in Brazil and Neighboring Countries* (eds A.P.A. Travassosda Rosa, P.F.C. Vasconcelos, and J.F.S. Traavassosda Rosa), Instituto

Evandro Chaggas, Belem, Brazil, pp. 219–232.

41. Maiztegui, J.I. (1975) Clinical and epidemiological patterns of Argentine haemorrhagic fever. *Bull. World Health Organ.*, **52**, 567–575.

42. Mackenzie, R.B., Beye, H.K., Valverde, L., and Garron, H. (1964) Epidemic hemorrhagic fever in Bolivia. I. A preliminary report of the epidemiologic and clinical findings in a new epidemic area in South America. *Am. J. Trop. Med. Hyg.*, **13**, 620–625.

43. Johnson, K.M., Wiebenga, N.H., Mackenzie, R.B., Kuns, M.L., Tauraso, N.M., Shelokov, A., Webb, P.A., Justines, G., and Beye, H.K. (1965) Virus Isolations from human cases of hemorrhagic fever in bolivia. *Proc. Soc. Exp. Biol. Med.*, **118**, 113–118.

44. Kuns, M.L. (1965) Epidemiology of machupo virus infection II. Ecological and control studies of hemorrhagic fever. *Am. J. Trop. Med. Hyg.*, **14**, 813–816.

45. Aguilar, P.V., Camargo, W., Vargas, J., Guevara, C., Roca, Y., Felices, V., Laguna-Torres, V.A., Tesh, R., Ksiazek, T.G., and Kochel, T.J. (2009) Reemergence of Bolivian hemorrhagic fever, 2007–2008. *Emerging Infect. Dis.*, **15**, 1526–1528.

46. Delgado, S., Erickson, B.R., Agudo, R., Blair, P.J., Vallejo, E., Albarino, C.G., Vargas, J., Comer, J.A., Rollin, P.E., Ksiazek, T.G., Olson, J.G., and Nichol, S.T. (2008) Chapare virus, a newly discovered arenavirus isolated from a fatal hemorrhagic fever case in Bolivia. *PLoS Pathog.*, **4**, e1000047.

47. Salas, R., de Manzione, N., Tesh, R.B., Rico-Hesse, R., Shope, R.E., Betancourt, A., Godoy, O., Bruzual, R., Pacheco, M.E., Ramos, B. *et al.* (1991) Venezuelan haemorrhagic fever. *Lancet*, **338**, 1033–1036.

48. Tesh, R.B., Jahrling, P.B., Salas, R., and Shope, R.E. (1994) Description of Guanarito virus (Arenaviridae: Arenavirus), the etiologic agent of Venezuelan hemorrhagic fever. *Am. J. Trop. Med. Hyg.*, **50**, 452–459.

49. de Manzione, N., Salas, R.A., Paredes, H., Godoy, O., Rojas, L., Araoz, F., Fulhorst, C.F., Ksiazek, T.G., Mills, J.N., Ellis, B.A., Peters, C.J., and Tesh, R.B. (1998) Venezuelan hemorrhagic fever: clinical and epidemiological studies of 165 cases. *Clin. Infect. Dis.*, **26**, 308–313.

50. Lisieux, T., Coimbra, M., Nassar, E.S., Burattini, M.N., de Souza, L.T., Ferreira, I., Rocco, I.M., da Rosa, A.P., Vasconcelos, P.F., Pinheiro, F.P. *et al.* (1994) New arenavirus isolated in Brazil. *Lancet*, **343**, 391–392.

51. Barry, M., Russi, M., Armstrong, L., Geller, D., Tesh, R., Dembry, L., Gonzalez, J.P., Khan, A.S., and Peters, C.J. (1995) Brief report: treatment of a laboratory-acquired Sabia virus infection. *N. Engl. J. Med.*, **333**, 294–296.

52. McCormick, J.B., King, I.J., Webb, P.A., Johnson, K.M., O'Sullivan, R., Smith, E.S., Trippel, S., and Tong, T.C. (1987) A case-control study of the clinical diagnosis and course of Lassa fever. *J. Infect. Dis.*, **155**, 445–455.

53. Cummins, D., Molinas, F.C., Lerer, G., Maiztegui, J.I., Faint, R., and Machin, S.J. (1990) A plasma inhibitor of platelet aggregation in patients with Argentine hemorrhagic fever. *Am. J. Trop. Med. Hyg.*, **42**, 470–475.

54. Harrison, L.H., Halsey, N.A., McKee, K.T. Jr., Peters, C.J., Barrera Oro, J.G., Briggiler, A.M., Feuillade, M.R., and Maiztegui, J.I. (1999) Clinical case definitions for Argentine hemorrhagic fever. *Clin. Infect. Dis.*, **28**, 1091–1094.

55. Molteni, H.D., Guarinos, H.C., Petrillo, C.O., and Jaschek, F. (1961) Clinico-statistical study of 338 patients with epidemic hemorrhagic fever in the northwest of the province of Buenos Aires. *Sem. Med.*, **118**, 839–855.

56. Rugiero, H.R., Ruggiero, H., Gonzalezcambaceres, C., Cintora, F.A., Maglio, F., Magnoni, C., Astarloa, L., Squassi, G., Giacosa, A., and Fernandez, D. (1964) Argentine hemorrhagic fever. II. Descriptive clinical study. *Rev. Asoc. Med. Argent.*, **78**, 281–294.

57. Stinebaugh, B.J., Schloeder, F.X., Johnson, K.M., Mackenzie, R.B., Entwisle, G., and De Alba, E. (1966)

Bolivian hemorrhagic fever. A report of four cases. *Am. J. Med.*, **40**, 217–230.
58. Peters, C.J. (2002) Human infection with arenaviruses in the Americas. *Curr. Top. Microbiol. Immunol.*, **262**, 65–74.
59. Vainrub, B. and Salas, R. (1994) Latin American hemorrhagic fever. *Infect. Dis. Clin. North Am.*, **8**, 47–59.
60. Walker, D.H., Johnson, K.M., Lange, J.V., Gardner, J.J., Kiley, M.P., and McCormick, J.B. (1982) Experimental infection of rhesus monkeys with Lassa virus and a closely related arenavirus, Mozambique virus. *J. Infect. Dis.*, **146**, 360–368.
61. Walker, D.H., Wulff, H., and Murphy, F.A. (1975) Experimental Lassa virus infection in the squirrel monkey. *Am. J. Pathol.*, **80**, 261–278.
62. Winn, W.C. Jr. and Walker, D.H. (1975) The pathology of human Lassa fever. *Bull. World Health Organ.*, **52**, 535–545.
63. Gunther, S. and Lenz, O. (2004) Lassa virus. *Crit. Rev. Clin. Lab. Sci.*, **41**, 339–390.
64. Cao, W., Henry, M.D., Borrow, P., Yamada, H., Elder, J.H., Ravkov, E.V., Nichol, S.T., Compans, R.W., Campbell, K.P., and Oldstone, M.B. (1998) Identification of alpha-dystroglycan as a receptor for lymphocytic choriomeningitis virus and Lassa fever virus. *Science*, **282**, 2079–2081.
65. Jahrling, P.B., Smith, S., Hesse, R.A., and Rhoderick, J.B. (1982) Pathogenesis of Lassa virus infection in guinea pigs. *Infect. Immun.*, **37**, 771–778.
66. Carrion, R. Jr., Brasky, K., Mansfield, K., Johnson, C., Gonzales, M., Ticer, A., Lukashevich, I., Tardif, S., and Patterson, J. (2007) Lassa virus infection in experimentally infected marmosets: liver pathology and immunophenotypic alterations in target tissues. *J. Virol.*, **81**, 6482–6490.
67. Avila, M.M., Frigerio, M.J., Weber, E.L., Rondinone, S., Samoilovich, S.R., Laguens, R.P., de Guerrero, L.B., and Weissenbacher, M.C. (1985) Attenuated Junin virus infection in Callithrix jacchus. *J. Med. Virol.*, **15**, 93–100.
68. Peters, C.J., Liu, C.T., Anderson, G.W. Jr., Morrill, J.C., and Jahrling, P.B. (1989) Pathogenesis of viral hemorrhagic fevers: Rift Valley fever and Lassa fever contrasted. *Rev. Infect. Dis.*, **11** (Suppl. 4), S743–S749.
69. Jahrling, P.B., Hesse, R.A., Eddy, G.A., Johnson, K.M., Callis, R.T., and Stephen, E.L. (1980) Lassa virus infection of rhesus monkeys: pathogenesis and treatment with ribavirin. *J. Infect. Dis.*, **141**, 580–589.
70. Johnson, K.M., McCormick, J.B., Webb, P.A., Smith, E.S., Elliott, L.H., and King, I.J. (1987) Clinical virology of Lassa fever in hospitalized patients. *J. Infect. Dis.*, **155**, 456–464.
71. Kunz, S. (2009) The role of the vascular endothelium in arenavirus haemorrhagic fevers. *Thromb. Haemost.*, **102**, 1024–1029.
72. Jahrling, P.B., Hesse, R.A., Rhoderick, J.B., Elwell, M.A., and Moe, J.B. (1981) Pathogenesis of a pichinde virus strain adapted to produce lethal infections in guinea pigs. *Infect. Immun.*, **32**, 872–880.
73. Child, P.L., MacKenzie, R.B., Valverde, L.R., and Johnson, K.M. (1967) Bolivian hemorrhagic fever. A pathologic description. *Arch. Pathol.*, **83**, 434–445.
74. Elsner, B., Schwarz, E., Mando, O.G., Maiztegui, J., and Vilches, A. (1973) Pathology of 12 fatal cases of Argentine hemorrhagic fever. *Am. J. Trop. Med. Hyg.*, **22**, 229–236.
75. Cossio, P., Laguens, R., Arana, R., Segal, A., and Maiztegui, J. (1975) Ultrastructural and immunohistochemical study of the human kidney in Argentine haemorrhagic fever. *Virchows Arch. A Pathol. Anat. Histol.*, **368**, 1–9.
76. Gonzalez, P.H., Cossio, P.M., Arana, R., Maiztegui, J.I., and Laguens, R.P. (1980) Lymphatic tissue in Argentine hemorrhagic fever. Pathologic features. *Arch. Pathol. Lab. Med.*, **104**, 250–254.
77. Johnson, K.M. (1965) Epidemiology of Machupo virus infection. 3. Significance of virological observations in man and animals. *Am. J. Trop. Med. Hyg.*, **14**, 816–818.
78. McLeod, C.G. Jr., Stookey, J.L., White, J.D., Eddy, G.A., and Fry, G.A. (1978)

Pathology of Bolivian hemorrhagic fever in the African green monkey. *Am. J. Trop. Med. Hyg.*, **27**, 822–826.

79. McLeod, C.G., Stookey, J.L., Eddy, G.A., and Scott, K. (1976) Pathology of chronic Bolivian hemorrhagic fever in the rhesus monkey. *Am. J. Pathol.*, **84**, 211–224.

80. Terrell, T.G., Stookey, J.L., Eddy, G.A., and Kastello, M.D. (1973) Pathology of Bolivian hemorrhagic fever in the rhesus monkey. *Am. J. Pathol.*, **73**, 477–494.

81. Kenyon, R.H., Green, D.E., Maiztegui, J.I., and Peters, C.J. (1988) Viral strain dependent differences in experimental Argentine hemorrhagic fever (Junin virus) infection of guinea pigs. *Intervirology*, **29**, 133–143.

82. McKee, K.T. Jr., Mahlandt, B.G., Maiztegui, J.I., Eddy, G.A., and Peters, C.J. (1985) Experimental Argentine hemorrhagic fever in rhesus macaques: viral strain-dependent clinical response. *J. Infect. Dis.*, **152**, 218–221.

83. McKee, K.T. Jr., Mahlandt, B.G., Maiztegui, J.I., Green, D.E., and Peters, C.J. (1987) Virus-specific factors in experimental Argentine hemorrhagic fever in rhesus macaques. *J. Med. Virol.*, **22**, 99–111.

84. Hall, W.C., Geisbert, T.W., Huggins, J.W., and Jahrling, P.B. (1996) Experimental infection of guinea pigs with Venezuelan hemorrhagic fever virus (Guanarito): a model of human disease. *Am. J. Trop. Med. Hyg.*, **55**, 81–88.

85. Kenyon, R.H., McKee, K.T. Jr., Zack, P.M., Rippy, M.K., Vogel, A.P., York, C., Meegan, J., Crabbs, C., and Peters, C.J. (1992) Aerosol infection of rhesus macaques with Junin virus. *Intervirology*, **33**, 23–31.

86. Peters, C.J., Jahrling, P.B., Liu, C.T., Kenyon, R.H., McKee, K.T. Jr., and Barrera Oro, J.G. (1987) Experimental studies of arenaviral hemorrhagic fevers. *Curr. Top. Microbiol. Immunol.*, **134**, 5–68.

87. Andrews, B.S., Theofilopoulos, A.N., Peters, C.J., Loskutoff, D.J., Brandt, W.E., and Dixon, F.J. (1978) Replication of dengue and Junin viruses in cultured rabbit and human endothelial cells. *Infect. Immun.*, **20**, 776–781.

88. Gomez, R.M., Pozner, R.G., Lazzari, M.A., D'Atri, L.P., Negrotto, S., Chudzinski-Tavassi, A.M., Berria, M.I., and Schattner, M. (2003) Endothelial cell function alteration after Junin virus infection. *Thromb. Haemost.*, **90**, 326–333.

89. Lukashevich, I.S., Maryankova, R., Vladyko, A.S., Nashkevich, N., Koleda, S., Djavani, M., Horejsh, D., Voitenok, N.N., and Salvato, M.S. (1999) Lassa and Mopeia virus replication in human monocytes/macrophages and in endothelial cells: different effects on IL-8 and TNF-alpha gene expression. *J. Med. Virol.*, **59**, 552–560.

90. Molinas, F.C., de Bracco, M.M., and Maiztegui, J.I. (1989) Hemostasis and the complement system in Argentine hemorrhagic fever. *Rev. Infect. Dis.*, **11** (Suppl. 4), S762–S770.

91. Molinas, F.C., Giavedoni, E., Frigerio, M.J., Calello, M.A., Barcat, J.A., and Weissenbacher, M.C. (1983) Alteration of blood coagulation and complement system in neotropical primates infected with Junin virus. *J. Med. Virol.*, **12**, 281–292.

92. Molinas, F.C. and Maiztegui, J.I. (1981) Factor VIII: C and factor VIII R: Ag in Argentine hemorrhagic fever. *Thromb. Haemost.*, **46**, 525–527.

93. Molinas, F.C., Paz, R.A., Rimoldi, M.T., and de Bracco, M.M. (1978) Studies of blood coagulation and pathology in experimental infection of guinea pigs with Junin virus. *J. Infect. Dis.*, **137**, 740–746.

94. Scott, S.K., Hickman, R.L., Lang, C.M., Eddy, G.A., Hilmas, D.E., and Spertzel, R.O. (1978) Studies of the coagulation system and blood pressure during experimental Bolivian hemorrhagic fever in rhesus monkeys. *Am. J. Trop. Med. Hyg.*, **27**, 1232–1239.

95. Fisher-Hoch, S.P., Mitchell, S.W., Sasso, D.R., Lange, J.V., Ramsey, R., and McCormick, J.B. (1987) Physiological and immunologic disturbances associated with shock in a primate model of Lassa fever. *J. Infect. Dis.*, **155**, 465–474.

96. Green, D.E., Mahlandt, B.G., and McKee, K.T. Jr. (1987) Experimental Argentine hemorrhagic fever in rhesus macaques: virus-specific variations in pathology. *J. Med. Virol.*, **22**, 113–133.
97. Marta, R., Heller, V., Maiztegui, J.I., and Molinas, F.C. (1990) Normal platelet aggregation and release are inhibited by plasma from patients with Argentine haemorrhagic fever. *Blood*, **76**, 46.
98. Marta, R.F., Heller, M.V., Maiztegui, J.I., and Molinas, F.C. (1993) Further studies on the plasma inhibitor of platelet activation in Argentine hemorrhagic fever. *Thromb. Haemost.*, **69**, 526–527.
99. Pozner, R.G., Ure, A.E., Jaquenod de Giusti, C., D'Atri, L.P., Italiano, J.E., Torres, O., Romanowski, V., Schattner, M., and Gomez, R.M. Junin virus infection of human hematopoietic progenitors impairs in vitro proplatelet formation and platelet release via a bystander effect involving type I IFN signaling. *PLoS Pathog.*, **6**, e1000847.
100. Agrest, A., Sanchez Avalos, J.C., Arce, M., and Slepoy, A. (1969) Argentine hemorrhagic fever and comsumption coagulation disorders. *Medicina (B Aires)*, **29**, 194–201.
101. Gallardo, F. (1970) Argentine hemorrhagic fever. Anatomo-pathological findings in 10 necropsies. *Medicina (B Aires)*, **1970**, (Suppl. 1), 77–84.
102. Gonzalez, P.H., Laguens, R.P., Frigerio, M.J., Calello, M.A., and Weissenbacher, M.C. (1983) Junin virus infection of Callithrix jacchus: pathologic features. *Am. J. Trop. Med. Hyg.*, **32**, 417–423.
103. Heller, M.V., Saavedra, M.C., Falcoff, R., Maiztegui, J.I., and Molinas, F.C. (1992) Increased tumor necrosis factor-alpha levels in Argentine hemorrhagic fever. *J. Infect. Dis.*, **166**, 1203–1204.
104. Molinas, F.C., de Bracco, M.M., and Maiztegui, J.I. (1981) Coagulation studies in Argentine hemorrhagic fever. *J. Infect. Dis.*, **143**, 1–6.
105. Heller, M.V., Marta, R.F., Sturk, A., Maiztegui, J.I., Hack, C.E., Cate, J.W., and Molinas, F.C. (1995) Early markers of blood coagulation and fibrinolysis activation in Argentine hemorrhagic fever. *Thromb. Haemost.*, **73**, 368–373.
106. Molinas, F.C., Kordich, L., Porterie, P., Lerer, G., and Maiztegui, J.I. (1987) Plasminogen abnormalities in patients with Argentine hemorrhagic fever. *Thromb. Res.*, **48**, 713–720.
107. Budzko, D.B., Rimoldi, M.T., Acobettro, R.I., and De Bracco, M.M. (1975) Mecanismo de la activación del complemento sérico en la fiebre hemorrágica argentina experimental en el cobayé. *Medicina (B Aires)*, **35**, 588–589.
108. de Bracco, M.M., Rimoldi, M.T., Cossio, P.M., Rabinovich, A., Maiztegui, J.I., Carballal, G., and Arana, R.M. (1978) Argentine hemorrhagic fever. Alterations of the complement system and anti-Junin-virus humoral response. *N. Engl. J. Med.*, **299**, 216–221.
109. Eddy, G.A., Wagner, F.S., Scott, S.K., and Mahlandt, B.J. (1975) Protection of monkeys against Machupo virus by the passive administration of Bolivian haemorrhagic fever immunoglobulin (human origin). *Bull. World Health Organ.*, **52**, 723–727.
110. Kenyon, R.H. and Peters, C.J. (1989) Actions of complement on Junin virus. *Rev. Infect. Dis.*, **11** (Suppl. 4), S771–S776.
111. Laguens, M., Chambo, J.G., and Laguens, R.P. (1983) In vivo replication of pathogenic and attenuated strains of Junin virus in different cell populations of lymphatic tissue. *Infect. Immun.*, **41**, 1279–1283.
112. Laguens, R.P., Gonzalez, P.H., Ponzinibbio, C., and Chambo, J. (1986) Damage of human polymorphonuclear leukocytes by Junin virus. *Med. Microbiol. Immunol.*, **175**, 177–180.
113. Carballal, G., Rodriguez, M., Frigerio, M.J., and Vasquez, C. (1977) Junin virus infection of guinea pigs: electron microscopic studies of peripheral blood and bone marrow. *J. Infect. Dis.*, **135**, 367–373.
114. Schwarz, E.R., Mando, O.G., Maiztegui, J.I., and Vilches, A.M. (1970) Symptoms and early signs of major

diagnostic value in Argentine hemorrhage fever. *Medicina (B Aires)*, **1970**, (Suppl. 1), 8–14.
115. Arana, R.M., Ritacco, G.V., and de la Vega, M.T. (1977) Immunological studiens in Argentine haemorrhagic fever. *Medicina (B Aires)*, **37**, 186–189.
116. Carballal, G., Oubina, J.R., Rondinone, S.N., Elsner, B., and Frigerio, M.J. (1981) Cell-mediated immunity and lymphocyte populations in experimental Argentine hemorrhagic fever (Junin Virus). *Infect. Immun.*, **34**, 323–327.
117. Enria, D., Franco, S.G., Ambrosio, A., Vallejos, D., Levis, S., and Maiztegui, J. (1986) Current status of the treatment of Argentine hemorrhagic fever. *Med. Microbiol. Immunol.*, **175**, 173–176.
118. Vallejos, D.A., Ambrosio, A.M., Feuillade, M.R., and Maiztegui, J.I. (1989) Lymphocyte subsets alteration in patients with Argentine hemorrhagic fever. *J. Med. Virol.*, **27**, 160–163.
119. Carballal, G., Cossio, P.M., Laguens, R.P., Ponzinibbio, C., Oubina, J.R., Meckert, P.C., Rabinovich, A., and Arana, R.M. (1981) Junin virus infection of guinea pigs: immunohistochemical and ultrastructural studies of hemopoietic tissue. *J. Infect. Dis.*, **143**, 7–14.
120. de Guerrero, L.B., Boxaca, M., Weissenbacher, M., and Frigerio, M.J. (1977) Experimental infection of the guinea pig with Junin virus. Clinical picture, dissemination, and elimination of the virus. *Medicina (B Aires)*, **37**, 271–278.
121. Frigerio, J. (1977) Immunologic aspects of guinea pigs infected with Junin virus. *Medicina (B Aires)*, **37**, 96–100.
122. Maiztegui, J.I., Laguens, R.P., Cossio, P.M., Casanova, M.B., de la Vega, M.T., Ritacco, V., Segal, A., Fernandez, N.J., and Arana, R.M. (1975) Ultrastructural and immunohistochemical studies in five cases of Argentine hemorrhagic fever. *J. Infect. Dis.*, **132**, 35–53.
123. Ambrosio, M., Vallejos, A., Saavedra, C., and Maiztegui, J.I. (1990) Junin virus replication in peripheral blood mononuclear cells of patients with Argentine haemorrhagic fever. *Acta Virol.*, **34**, 58–63.
124. Peters, C.J., Webb, P.A., and Johnson, K.M. (1973) Measurement of antibodies to Machupo virus by the indirect fluorescent technique. *Proc. Soc. Exp. Biol. Med.*, **142**, 526–531.
125. Webb, P.A., Johnson, K.M., and Mackenzie, R.B. (1969) The measurement of specific antibodies in Bolivian hemorrhagic fever by neutralization of virus plaques. *Proc. Soc. Exp. Biol. Med.*, **130**, 1013–1019.
126. Fisher-Hoch, S.P., Hutwagner, L., Brown, B., and McCormick, J.B. (2000) Effective vaccine for Lassa fever. *J. Virol.*, **74**, 6777–6783.
127. Fisher-Hoch, S.P., McCormick, J.B., Auperin, D., Brown, B.G., Castor, M., Perez, G., Ruo, S., Conaty, A., Brammer, L., and Bauer, S. (1989) Protection of rhesus monkeys from fatal Lassa fever by vaccination with a recombinant vaccinia virus containing the Lassa virus glycoprotein gene. *Proc. Natl. Acad. Sci. U.S.A.*, **86**, 317–321.
128. Geisbert, T.W., Jones, S., Fritz, E.A., Shurtleff, A.C., Geisbert, J.B., Liebscher, R., Grolla, A., Stroher, U., Fernando, L., Daddario, K.M., Guttieri, M.C., Mothe, B.R., Larsen, T., Hensley, L.E., Jahrling, P.B., and Feldmann, H. (2005) Development of a new vaccine for the prevention of Lassa fever. *PLoS Med.*, **2**, e183.
129. Lukashevich, I.S., Carrion, R. Jr., Salvato, M.S., Mansfield, K., Brasky, K., Zapata, J., Cairo, C., Goicochea, M., Hoosien, G.E., Ticer, A., Bryant, J., Davis, H., Hammamieh, R., Mayda, M., Jett, M., and Patterson, J. (2008) Safety, immunogenicity, and efficacy of the ML29 reassortant vaccine for Lassa fever in small non-human primates. *Vaccine*, **26**, 5246–5254.
130. Clayton, A.J. (1977) Lassa immune serum. *Bull. World Health Organ.*, **55**, 435–439.
131. McCormick, J.B. (1986) Clinical, epidemiologic, and therapeutic aspects of Lassa fever. *Med. Microbiol. Immunol.*, **175**, 153–155.
132. Jahrling, P.B., Peters, C.J., and Stephen, E.L. (1984) Enhanced treatment of Lassa fever by immune plasma

combined with ribavirin in cynomolgus monkeys. *J. Infect. Dis.*, **149**, 420–427.

133. Maiztegui, J.I., Fernandez, N.J., and de Damilano, A.J. (1979) Efficacy of immune plasma in treatment of Argentine haemorrhagic fever and association between treatment and a late neurological syndrome. *Lancet*, **2**, 1216–1217.

134. Enria, D.A., Briggiler, A.M., Fernandez, N.J., Levis, S.C., and Maiztegui, J.I. (1984) Importance of dose of neutralising antibodies in treatment of Argentine haemorrhagic fever with immune plasma. *Lancet*, **2**, 255–256.

135. Enria, D.A., Briggiler, A.M., and Sanchez, Z. (2008) Treatment of Argentine hemorrhagic fever. *Antiviral Res.*, **78**, 132–139.

136. Jahrling, P.B. (1983) Protection of Lassa virus-infected guinea pigs with Lassa-immune plasma of guinea pig, primate, and human origin. *J. Med. Virol.*, **12**, 93–102.

137. Jahrling, P.B., Frame, J.D., Rhoderick, J.B., and Monson, M.H. (1985) Endemic Lassa fever in Liberia. IV. Selection of optimally effective plasma for treatment by passive immunization. *Trans. R. Soc. Trop. Med. Hyg.*, **79**, 380–384.

138. Baize, S., Marianneau, P., Loth, P., Reynard, S., Journeaux, A., Chevallier, M., Tordo, N., Deubel, V., and Contamin, H. (2009) Early and strong immune responses are associated with control of viral replication and recovery in lassa virus-infected cynomolgus monkeys. *J. Virol.*, **83**, 5890–5903.

139. Flatz, L., Rieger, T., Merkler, D., Bergthaler, A., Regen, T., Schedensack, M., Bestmann, L., Verschoor, A., Kreutzfeldt, M., Bruck, W., Hanisch, U.K., Gunther, S., and Pinschewer D.D. (2007) T cell-dependence of Lassa fever pathogenesis. *PLoS Pathog.*, **6**, e1000836.

140. Botten, J., Alexander, J., Pasquetto, V., Sidney, J., Barrowman, P., Ting, J., Peters, B., Southwood, S., Stewart, B., Rodriguez-Carreno, M.P., Mothe, B., Whitton, J.L., Sette, A., and Buchmeier, M.J. (2006) Identification of protective Lassa virus epitopes that are restricted by HLA-A2. *J. Virol.*, **80**, 8351–8361.

141. Botten, J., Whitton, J.L., Barrowman, P., Sidney, J., Whitmire, J.K., Alexander, J., Kotturi, M.F., Sette, A., and Buchmeier M.J. (2007) A multivalent vaccination strategy for the prevention of old world arenavirus infection in humans. *J. Virol.* **84**, 9947–9956.

142. Kotturi, M.F., Botten, J., Sidney, J., Bui, H.H., Giancola, L., Maybeno, M., Babin, J., Oseroff, C., Pasquetto, V., Greenbaum, J.A., Peters, B., Ting, J., Do, D., Vang, L., Alexander, J., Grey, H., Buchmeier, M.J., and Sette, A. (2009) A multivalent and cross-protective vaccine strategy against arenaviruses associated with human disease. *PLoS Pathog.*, **5**, e1000695.

143. Oldstone, M.B., Lewicki, H., Homann, D., Nguyen, C., Julien, S., and Gairin, J.E. (2001) Common antiviral cytotoxic t-lymphocyte epitope for diverse arenaviruses. *J. Virol.*, **75**, 6273–6278.

144. McKee, K.T. Jr., Oro, J.G., Kuehne, A.I., Spisso, J.A., and Mahlandt, B.G. (1992) Candid No. 1 Argentine hemorrhagic fever vaccine protects against lethal Junin virus challenge in rhesus macaques. *Intervirology*, **34**, 154–163.

145. McKee, K.T. Jr., Oro, J.G., Kuehne, A.I., Spisso, J.A., and Mahlandt, B.G. (1993) Safety and immunogenicity of a live-attenuated Junin (Argentine hemorrhagic fever) vaccine in rhesus macaques. *Am. J. Trop. Med. Hyg.*, **48**, 403–411.

146. Maiztegui, J.I., McKee, K.T. Jr., Barrera Oro, J.G., Harrison, L.H., Gibbs, P.H., Feuillade, M.R., Enria, D.A., Briggiler, A.M., Levis, S.C., Ambrosio, A.M., Halsey, N.A., and Peters, C.J., AHF Study Group (1998) Protective efficacy of a live attenuated vaccine against Argentine hemorrhagic fever. *J. Infect. Dis.*, **177**, 277–283.

147. Ambrosio, A.M., Riera, L.M., Saavedra Mdel, C., and Sabattini, M.S. (2006) Immune response to vaccination against Argentine hemorrhagic fever in an area where different arenaviruses coexist. *Viral Immunol.*, **19**, 196–201.

148. Kilgore, P.E., Ksiazek, T.G., Rollin, P.E., Mills, J.N., Villagra, M.R., Montenegro, M.J., Costales, M.A., Paredes, L.C., and Peters, C.J. (1997) Treatment of Bolivian hemorrhagic fever with intravenous ribavirin. *Clin. Infect. Dis.*, **24**, 718–722.

149. Artuso, M.C., Ellenberg, P.C., Scolaro, L.A., Damonte, E.B., and Garcia, C.C. (2009) Inhibition of Junin virus replication by small interfering RNAs. *Antiviral Res.*, **84**, 31–37.

150. Bolken, T.C., Laquerre, S., Zhang, Y., Bailey, T.R., Pevear, D.C., Kickner, S.S., Sperzel, L.E., Jones, K.F., Warren, T.K., Amanda Lund, S., Kirkwood-Watts, D.L., King, D.S., Shurtleff, A.C., Guttieri, M.C., Deng, Y., Bleam, M., and Hruby, D.E. (2006) Identification and characterization of potent small molecule inhibitor of hemorrhagic fever New World arenaviruses. *Antiviral Res.*, **69**, 86–97.

151. de la Torre, J.C. (2008) Reverse genetics approaches to combat pathogenic arenaviruses. *Antiviral Res.*, **80**, 239–250.

152. Gowen, B.B., Smee, D.F., Wong, M.H., Hall, J.O., Jung, K.H., Bailey, K.W., Stevens, J.R., Furuta, Y., and Morrey, J.D. (2008) Treatment of late stage disease in a model of arenaviral hemorrhagic fever: T-705 efficacy and reduced toxicity suggests an alternative to ribavirin. *PLoS One*, **3**, e3725.

153. Larson, R.A., Dai, D., Hosack, V.T., Tan, Y., Bolken, T.C., Hruby, D.E., and Amberg, S.M. (2008) Identification of a broad-spectrum arenavirus entry inhibitor. *J. Virol.*, **82**, 10768–10775.

154. Rojek, J.M., Pasqual, G., Sanchez, A.B., Nguyen, N.T., de la Torre, J.C., and Kunz, S. (2007) Targeting the proteolytic processing of the viral glycoprotein precursor is a promising novel antiviral strategy against arenaviruses. *J. Virol.*, **84**, 573–584.

155. Syromiatnikova, S.I., Khmelev, A.L., Pantiukhov, V.B., Shatokhina, I.V., Pirozhkov, A.P., Khamitov, R.A., Markov, V.I., Birisevich, I.B., and Bondarev, V.P. (2009) Chemotherapy for Bolivian hemorrhagic fever in experimentally infected guinea pigs. *Vopr. Virusol.*, **54**, 37–40.

156. Uckun, F.M., Petkevich, A.S., Vassilev, A.O., Tibbles, H.E., and Titov, L. (2004) Stampidine prevents mortality in an experimental mouse model of viral hemorrhagic fever caused by lassa virus. *BMC Infect. Dis.*, **4**, 1.

157. Levis, S.C., Saavedra, M.C., Ceccoli, C., Falcoff, E., Feuillade, M.R., Enria, D.A., Maiztegui, J.I., and Falcoff, R. (1984) Endogenous interferon in Argentine hemorrhagic fever. *J. Infect. Dis.*, **149**, 428–433.

158. Levis, S.C., Saavedra, M.C., Ceccoli, C., Feuillade, M.R., Enria, D.A., Maiztegui, J.I., and Falcoff, R. (1985) Correlation between endogenous interferon and the clinical evolution of patients with Argentine hemorrhagic fever. *J. Interferon Res.*, **5**, 383–389.

159. Marta, R.F., Montero, V.S., Hack, C.E., Sturk, A., Maiztegui, J.I., and Molinas, F.C. (1999) Proinflammatory cytokines and elastase-alpha-1-antitrypsin in Argentine hemorrhagic fever. *Am. J. Trop. Med. Hyg.*, **60**, 85–89.

160. Asper, M., Sternsdorf, T., Hass, M., Drosten, C., Rhode, A., Schmitz, H., and Gunther, S. (2004) Inhibition of different Lassa virus strains by alpha and gamma interferons and comparison with a less pathogenic arenavirus. *J. Virol.*, **78**, 3162–3169.

161. Fan, L., Briese, T., and Lipkin W.I. (2007) Z proteins of new world arenaviruses bind RIG-I and interfere with type I interferon induction. *J. Virol.*, **84**, 1785–1791.

162. Mahanty, S., Bausch, D.G., Thomas, R.L., Goba, A., Bah, A., Peters, C.J., and Rollin, P.E. (2001) Low levels of interleukin-8 and interferon-inducible protein-10 in serum are associated with fatal infections in acute Lassa fever. *J. Infect. Dis.*, **183**, 1713–1721.

163. Lukashevich, I.S., Tikhonov, I., Rodas, J.D., Zapata, J.C., Yang, Y., Djavani, M., and Salvato, M.S. (2003) Arenavirus-mediated liver pathology: acute lymphocytic choriomeningitis virus infection of rhesus macaques is characterized by high-level interleukin-6 expression and hepatocyte proliferation. *J. Virol.*, **77**, 1727–1737.

164. Qian, C., Jahrling, P.B., Peters, C.J., and Liu, C.T. (1994) Cardiovascular and pulmonary responses to Pichinde virus infection in strain 13 guinea pigs. *Lab. Anim. Sci.*, **44**, 600–607.

165. Bausch, D.G., Rollin, P.E., Demby, A.H., Coulibaly, M., Kanu, J., Conteh, A.S., Wagoner, K.D., McMullan, L.K., Bowen, M.D., Peters, C.J., and Ksiazek, T.G. (2000) Diagnosis and clinical virology of Lassa fever as evaluated by enzyme-linked immunosorbent assay, indirect fluorescent-antibody test, and virus isolation. *J. Clin. Microbiol.*, **38**, 2670–2677.

166. Ivanov, A.P., Bashkirtsev, V.N., and Tkachenko, E.A. (1981) Enzyme-linked immunosorbent assay for detection of arenaviruses. *Arch. Virol.*, **67**, 71–74.

167. Ter Meulen, J., Koulemou, K., Wittekindt, T., Windisch, K., Strigl, S., Conde, S., and Schmitz, H. (1998) Detection of Lassa virus antinucleoprotein immunoglobulin G (IgG) and IgM antibodies by a simple recombinant immunoblot assay for field use. *J. Clin. Microbiol.*, **36**, 3143–3148.

168. Wulff, H. and Lange, J.V. (1975) Indirect immunofluorescence for the diagnosis of Lassa fever infection. *Bull. World Health Organ.*, **52**, 429–436.

169. Jahrling, P.B., Niklasson, B.S., and McCormick, J.B. (1985) Early diagnosis of human Lassa fever by ELISA detection of antigen and antibody. *Lancet*, **1**, 250–252.

170. Niklasson, B.S., Jahrling, P.B., and Peters, C.J. (1984) Detection of Lassa virus antigens and Lassa virus-specific immunoglobulins G and M by enzyme-linked immunosorbent assay. *J. Clin. Microbiol.*, **20**, 239–244.

171. Demby, A.H., Chamberlain, J., Brown, D.W., and Clegg, C.S. (1994) Early diagnosis of Lassa fever by reverse transcription-PCR. *J. Clin. Microbiol.*, **32**, 2898–2903.

172. Drosten, C., Gottig, S., Schilling, S., Asper, M., Panning, M., Schmitz, H., and Gunther, S. (2002) Rapid detection and quantification of RNA of Ebola and Marburg viruses, Lassa virus, Crimean-Congo hemorrhagic fever virus, Rift Valley fever virus, dengue virus, and yellow fever virus by real-time reverse transcription-PCR. *J. Clin. Microbiol.*, **40**, 2323–2330.

173. Gunther, S., Emmerich, P., Laue, T., Kuhle, O., Asper, M., Jung, A., Grewing, T., ter Meulen, J., and Schmitz, H. (2000) Imported Lassa fever in Germany: molecular characterization of a new lassa virus strain. *Emerging Infect. Dis.*, **6**, 466–476.

174. Lozano, M.E., Enria, D., Maiztegui, J.I., Grau, O., and Romanowski, V. (1995) Rapid diagnosis of Argentine hemorrhagic fever by reverse transcriptase PCR-based assay. *J. Clin. Microbiol.*, **33**, 1327–1332.

175. Trappier, S.G., Conaty, A.L., Farrar, B.B., Auperin, D.D., McCormick, J.B., and Fisher-Hoch, S.P. (1993) Evaluation of the polymerase chain reaction for diagnosis of Lassa virus infection. *Am. J. Trop. Med. Hyg.*, **49**, 214–221.

176. Trombley, A.R., Wachter, L., Garrison, J., Buckley-Beason, V.A., Jahrling, J., Hensley, L.E., Schoepp, R.J., Norwood, D.A., Goba, A., Fair, J.N., and Kulesh, D.A. (2007) Comprehensive panel of real-time TaqMan polymerase chain reaction assays for detection and absolute quantification of filoviruses, arenaviruses, and New World hantaviruses. *Am. J. Trop. Med. Hyg.*, **82**, 954–960.

177. Vieth, S., Drosten, C., Charrel, R., Feldmann, H., and Gunther, S. (2005) Establishment of conventional and fluorescence resonance energy transfer-based real-time PCR assays for detection of pathogenic New World arenaviruses. *J. Clin. Virol.*, **32**, 229–235.

178. Drosten, C., Kummerer, B.M., Schmitz, H., and Gunther, S. (2003) Molecular diagnostics of viral hemorrhagic fevers. *Antiviral Res.*, **57**, 61–87.

179. Palacios, G., Druce, J., Du, L., Tran, T., Birch, C., Briese, T., Conlan, S., Quan, P.L., Hui, J., Marshall, J., Simons, J.F., Egholm, M., Paddock, C.D., Shieh, W.J., Goldsmith, C.S., Zaki, S.R., Catton, M., and Lipkin, W.I. (2008) A new arenavirus in a cluster of fatal transplant-associated diseases. *N. Engl. J. Med.*, **358**, 991–998.

180. Vieth, S., Drosten, C., Lenz, O., Vincent, M., Omilabu, S., Hass, M., Becker-Ziaja, B., ter Meulen, J., Nichol, S.T., Schmitz, H., and Gunther, S. (2007) RT-PCR assay for detection of Lassa virus and related Old World arenaviruses targeting the L gene. *Trans. R. Soc. Trop. Med. Hyg.*, **101**, 1253–1264.

13
Filoviruses: Hemorrhagic Fevers

Victoria Wahl-Jensen, Sheli R. Radoshitzky, Sina Bavari, Peter B. Jahrling, and Jens H. Kuhn

13.1
Characteristics

The family *Filoviridae* (order *Mononegavirales*) contains three genera. Marburg virus (MARV) and Ravn virus (RAVV) are members of the genus *Marburgvirus* and are assigned to the same species (*Marburg marburgvirus*). Bundibugyo virus (BDBV), Ebola virus (EBOV), Reston virus (RESTV), Sudan virus (SUDV), and Taï Forest virus (TAFV) are members of individual species (*Bundibugyo ebolavirus, Zaire ebolavirus, Reston ebolavirus, Sudan ebolavirus,* and *Taï Forest ebolavirus,* respectively) in the genus *Ebolavirus*. Finally, Lloviu virus (LLOV) is the sole member of the tentative species "Lloviu cuevavirus" in the tentative genus "Cuevavirus" (Table 13.1) [1]. The genomic sequences of the individual filoviruses are rather divergent. However, their genomic organization and the morphology of their virions are strikingly similar. A filovirus genome is a linear nonsegmented single-stranded negative-sense RNA molecule approximately 19 kb in length. It contains seven genes in the order *3′-NP-VP35-VP40-GP-VP30-VP24-L-5′*. One distinguishing feature of marburgviruses, ebolaviruses, and "cuevaviruses" is the number and location of gene overlaps. The genes, respectively, encode seven structural proteins, the nucleoprotein (NP), RNA-dependent RNA polymerase cofactor (VP35), matrix protein (VP40), spike glycoprotein ($GP_{1,2}$), transcriptional activator (VP30), potential secondary matrix protein (VP24), and the RNA-dependent RNA polymerase (L) [2–4]. NP helically encapsidates the genome and associates with VP30, VP35, and L to form the tubule-like ribonucleoprotein (RNP) complex [5–7]. VP40 and VP24 surround the RNP and are surrounded by a lipid envelope derived from the host cell during virion egress [8, 9]. $GP_{1,2}$ is found inserted into the membrane, protruding from the virion surface [10]. Mature virions are 80 nm in diameter but differ in average length ($\approx$800–1100 nm). The filaments can branch, or form shapes resembling Us, 6s, or toroids [11]. Ebolavirus and "cuevavirus", but not marburgvirus, *GP* genes encode additional, secreted glycoproteins (sGPs, ssGP, Δ-peptide) of unknown function [12, 13].

BSL3 and BSL4 Agents: Epidemiology, Microbiology, and Practical Guidelines, First Edition.
Edited by Mandy C. Elschner, Sally J. Cutler, Manfred Weidmann, and Patrick Butaye.

Table 13.1 Filovirus taxonomy [1, 14].

Order *Mononegavirales*
Family *Filoviridae*
Genus *Marburgvirus*
Species *Marburg marburgvirus*
Virus 1: Marburg virus (MARV)
Virus 2: Ravn virus (RAVV)
Genus *Ebolavirus*
Species *Taï Forest ebolavirus*
Virus: Taï Forest virus (TAFV)
Species *Reston ebolavirus*
Virus: Reston virus (RESTV)
Species *Sudan ebolavirus*
Virus: Sudan virus (SUDV)
Species *Zaire ebolavirus*
Virus: Ebola virus (EBOV)
Species *Bundibugyo ebolavirus*
Virus: Bundibugyo virus (BDBV)
Genus "Cuevavirus" (tentative)
Species "Lloviu cuevavirus" (tentative)
Virus: Lloviu virus (LLOV)

The filovirus life cycle begins with cell entry. As in all classical class I fusion proteins, the ectodomain of $GP_{1,2}$ (GP_1) mediates virion binding to an unidentified cell surface receptor [15], followed by cellular uptake of the particle into lysosome. There, GP1 binds to the protein NPC-1 [16, 17], upon which the transmembrane domain (GP_2) mediates fusion of the virion envelope with a cellular membrane and the release of the virion RNP into the cytosol [18, 19]. The host cell cytoplasm is the location for replication of the filovirus genome. Once released into the cytoplasm, the negative-sense genome must be transcribed. Encapsidated RNA acts as template for the generation of polyadenylated, monocistronic mRNAs that are transcribed from the genes in a 3′ to 5′ manner. As early as 7 h post infection, NP mRNA can be detected and transcription peaks approximately 11 h later [20]. The host cellular machinery functions to transcribe and subsequently translate all the virus genes leading to their accumulation within cells. After the translation of viral proteins, there is a switch from transcription to replication that leads to the synthesis and encapsidation of full-length positive-sense RNA (antigenome). Antigenome copies serve as templates for the synthesis of full-length genomic RNA that is rapidly encapsidated by RNP complex proteins. As newly transcribed RNP proteins are sequestered for encapsidation, their depletion is thought to signal a switch back to transcription, eventually leading to a homeostatic balance between transcription and replication. Viral egress occurs when nucleocapsids in the cytoplasm and membrane-bound proteins (VP24, VP40, $GP_{1,2}$) accumulate in the cytoplasm of the cell and eventually amalgamate at the plasma membrane [21, 22].

In addition to playing a critical role in virus transcription/replication, several filovirus proteins are multifunctional. VP35 functions as a type 1 interferon antagonist [23]. Specifically, VP35 can inhibit double-stranded RNA-mediated and virus-mediated induction of an IFN-stimulated response element reporter gene and inhibit double-stranded RNA- and virus-mediated induction of the IFN-β promoter by inhibiting phosphorylation and subsequent dimerization and nuclear translocation of IRF-3, a cellular transcription factor of central importance in initiating the host cell IFN response [24]. In addition, VP35 possesses double-stranded RNA (dsRNA)-binding activity and blocks activation of IRF-3 induced by overexpression of RIG-I [25]. VP35 is also able to inhibit the antiviral response induced by interferon IFN-α. This depends on a VP35 function that interferes with the pathway regulated by double-stranded RNA-dependent protein kinase PKR [26]. Taken together, these data suggest that VP35 may act in multiple ways to interfere with the host interferon response. Likewise, EBOV VP24 inhibits IFN-$\alpha/\beta/\gamma$ signaling [27], a function that in the case of MARV is overtaken by VP40 [28]. This suppression of the innate cellular antiviral response allows the structural proteins to be assembled into progeny virions and to leave the cell in large numbers.

13.2 Epidemiology

Filoviruses are emerging zoonotic pathogens. Most of them cause severe viral hemorrhagic fevers (VHFs) in humans characterized by high lethality. MARV, BDBV, EBOV, and SUDV cause the majority of human infections, whereas RAVV and TAFV have only been isolated from single cases of VHF. RESTV and LLOV have thus far not been isolated from humans and their ecology and importance remains unclear [29]. Experimental infections demonstrated that filoviruses are highly infectious and lethal for a variety of both Old World and New World nonhuman primates. In the wild, filoviruses have not been isolated from monkeys or apes. There is, however, evidence that at least EBOV infects large numbers of nonhuman primates. For one, group decimation of Western lowland gorillas (*Gorilla gorilla gorilla*) and central chimpanzees (*Pan troglodytes verus*) often coincides with Ebola virus disease (EVD) outbreaks in humans in the same area. It is currently thought that these ape infections contribute to the dramatic ape population decline observed in Africa [30]. Second, EBOV-like RNA sequences could be detected in ape carcasses [31]. Third, most recent EVD outbreaks could be traced back to index cases hunting nonhuman primates or finding and consuming deceased ones [29]. Likewise, TAFV was isolated from a human who performed necropsies on western chimpanzees (*Pan troglodytes troglodytes*) that had died naturally of VHF [32]. Laboratory tests other than virus isolation confirmed TAFV infection of these animals, indicating that the chimpanzees had been in contact with the yet elusive EBOV reservoir host [33]. However, not all other filovirus infections have yet been associated with natural nonhuman primate infections. MARV and RAVV have been isolated from low numbers of frugivorous bats (Egyptian rousettes; *Rousettus aegyptiacus*) [34], suggesting that human infections occur after direct contact with or consumption of bats or their secretions. Interestingly, MARV and RAVV infections have almost always been associated with visits or mining in caves [29]. Even more interesting is that the yet to be isolated LLOV was discovered in Schreiber's long-fingered bats (*Miniopterus schreibersii*) and that antibodies against and/or genomic RNA fragments of EBOV could be detected in hammer-headed fruit bats (*Hypsignathus monstrosus*), Franquet's epauletted fruit bats (*Epomops franqueti*), little collared fruit bats (*Myonycteris* (*Myonycteris*) *torquata*) [35], and African straw-colored fruit bats (*Eidolon helvum*) [36]. It remains to be determined whether bats are the true host reservoirs of filoviruses or whether the bats themselves are in contact with a reservoir species, such as ectoparasites. Laboratory infections of arthropods or arthropod-derived cell lines with filoviruses have thus far failed [37, 38].

Judging by virus isolation (Table 13.2), EBOV is endemic in areas of humid rain forest in central and western Africa (Democratic Republic of the Congo, Gabon, Republic of the Congo, Uganda). MARV and RAVV are endemic in the arid woodlands of east, south-central, and western Africa (Angola, Democratic Republic of the Congo, Kenya, Uganda, Zimbabwe). SUDV has thus far only been detected in (South) Sudan, whereas TAFV and RESTV have thus far been found only in Côte d'Ivoire and the Philippines, respectively. The epidemiology of LLOV is unclear due to its recent discovery in Spain [39]. The circulation of RESTV in domestic pigs (*Sus scrofa*) [40] in the Philippines and the detection of LLOV in Spain indicate that filoviruses are not restricted to the African continent as was commonly thought. Indeed, serological surveys, performed with human or animal sera predominantly in the 1980s, suggest an almost worldwide distribution. In general, most of these studies were disregarded due to the experimental methods employed [primarily immunofluorescence assay (IFA), which is plagued by subjective interpretation and false-positives] and due to the fact that VHF-like cases were absent from most areas where positive sera were collected [29]. They may have to be revisited and repeated with more specific methods such as ELISA and Western blot now that filoviruses not pathogenic for humans have been discovered.

Most filovirus disease outbreaks trace back to a single infected index patient. Due to the unspecific symptoms, additional people usually become infected after being in contact with this individual. This includes family members who care for the index patient and cohabitants (sexual transmission) [29, 41]. Other than caves in the case of MARV/RAVV and contact with nonhuman primates in the case of EBOV, no common denominator has been found among the index cases of the various filovirus disease outbreaks. Fortunately, filoviruses are not especially contagious. Indeed, aerosol transmission has never been observed in natural settings. While transmission chains readily get started from unsuspecting index cases, transmission usually is limited to a few people only. This is the reason why even the largest outbreaks of filovirus disease have not involved more than 500 people and why the overall number of known filovirus disease cases over the last 43 years has not yet reached 3000 [29]. The most important amplifier of filovirus epidemics are underequipped African hospitals, in which doctors and predominantly nurses have to work without proper barrier safety measures and where the reuse of

Table 13.2 Filovirus disease outbreaks.

Year	Filovirus	Location	Human cases/deaths (% case-fatality rate)
1967	MARV	Marburg an der Lahn, Frankfurt am Main (West Germany); Belgrade (Yugoslavia) ex Uganda	Humans: 31/7 (23%)
1975	MARV	Rhodesia	Humans: 3/1 (33%)
1976	SUDV	Nzara, Maridi, Tembura (Sudan)	Humans: 151/284 (53%)
1976	EBOV	Yambuku (Zaire)	Humans: 318/280 (88%)
1977	EBOV	Bonduni (Zaire)	Humans: 1/1 (100%)
1979	SUDV	Nzara (Sudan)	Humans: 34/22 (65%)
1980	MARV	Mt. Elgon/Nzoia (Kenya)	Humans: 2/1 (50%)
1987	RAVV	Mt. Elgon/Mombasa (Kenya Elgon)	Humans: 1/1 (100%)
1989–1990	RESTV	Alice, Philadelphia, Reston (USA) ex Luzon (Philippines)	Captive crab-eating macaques
1992	RESTV	Siena (Italy) ex Luzon (Philippines)	Captive crab-eating macaques
1994	TAFV	Guiglot (Côte d'Ivoire)	Humans: 1/0 (0%)
1994–1995	EBOV	Andok, Mékouka, Minkébé, Mayéla-Mbeza, Ovan, Etakangaye (Gabon)	Humans : 52/32 (62%)
1995	EBOV	Kikwit (Zaire)	Humans: 317/245 (77%)
1996	EBOV	Mayibout II, Makokou (Gabon)	Humans: 31/21 (68%)
1996	RESTV	Alice (US) ex Luzon (Philippines)	Captive crab-eating macaques
1996–1997	EBOV	Balimba, Bouée, Lastoursville, Libreville, Lolo (Gabon)	Humans: 62/46 (74%)
1998–2000	MARV/ RAVV	Durba, Watsa [Democratic Republic of the (DR) Congo]	Humans: 154/128 (83%)
2000–2001	SUDV	Gulu, Masindi, Mbarara Districts (Uganda)	Humans: 425/224 (53%)
2001–2002	EBOV	Ekata, Etakangaye, Franceville, Grand Etoumbi, Ilahounene, Imbong, Makokou, Mékambo, Mendema, Ntolo (Gabon); Abolo, Ambomi, Entsiami, Kéllé, Olloba (Republic of the Congo)	Humans: 124/97 (78%)
2002	EBOV	Olloba (Republic of the Congo); Ekata (Gabon)	Humans: 11/10 (91%)
2002	LLOV	Spain	Schreiber's long-fingered bats
2002–2003	EBOV	Yembelengoye, Mvoula (Republic of the Congo)	Humans: 143/128 (90%)
2003–2004	EBOV	Mbomo, Mbanza (Republic of the Congo)	Humans: 35/29 (83%)
2004	SUDV	Yambio (Sudan)	Humans: 17/7 (41%)
2004–2005	MARV	Uíge Province (Angola)	Humans: 252/227 (90%)
2005	EBOV	Etoumbi, Mbomo (Republic of the Congo)	Humans: 11/9 (82%)
2007	MARV/ RAVV	Kakasi Forest Reserve (Uganda)	Humans: 4/1 (25%)
2007	EBOV	Kampungu, Mweka, Mwene-Ditu (DR Congo)	Humans: 264/186 (71%)
2007	BDBV	Bundibugyo District (Uganda)	Humans: 116/39 (34%)
2008	EBOV	Kampungu, Mweka, Mwene-Ditu (DR Congo)	Humans: 32/15 (47%)
2008	RESTV	Philippines	Domestic pigs
2008	MARV	Leiden (Holland) ex Uganda	Humans: 1/1 (100%)
2011	SUDV	Sudan	Humans: 1/1 (100%)

[a] Updated from Ref. 29.

needles is not uncommon. In fact, a considerable percentage of fatalities during a filovirus disease outbreak reflect secondary and tertiary spread among hospital staff (close to one-third in Kikwit in 1995) [29]. Filovirus disease is therefore a bona fide nosocomial disease. The second most important amplifier are local customs, such as greeting rituals (hand shaking) and burial ceremonies (ritual hand washing, embalming of the dead). Outbreaks usually come to an end once proper equipment arrives in hospitals via humanitarian efforts, staff is trained properly, and the local population is educated regarding filovirus transmission routes. The latter may be more successful when medical anthropologists are involved [29, 42].

The infectious dose of filoviruses for humans is unclear, but it is suspected to be extremely low because doses as low as one plaque-forming unit sufficed to establish infections, at least in experimental animal models. The genomes of the viruses remain stable over time even in secondary and tertiary cases, suggesting that little or no adaptation of the virus to the human host occurs [43, 44].

13.3 Clinical Signs

Marburgviruses and ebolaviruses are the etiological agents of severe VHFs in primates. Marburg virus disease (MVD), the disease caused by all marburgviruses, has thus far only been observed in humans (ICD-10: A98-3). In 1967, MARV was imported from Uganda into Germany via lethally infected vervet monkeys (*Chlorocebus aethiops*) [45], but it is unclear whether these animals had been infected naturally or in captivity after being held in proximity with a natural MARV reservoir [29]. Ebolaviruses infect mammals of several species in nature. TAFV thus far caused one lethal VHF epizootic among western chimpanzees and a single case of EVD in humans (ICD-10: A98.4) [32]. EBOV, the prototype ebolavirus thus far causing the majority of EVD outbreaks in humans is suspected to also fatally infect central chimpanzees and western lowland gorillas and to contribute to the dramatic decline of their populations [30]. BDBV and SUDV have only been directly associated with human infections, whereas RESTV is thought to be apathogenic for humans [29]. RESTV has been imported from the Philippines into the United States and Italy via lethally infected crab-eating macaques (*Macaca fascicularis*) from primate-holding facilities [46], but as in the case of MARV it is unclear whether these primates became naturally infected or after being placed in proximity with an infected animal they would not have encountered in their natural habitat [29]. Recently, RESTV was also detected in sick domestic pigs [40]. However, because of coinfection with at least two other viral agents (circoviruses, arteriviruses) it is unclear which agent caused the disease. The "cuevavirus" LLOV is not yet known to cause disease in any primate but is implicated in lethal disease of Schreiber's long-fingered bats. With the exception of humans, none of the animals mentioned above could be observed or monitored while being sick. Instead they were usually found dead and in advanced stages of decay. Consequently, clinical signs of MVD and EVD are only known from humans or from experimentally infected animals.

Clinically, MVD and EVD of humans cannot be differentiated and also reflect the symptomatology of other VHFs (Table 13.3) [47, 48]. Both diseases begin after an incubation period of 3–19 days with a sudden onset of influenza-like symptoms (fever, headache, myalgia, nausea, malaise), followed by clinical signs and symptoms indicating impairment of the gastrointestinal tract (anorexia, nausea, diarrhea, vomiting), CNS (headaches, tremors, confusion, coma), vasculature (edema, hypotension, shock), and respiratory system (chest pains, cough). Hemorrhagic manifestations, which usually occur in less than 50% of patients, include maculopapular rash, melena, hemoptysis, epistaxis, bleeding from anus or vagina, and internal bleeding. Overall blood loss is minimal and death is rather the result of fulminant shock and multiorgan failure [29, 47, 48].

Subclinical or mild filovirus infections in humans are either extremely rare or do not occur. There are, however, survivors. The lethality of MVD/EVD varies depending on the agent involved. MARV and EBOV usually cause disease with case-fatality rates surpassing 89%, whereas SUDV and BDBV infections results in case-fatality rates of 50–60 and 37%, respectively (Table 13.3) [29]. Whether these numbers are a true reflection of the disease or rather of the availability or absence of modern hospital equipment and care in endemic areas is unclear. For instance, MARV infections in West Germany in 1967, which were a result of contact with imported vervet monkeys from Uganda, resulted in death in only 25% of the cases, suggesting that proper and modern supportive care (for that time) can have a significant effect on disease outcome [45, 49, 50]. Survivors usually have to endure a prolonged period of convalescence and may suffer of arthralgia, uveitis, orchitis, and psychological ailments that may or may not resolve with time. Importantly, filovirus infections may persist and reactivate months after apparent convalescence [41, 44, 51, 52].

13.4 Pathological Findings

Most human filovirus disease outbreaks occur in central Africa. Due to local customs and beliefs, but also due to biosafety concerns, autopsies have rarely been performed [29]. Consequently, the pathologic characterization of filovirus disease in humans is not fully elucidated. However, several animal models have been developed in maximum containment facilities primarily located in the United States and Russia. These include the very well characterized rhesus (*Macaca mulatta*) and crab-eating macaque models, hamadryas baboons (*Papio hamadryas*), and vervet monkeys for EBOV, RESTV, SUDV, MARV, and RAVV infections [29, 53–55], as well as laboratory mouse and guinea pig models for EBOV, MARV, and RAVV after their adaptation through sequential passage [29, 56–59]. Newer models employ common marmosets (*Callithrix jacchus*) for EBOV and MARV infections [60], hamsters for mouse-adapted EBOV, and domestic pigs for EBOV and RESTV [61, 62]. Although important differences exist among the various models, the

Table 13.3 Symptoms and clinical signs of filovirus disease survivors and fatally infected patients in two exemplary outbreaks (data are not available for all symptoms) [47, 48].

Symptom/clinical sign	Frequency observed in survivors MARV/ EBOV (%)	Frequency observed in fatal cases MARV/ EBOV (%)
Abdominal pain	59/68	57/62
Abortion	–/5	–/2
Anorexia	77/47	72/43
Anuria	–/0	–/7
Arthralgia or myalgia	55/79	55/50
Asthenia	–/95	?/85
Bleeding from puncture sites	0/5	7/8
Bleeding from the gums	23/0	36/15
Bleeding from any site	59/–	71/–
Bloody stools	–/5	–/7
Chest pain	18/5	4/10
Conjunctival injection	14/47	42/42
Convulsions	–/0	–/2
Cough	9/26	5/7
Diarrhea	59/84	56/86
Difficulty breathing	36/–	58/–
Dysesthesia	–/5	–/0
Epistaxis	18/0	34/2
Fever	100/95	92/93
Headaches	73/74	79/52
Hearing loss	–/11	–/5
Hematemesis	68/0	76/13
Hematoma	0/0	3/2
Hematuria	–/16	–/7
Hemoptysis	9/11	44/0
Hepatomegaly	–/5	–/2
Hiccups	18/5	44/17
Lumbar pain	5/26	8/12
Macupapular rash	–/16	–/14
Malaise or fatigue	86/–	83/–
Melena	41/16	58/8
Nausea and vomiting	77/68	76/73
Petechiae	9/0	7/8
Sore throat, odynophagia, or dysphagia	43/58	43/56
Splenomegaly	–/5	–/2
Tachypnea	–/0	–/31
Tinnitus	–/11	–/1

combined data that accumulated from necropsies permitted to establish a general filovirus disease pathology that can be extrapolated to humans.

The pathology of MVD and EVD is strikingly similar and does not allow the differentiation of both. There are also no pathognomic markers that allow a clear-cut differentiation of filovirus disease from other VHFs [63]. Typical findings are extensive visceral effusions and abundant petechiae and ecchymoses in the skin, mucous membranes, and internal organs. Disseminated intravascular coagulation (DIC) is the cause of myriads of microthrombi that clog blood vessels and thereby prevent the oxygenation of major organs. This results in massive focal necroses in primary target organs, such as liver, spleen, lymph nodes, and kidneys, characteristically with little to no inflammatory reactions. Lung interstitial edema and myocardial edema are typical gross findings. Histopathologically, tissue damage is often directly correlated with the presence of filoviral RNA or antigen. In fact, primary target cells of filoviruses, such as macrophages and their derivatives (dendritic cells, Kupffer cells, microglia), hepatocytes, and fibroblasts and secondary target cells, such as endothelial cells, often are filled with characteristic filovirus nucleocapsids in paracrystalline arrangements. These inclusion bodies are eosinophilic and can often be detected in H + E-stained sections using light microscopy [64–69].

13.5 Pathogenesis

Filovirus infections are usually initiated after direct person to person (skin to skin) contact or after contact with tissues or bodily fluids from sick or deceased individuals. Aerosol transmission has not been observed during natural outbreaks of filovirus disease. High concentrations of EBOV antigen have been detected in the skin, especially in the surrounding of sweat glands, in infected humans [63]. Consequently, it is thought that sick individuals secrete filovirions that infect others during, for instance, patient care through small skin lesions or direct contact with mucous membranes. The virus particles then gain access to the vascular or lymphatic system of the newly infected individual, which distributes them to distal areas of the body [70].

The primary cellular targets of filoviruses are macrophages, dendritic cells, and fibroblastic reticular cells located in various organs systems, in particular liver (Kupffer cells), spleen and lymph nodes, kidneys, lungs, pleura, peritoneum, and nervous system (microglia) [53, 71–74]. These cells then excrete progeny virions, leading to exponential increases of filovirion titers in blood (viremia) and further spread. Hepatocytes, endothelial cells, and most other cell types (with the exception of neurons, bone cells such as osteoblasts and osteclasts, and lymphocytes) are infected later during infection [75].

Infected macrophages release proinflammatory cytokines (in particular TNF-α, IL-1β, IL-6, and IL-8), leading not only to a wide array of detrimental effects but also to recruitment of additional macrophages to the site of infection, thereby furthering filovirus amplification [21, 76, 77]. Dendritic cells, in contrast, become impaired in function, leading to aberrant cytokine responses and partial suppression of MHC class II. The early disabling of these cells could explain the absence of detectable immune response to infection, and the massive loss of lymphocytes due to bystander apoptosis as infected dendritic cells secrete TNF-related apoptosis-inducing ligand (TRAIL) [78, 79]. Fatally infected patients are characterized by increased serum concentrations of IL-10, IL1-RA, and IL-1, whereas survivors usually have elevated concentrations of IL-1β and IL-6 [80–82].

The severe cytokine disarray not only impairs innate and adaptive immune response but is also suspected as the cause of DIC. Vascular leakage and hemorrhages occur once all clotting factors have been consumed, thereby leading to imbalances in blood pressure and visible edema. The cause of death of filovirus-infected individuals is usually due to multiorgan failure and shock either as a direct consequence of all these events or exacerbated by secondary bacterial or fungal infections [69].

13.6 Diagnostic Procedures

The clinical diagnosis of filovirus disease is difficult due to its nonspecific clinical signs. Not only other VHFs, but also much more common ailments such as falciparum malaria, Gram-negative septicemia, plague, typhoid fever, leptospirosis, and rickettsial fevers, mimic it in clinical presentation [29, 83]. However, the location where infection was most likely acquired can provide valuable clues. For instance, recent direct or indirect contact with bats in caves or with deceased nonhuman primates in central Africa should be sufficient for physicians to at least consider filovirus infection after a patient tested negative for plasmodia.

Definite diagnosis of filovirus infection must be established using molecular biological methods, such as antigen capture ELISA and or RT-PCR [84, 85], as well as, if available, isolation of agent in tissue culture with subsequent electron-microscopic confirmation. Viral antigen and virus genomic RNA are readily detectable during the acute phase of MVD and EVD with ELISA and RT-PCR, even in the field [86]. Serum and tissue samples should be frozen and, if possible, sent to a WHO reference laboratory possessing a maximum containment laboratory. There, filoviruses can easily be propagated in Vero cells and its subclones, as well as in MA-104, SW13, A549, or HeLa cells and be detected using relatively simple IFA [29]. Due to the unique filamentous morphology of filovirions, filovirus infection can quickly be identified using electron microscopy as long as infectious titers are high enough to make detection of virions on grid likely [11, 74]. EM, of course, may also be used directly on serum samples. Antibody capture ELISAs are commonly used to establish filovirus infection retrospectively, as IgG concentrations increase over ongoing convalescence [85].

13.7
Disclaimer

The content of this publication does not necessarily reflect the views or policies of the United States (US) Department of Defense, the US Department of the Army, thc US Department of Health and Human Services, or the institutions and companies affiliated with the authors.

References

1. Kuhn, J.H., Becker, S., Ebihara, H., Geisbert, T.W., Johnson, K.M., Kawaoka, Y., Lipkin, W.I., Negredo, A.I., Netesov, S.V., Nichol, S.T., Palacios, G., Peters, C.J., Tenorio, A., Volchkov, V.E., and Jahrling, P.B. (2010) Proposal for a revised taxonomy of the family *Filoviridae*: classification, names of taxa and viruses, and virus abbreviations. *Arch. Virol.*, **155** (12), 2083–2103.
2. Feldmann, H., Muhlberger, E., Randolf, A., Will, C., Kiley, M.P., Sanchez, A., and Klenk, H.D. (1992) Marburg virus, a filovirus: messenger RNAs, gene order, and regulatory elements of the replication cycle. *Virus Res.*, **24**, 1–19.
3. Ikegami, T., Calaor, A.B., Miranda, M.E., Niikura, M., Saijo, M., Kurane, I., Yoshikawa, Y., and Morikawa, S. (2001) Genome structure of Ebola virus subtype Reston: differences among Ebola subtypes brief report. *Arch. Virol.*, **146**, 2021–2027.
4. Sanchez, A., Kiley, M.P., Holloway, B.P., and Auperin, D.D. (1993) Sequence analysis of the Ebola virus genome: organization, genetic elements, and comparison with the genome of Marburg virus. *Virus Res.*, **29**, 215–240.
5. Becker, S., Rinne, C., Hofsass, U., Klenk, H.D., and Muhlberger, E. (1998) Interactions of marburg virus nucleocapsid proteins. *Virology*, **249**, 406–417.
6. Muhlberger, E., Lotfering, B., Klenk, H.D., and Becker, S. (1998) Three of the four nucleocapsid proteins of Marburg virus, NP, VP35, and L, are sufficient to mediate replication and transcription of Marburg virus-specific monocistronic minigenomes. *J. Virol.*, **72**, 8756–8764.
7. Muhlberger, E., Weik, M., Volchkov, V.E., Klenk, H.D., and Becker, S. (1999) Comparison of the transcription and replication strategies of marburg virus and Ebola virus by using artificial replication systems. *J. Virol.*, **73**, 2333–2342.
8. Han, Z., Boshra, H., Sunyer, J.O., Zwiers, S.H., Paragas, J., and Harty, R.N. (2003) Biochemical and functional characterization of the Ebola virus VP24 protein: implications for a role in virus assembly and budding. *J. Virol.*, **77**, 1793–1800.
9. Noda, T., Sagara, H., Suzuki, E., Takada, A., Kida, H., and Kawaoka, Y. (2002) Ebola virus VP40 drives the formation of virus-like filamentous particles along with GP. *J. Virol.*, **76**, 4855–4865.
10. Jeffers, S.A., Sanders, D.A., and Sanchez, A. (2002) Covalent modifications of the ebola virus glycoprotein. *J. Virol.*, **76**, 12463–12472.
11. Geisbert, T.W. and Jahrling, P.B. (1995) Differentiation of filoviruses by electron microscopy. *Virus Res.*, **39**, 129–150.
12. Volchkova, V.A., Feldmann, H., Klenk, H.D., and Volchkov, V.E. (1998) The nonstructural small glycoprotein sGP of Ebola virus is secreted as an antiparallel-orientated homodimer. *Virology*, **250**, 408–414.
13. Volchkova, V.A., Klenk, H.D., and Volchkov, V.E. (1999) Delta-peptide is the carboxy-terminal cleavage fragment of the nonstructural small glycoprotein sGP of Ebola virus. *Virology*, **265**, 164–171.
14. Kuhn, J.H., Becker, S., Ebihara, H., Geisbert, T.W., Jahrling, P.B., Kawaoka, Y., Netesov, S V., Nichol, S.T., Peters, C.J., Volchkov, V.E., and Ksiazek, T.G. (2011) Family Filoviridae. in (eds A.M.Q.

King, M.J. Adams, E.B. Carstens, and E.J. Lefkowitz) *Virus Taxonomy - Ninth Report of the International Committee on Taxonomy of Viruses.* Elsevier, Academic Press, London, UK, pp. 665–671.

15. Kuhn, J.H., Radoshitzky, S.R., Guth, A.C., Warfield, K.L., Li, W., Vincent, M.J., Towner, J.S., Nichol, S.T., Bavari, S., Choe, H., Aman, M.J., and Farzan, M. (2006) Conserved receptor-binding domains of Lake Victoria marburgvirus and Zaire ebolavirus bind a common receptor. *J. Biol. Chem.*, **281**, 15951–15958.

16. Cote, M., Misasi, J., Ren, T., Bruchez, A., Lee, K., Filone, C.M., Hensley, L., Li, Q., Ory, D., Chandran, K., and Cunningham, J. (2011) Small molecule inhibitors reveal Niemann-Pick C1 is essential for Ebola virus infection. *Nature*, **477** (7364), 344–348.

17. Carette, J.E., Raaben, M., Wong, A.C., Herbert, A.S., Obernosterer, G., Mulherkar, N., Kuehne, A.I., Kranzusch, P.J., Griffin, A.M., Ruthel, G., Dal Cin, P., Dye, J.M., Whelan, S.P., Chandran, K., and Brummelkamp, T.R. (2011) Ebola virus entry requires the cholesterol transporter Niemann-Pick C1. *Nature*, **477** (7364), 340–343.

18. Gallaher, W.R. (1996) Similar structural models of the transmembrane proteins of Ebola and avian sarcoma viruses. *Cell*, **85**, 477–478.

19. Ruiz-Arguello, M.B., Goni, F.M., Pereira, F.B., and Nieva, J.L. (1998) Phosphatidylinositol-dependent membrane fusion induced by a putative fusogenic sequence of Ebola virus. *J. Virol.*, **72**, 1775–1781.

20. Sanchez, A. and Kiley, M.P. (1987) Identification and analysis of Ebola virus messenger RNA. *Virology*, **157**, 414–420.

21. Feldmann, H., Bugany, H., Mahner, F., Klenk, H.D., Drenckhahn, D., and Schnittler, H.J. (1996) Filovirus-induced endothelial leakage triggered by infected monocytes/macrophages. *J. Virol.*, **70**, 2208–2214.

22. Mühlberger, E. (2007) Filovirus replication and transcription. *Future Virol.*, **2**, 205–215.

23. Basler, C.F., Wang, X., Muhlberger, E., Volchkov, V., Paragas, J., Klenk, H.D., Garcia-Sastre, A., and Palese, P. (2000) The Ebola virus VP35 protein functions as a type I IFN antagonist. *Proc. Natl. Acad. Sci. U.S.A.*, **97**, 12289–12294.

24. Basler, C.F., Mikulasova, A., Martinez-Sobrido, L., Paragas, J., Muhlberger, E., Bray, M., Klenk, H.D., Palese, P., and Garcia-Sastre, A. (2003) The Ebola virus VP35 protein inhibits activation of interferon regulatory factor 3. *J. Virol.*, **77**, 7945–7956.

25. Cardenas, W.B., Loo, Y.M., Gale, M. Jr., Hartman, A.L., Kimberlin, C.R., Martinez-Sobrido, L., Saphire, E.O., and Basler, C.F. (2006) Ebola virus VP35 protein binds double-stranded RNA and inhibits alpha/beta interferon production induced by RIG-I signaling. *J. Virol.*, **80**, 5168–5178.

26. Feng, Z., Cerveny, M., Yan, Z., and He, B. (2007) The VP35 protein of Ebola virus inhibits the antiviral effect mediated by double-stranded RNA-dependent protein kinase PKR. *J. Virol.*, **81**, 182–192.

27. Reid, S.P., Leung, L.W., Hartman, A.L., Martinez, O., Shaw, M.L., Carbonnelle, C., Volchkov, V.E., Nichol, S.T., and Basler, C.F. (2006) Ebola virus VP24 binds karyopherin alpha1 and blocks STAT1 nuclear accumulation. *J. Virol.*, **80**, 5156–5167.

28. Valmas, C., Grosch, M.N., Schumann, M., Olejnik, J., Martinez, O., Best, S.M., Krahling, V., Basler, C.F., and Muhlberger, E. (2010) Marburg virus evades interferon responses by a mechanism distinct from ebola virus. *PLoS Pathog.*, **6**, e1000721.

29. Kuhn, J.H. (2008) *Filoviruses – A Compendium of 40 Years of Epidemiological, Clinical, and Laboratory Studies*, SpringerWienNewYork, Vienna, Austria.

30. Bermejo, M., Rodriguez-Teijeiro, J.D., Illera, G., Barroso, A., Vila, C., and Walsh, P.D. (2006) Ebola outbreak killed 5000 gorillas. *Science*, **314**, 1564.

31. Leroy, E.M., Rouquet, P., Formenty, P., Souquiere, S., Kilbourne, A., Froment, J.M., Bermejo, M., Smit, S., Karesh, W., Swanepoel, R., Zaki, S.R., and Rollin,

P.E. (2004) Multiple Ebola virus transmission events and rapid decline of central African wildlife. *Science*, **303**, 387–390.

32. Le Guenno, B., Formenty, P., Wyers, M., Gounon, P., Walker, F., and Boesch, C. (1995) Isolation and partial characterisation of a new strain of Ebola virus. *Lancet*, **345**, 1271–1274.

33. Wyers, M., Formenty, P., Cherel, Y., Guigand, L., Fernandez, B., Boesch, C., and Le Guenno, B. (1999) Histopathological and immunohistochemical studies of lesions associated with Ebola virus in a naturally infected chimpanzee. *J. Infect. Dis.*, **179** (Suppl. 1), S54–S59.

34. Towner, J.S., Amman, B.R., Sealy, T.K., Carroll, S.A., Comer, J.A., Kemp, A., Swanepoel, R., Paddock, C.D., Balinandi, S., Khristova, M.L., Formenty, P.B., Albarino, C.G., Miller, D.M., Reed, Z.D., Kayiwa, J.T., Mills, J.N., Cannon, D.L., Greer, P.W., Byaruhanga, E., Farnon, E.C., Atimnedi, P., Okware, S., Katongole-Mbidde, E., Downing, R., Tappero, J.W., Zaki, S.R., Ksiazek, T.G., Nichol, S.T., and Rollin, P.E. (2009) Isolation of genetically diverse Marburg viruses from Egyptian fruit bats. *PLoS Pathog.*, **5**, e1000536.

35. Leroy, E.M., Kumulungui, B., Pourrut, X., Rouquet, P., Hassanin, A., Yaba, P., Delicat, A., Paweska, J.T., Gonzalez, J.P., and Swanepoel, R. (2005) Fruit bats as reservoirs of Ebola virus. *Nature*, **438**, 575–576.

36. Hayman, D.T., Emmerich, P., Yu, M., Wang, L.F., Suu-Ire, R., Fooks, A.R., Cunningham, A.A., and Wood, J.L. (2010) Long-term survival of an urban fruit bat seropositive for Ebola and Lagos bat viruses. *PLoS ONE*, **5** (8), 24–26.

37. Swanepoel, R., Leman, P.A., Burt, F.J., Zachariades, N.A., Braack, L.E., Ksiazek, T.G., Rollin, P.E., Zaki, S.R., and Peters, C.J. (1996) Experimental inoculation of plants and animals with Ebola virus. *Emerg. Infect. Dis.*, **2**, 321–325.

38. van der Groen, G. (1978) in *Ebola Virus Haemorrhagic Fever* (ed. S.R. Pattyn), Elsevier/North-Holland Biomedical Press, Amsterdam, pp. 255–260.

39. Negredo, A., Palacios, G., Vazquez-Moron, S., Gonzalez, F., Dopazo, H., Molero, F., Juste, J., Quetglas, J., Savji, N., de la Cruz Martinez, M., Herrera, J.E., Pizarro, M., Hutchison, S.K., Echevarria, J.E., Lipkin, W.I., and Tenorio, A. (2011) Discovery of an ebolavirus-like filovirus in europe. *PLoS Pathog* **7** (10), e1002304.

40. Barrette, R.W., Metwally, S.A., Rowland, J.M., Xu, L., Zaki, S.R., Nichol, S.T., Rollin, P.E., Towner, J.S., Shieh, W.J., Batten, B., Sealy, T.K., Carrillo, C., Moran, K.E., Bracht, A.J., Mayr, G.A., Sirios-Cruz, M., Catbagan, D.P., Lautner, E.A., Ksiazek, T.G., White, W.R., and McIntosh, M.T. (2009) Discovery of swine as a host for the reston ebolavirus. *Science*, **325**, 204–206.

41. Martini, G.A. and Schmidt, H.A. (1968) Spermatogene Übertragung des "Virus Marburg" (Erreger der "Marburger Affenkrankheit"). *Klin. Wochenschr.*, **46**, 398–400.

42. Hewlett, B.S. and Hewlett, B.L. (2007) *Ebola, Culture and Politics – The Anthropology of Emerging Disease*, Thomson Wadsworth, Belmont, CA.

43. Dowell, S.F., Mukunu, R., Ksiazek, T.G., Khan, A.S., Rollin, P.E., and Peters, C.J. (1999) Transmission of Ebola hemorrhagic fever: a study of risk factors in family members, Kikwit, Democratic Republic of the Congo, 1995. Commission de Lutte contre les Epidemies a Kikwit. *J. Infect. Dis.*, **179** (Suppl. 1), S87–S91.

44. Rodriguez, L.L., De Roo, A., Guimard, Y., Trappier, S.G., Sanchez, A., Bressler, D., Williams, A.J., Rowe, A.K., Bertolli, J., Khan, A.S., Ksiazek, T.G., Peters, C.J., and Nichol, S.T. (1999) Persistence and genetic stability of Ebola virus during the outbreak in Kikwit, Democratic Republic of the Congo, 1995. *J. Infect. Dis.*, **179** (Suppl. 1), S170–S176.

45. Siegert, R., Shu, H.-L., Slenczka, W., Peters, D., and Müller, G. (1967) Zur ätiologie einer unbekannten, von affen ausgegangenen menschlichen infektionskrankheit. *Dtsch. Med. Wochenschr.*, **92**, 2341–2343.

46. Jahrling, P.B., Geisbert, T.W., Dalgard, D.W., Johnson, E.D., Ksiazek, T.G., Hall, W.C., and Peters, C.J. (1990) Preliminary report: isolation of Ebola virus from monkeys imported to USA. *Lancet*, **335**, 502–505.
47. Bausch, D.G., Nichol, S.T., Muyembe-Tamfum, J.J., Borchert, M., Rollin, P.E., Sleurs, H., Campbell, P., Tshioko, F.K., Roth, C., Colebunders, R., Pirard, P., Mardel, S., Olinda, L.A., Zeller, H., Tshomba, A., Kulidri, A., Libande, M.L., Mulangu, S., Formenty, P., Grein, T., Leirs, H., Braack, L., Ksiazek, T., Zaki, S., Bowen, M.D., Smit, S.B., Leman, P.A., Burt, F.J., Kemp, A., and Swanepoel, R. (2006) Marburg hemorrhagic fever associated with multiple genetic lineages of virus. *N. Engl. J. Med.*, **355**, 909–919.
48. Bwaka, M.A., Bonnet, M.J., Calain, P., Colebunders, R., De Roo, A., Guimard, Y., Katwiki, K.R., Kibadi, K., Kipasa, M.A., Kuvula, K.J., Mapanda, B.B., Massamba, M., Mupapa, K.D., Muyembe-Tamfum, J.J., Ndaberey, E., Peters, C.J., Rollin, P.E., and Van den Enden, E. (1999) Ebola hemorrhagic fever in Kikwit, Democratic Republic of the Congo: clinical observations in 103 patients. *J. Infect. Dis.*, **179** (Suppl. 1), S1–S7.
49. Stille, W., Böhle, E., Helm, E., van Rey, W., and Siede, W. (1968) Über eine durch Cercopithecus aethiops übertragene infektionskrankheit ("Grüne-Meerkatzen-Krankheit", "Green Monkey Disease"). *Dtsch. Med. Wochenschr.*, **93**, 572–582 [PMID: 4966281].
50. Todorović, K., Mocić, M., Klašnja, R., Stojković, L., Bordjoški, M., Gligić, A., and Stefanović, Ž. (1969) Nepoznato virusno oboljenje preneto sa infitsiranikh-obolelikh majmuna choveka. *Glas. Srpske Akad. Nauka Umetnosti, Odeljenje Med. Nauka*, **CCLXXV**, 91–101.
51. Kalongi, Y., Mwanza, K., Tshisuaka, M., Lusiama, N., Ntando, E., Kanzake, L., Shieh, W.J., Zaki, S.R., Lloyd, E.S., Ksiazek, T.G., and Rollin, P.E. (1999) Isolated case of Ebola hemorrhagic fever with mucormycosis complications, Kinshasa, Democratic Republic of the Congo. *J. Infect. Dis.*, **179** (Suppl. 1), S15–S17.
52. Nikiforov, V.V., Turovskii, Yu.I., Kalinin, P.P., Akinfeeva, L.A., Katkova, L.R., Barmin, V.S., Ryabchikova, Ye.I., Popkova, N.I., Shestopalov, A.M., Nazarov, V.P., Vedishchev, S.V. and Netesov, S.V. (1994) A case of Marburg virus laboratory infection. *Zh. Microbiol. Epidemiol. Immunobiol.*, (3), 104–106.
53. Geisbert, T.W., Hensley, L.E., Larsen, T., Young, H.A., Reed, D.S., Geisbert, J.B., Scott, D.P., Kagan, E., Jahrling, P.B., and Davis, K.J. (2003) Pathogenesis of Ebola hemorrhagic fever in cynomolgus macaques: evidence that dendritic cells are early and sustained targets of infection. *Am. J. Pathol.*, **163**, 2347–2370.
54. Gonchar, N.I., Pshenichnov, V.A., Pokhodyaev, V.A., Lopatov, K.L., and Firsova, I.V. (1991) Sensitivity of different experimental animals to Marburg virus. *Vopr. Virusol.*, **36** (5), 435–437.
55. Ryabchikova, Ye.I., Kolesnikova, L.V., and Rassadkin, Yu.N. (1998) Microscopic study of the species-specific features of hemostatic impairment in Ebola virus-infected monkeys. *Vestn. Ross. Akad. Med. Nauk*, (3), 51–55.
56. Bray, M., Davis, K., Geisbert, T., Schmaljohn, C., and Huggins, J. (1998) A mouse model for evaluation of prophylaxis and therapy of Ebola hemorrhagic fever. *J. Infect. Dis.*, **178**, 651–661.
57. Connolly, B.M., Steele, K.E., Davis, K.J., Geisbert, T.W., Kell, W.M., Jaax, N.K., and Jahrling, P.B. (1999) Pathogenesis of experimental Ebola virus infection in guinea pigs. *J. Infect. Dis.*, **179** (Suppl. 1), S203–S217.
58. Volchkov, V.E., Chepurnov, A.A., Volchkova, V.A., Ternovoj, V.A., and Klenk, H.D. (2000) Molecular characterization of guinea pig-adapted variants of Ebola virus. *Virology*, **277**, 147–155.
59. Warfield, K.L., Bradfute, S.B., Wells, J., Lofts, L., Cooper, M.T., Alves, D.A., Reed, D.K., VanTongeren, S.A., Mech, C.A., and Bavari, S. (2009) Development and characterization of a mouse model for Marburg hemorrhagic fever. *J. Virol.*, **83**, 6404–6415.

60. Carrion, R., Jr., Ro, Y., Hoosien, K., Ticer, A., Brasky, K., de la Garza, M., Mansfield, K., and Patterson, J.L. (2011) A small nonhuman primate model for filovirus-induced disease. *Virology*, **420** (2), 117–124.

61. Kobinger, G.P., Leung, A., Neufeld, J., Richardson, J.S., Falzarano, D., Smith, G., Tierney, K., Patel, A., and Weingartl, H.M. (2011) Replication, pathogenicity, shedding, and transmission of Zaire ebolavirus in pigs. *J. Infect. Dis.*, **204** (2), 200–208.

62. Marsh, G.A., Haining, J., Robinson, R., Foord, A., Yamada, M., Barr, J.A., Payne, J., White, J., Yu, M., Bingham, J., Rollin, P.E., Nichol, S.T., Wang, L.F., and Middleton, D. (2011) Ebola reston virus infection of pigs: clinical significance and transmission potential. *J. Infect. Dis.*, **204** (Suppl. 3), S804–S809.

63. Zaki, S.R., Shieh, W.J., Greer, P.W., Goldsmith, C.S., Ferebee, T., Katshitshi, J., Tshioko, F.K., Bwaka, M.A., Swanepoel, R., Calain, P., Khan, A.S., Lloyd, E., Rollin, P.E., Ksiazek, T.G., and Peters, C.J. (1999) A novel immunohistochemical assay for the detection of Ebola virus in skin: implications for diagnosis, spread, and surveillance of Ebola hemorrhagic fever. Commission de Lutte contre les Epidemies a Kikwit. *J. Infect. Dis.*, **179** (Suppl. 1), S36–S47.

64. Bechtelsheimer, H., Korb, G., and Gedigk, P. (1970) Die "Marburg-Virus"-Hepatitis – Untersuchungen bei Menschen und Meerschweinchen. *Virchows Arch. Abt. A Pathol. Pathol. Anat.*, **351**, 273–290.

65. Dietrich, M., Schumacher, H.H., Peters, D., and Knobloch, J. (1978) in *Ebola Virus Haemorrhagic Fever* (ed. S.R. Pattyn), Elsevier/North-Holland Biomedical Press, Amsterdam, pp. 37–41.

66. Gedigk, P., Korb, G., and Bechtelsheimer, H. (1968) in *Verhandlungen der Deutschen Gesellschaft für Pathologie*, vol. 52 (ed. G. Seifert), Gustav Fischer Verlag, Stuttgart, pp. 317–322.

67. Geisbert, T.W. and Jaax, N.K. (1998) Marburg hemorrhagic fever: report of a case studied by immunohistochemistry and electron microscopy. *Ultrastruct. Pathol.*, **22**, 3–17.

68. Murphy, F.A. (1978) in *Ebola Virus Haemorrhagic Fever* (ed. S.R. Pattyn), Elsevier/North-Holland Biomedical Press, Amsterdam, pp. 43–59.

69. Zaki, S.R. and Goldsmith, C.S. (1999) in *Marburg and Ebola Viruses. Current Topics in Microbiology and Immunology*, vol. 235 (ed. H.-D. Klenk), Springer-Verlag, Berlin, pp. 97–116.

70. Schnittler, H.J. and Feldmann H. (1999) in *Marburg and Ebola Viruses*, Current Topics in Microbiology and Immunology, Vol. 235 (ed. H.-D. Klenk), Springer-Verlag, Berlin, pp. 175–204.

71. Bray, M. and Geisbert, T.W. (2005) Ebola virus: the role of macrophages and dendritic cells in the pathogenesis of Ebola hemorrhagic fever. *Int. J. Biochem. Cell Biol.*, **37**, 1560–1566.

72. Davis, K.J., Anderson, A.O., Geisbert, T.W., Steele, K.E., Geisbert, J.B., Vogel, P., Connolly, B.M., Huggins, J.W., Jahrling, P.B., and Jaax, N.K. (1997) Pathology of experimental Ebola virus infection in African green monkeys. Involvement of fibroblastic reticular cells. *Arch. Pathol. Lab. Med.*, **121**, 805–819.

73. Geisbert, T.W., Jahrling, P.B., Hanes, M.A., and Zack, P.M. (1992) Association of Ebola-related Reston virus particles and antigen with tissue lesions of monkeys imported to the United States. *J. Comp. Pathol.*, **106**, 137–152.

74. Ryabchikova, E.I. and Price, B.B.S. (2004) *Ebola and Marburg Viruses – A View of Infection Using Electron Microscopy*, Battelle Press, Columbus, OH.

75. Schnittler, H.-J., Ströher, U., Afanasieva, T., and Feldmann, H. (2004) in *Ebola and Marburg Viruses – Molecular and Cellular Biology* (eds H.-D. Klenk and H. Feldmann) Horizon Bioscience, Wymondham, Norfolk, pp. 279–303.

76. Rubins, K.H., Hensley, L.E., Wahl-Jensen, V., Daddario DiCaprio, K.M., Young, H.A., Reed, D.S., Jahrling, P.B., Brown, P.O., Relman, D.A., and Geisbert, T.W. (2007) The temporal program of peripheral blood gene

expression in the response of nonhuman primates to Ebola hemorrhagic fever. *Genome. Biol.*, **8**, R174.
77. Stroher, U., West, E., Bugany, H., Klenk, H.D., Schnittler, H.J., and Feldmann, H. (2001) Infection and activation of monocytes by Marburg and Ebola viruses. *J. Virol.*, **75**, 11025–11033.
78. Geisbert, T.W., Hensley, L.E., Gibb, T.R., Steele, K.E., Jaax, N.K., and Jahrling, P.B. (2000) Apoptosis induced in vitro and in vivo during infection by Ebola and Marburg viruses. *Lab. Invest.*, **80**, 171–186.
79. Hensley, L.E., Young, H.A., Jahrling, P.B., and Geisbert, T.W. (2002) Proinflammatory response during Ebola virus infection of primate models: possible involvement of the tumor necrosis factor receptor superfamily. *Immunol. Lett.*, **80**, 169–179.
80. Baize, S., Leroy, E.M., Georges-Courbot, M.C., Capron, M., Lansoud-Soukate, J., Debre, P., Fisher-Hoch, S.P., McCormick, J.B., and Georges, A.J. (1999) Defective humoral responses and extensive intravascular apoptosis are associated with fatal outcome in Ebola virus-infected patients. *Nat. Med.*, **5**, 423–426.
81. Leroy, E.M., Baize, S., Volchkov, V.E., Fisher-Hoch, S.P., Georges-Courbot, M.C., Lansoud-Soukate, J., Capron, M., Debre, P., McCormick, J.B., and Georges, A.J. (2000) Human asymptomatic Ebola infection and strong inflammatory response. *Lancet*, **355**, 2210–2215.
82. Villinger, F., Rollin, P.E., Brar, S.S., Chikkala, N.F., Winter, J., Sundstrom, J.B., Zaki, S.R., Swanepoel, R., Ansari, A.A., and Peters, C.J. (1999) Markedly elevated levels of interferon (IFN)-gamma, IFN-alpha, interleukin (IL)-2, IL-10, and tumor necrosis factor-alpha associated with fatal Ebola virus infection. *J. Infect. Dis.*, **179** (Suppl. 1), S188–S191.
83. Peters, C.J. and Khan A.S. (1999) in *Marburg and Ebola Viruses*, Current Topics in Microbiology and Immunology, Vol. 235 (ed. H.-D. Klenk), Springer-Verlag, Berlin, pp. 85–95.
84. Formenty, P., Leroy, E.M., Epelboin, A., Libama, F., Lenzi, M., Sudeck, H., Yaba, P., Allarangar, Y., Boumandouki, P., Nkounkou, V.B., Drosten, C., Grolla, A., Feldmann, H., and Roth, C. (2006) Detection of Ebola virus in oral fluid specimens during outbreaks of Ebola virus hemorrhagic fever in the Republic of Congo. *Clin. Infect. Dis.*, **42**, 1521–1526.
85. Towner, J.S., Rollin, P.E., Bausch, D.G., Sanchez, A., Crary, S.M., Vincent, M., Lee, W.F., Spiropoulou, C.F., Ksiazek, T.G., Lukwiya, M., Kaducu, F., Downing, R., and Nichol, S.T. (2004) Rapid diagnosis of Ebola hemorrhagic fever by reverse transcription-PCR in an outbreak setting and assessment of patient viral load as a predictor of outcome. *J. Virol.*, **78**, 4330–4341.
86. Towner, J.S., Sealy, T.K., Ksiazek, T.G., and Nichol, S.T. (2007) High-throughput molecular detection of hemorrhagic fever virus threats with applications for outbreak settings. *J. Infect. Dis.*, **196** (Suppl. 2), S205–S212.

14
Bunyavirus: Hemorrhagic Fevers

Introduction

Viral hemorrhagic fever (VHF) diseases are caused by a diverse set of pathogens belonging to different virus families; *Bunyaviridae, Filoviridae, Arenaviridae,* and *Flaviviridae.*

VHF diseases differ in their pathogenesis, and the causative viruses differ in their replication, maintenance, epidemiology, and host interactions. All these viruses are maintained in nature, and infections in humans are an accidental consequence of the virus strategies for survival.

Some of these viruses cause very severe diseases with very high case fatalities. Knowledge regarding the molecular biology and pathogenesis of these viruses is very limited and in some cases completely unknown, due to the fact [1] that handling of these viruses requires a BSL (biosafety level)-3 or BSL-4 facility [2], specialized laboratory personnel, and [3] outbreaks happen sporadically.

BSL3 and BSL4 Agents: Epidemiology, Microbiology, and Practical Guidelines, First Edition.
Edited by Mandy C. Elschner, Sally J. Cutler, Manfred Weidmann, and Patrick Butaye.

14.1 Crimean Congo Hemorrhagic Fever Virus: an Enzootic Tick Borne Virus Causing Severe Disease in Man

Ali Mirazimi

14.1.1 Introduction

In the twelfth century a hemorrhagic disease, today considered to be Crimean Congo hemorrhagic fever, was described in region corresponding to contemporary Tadzhikistan. In modern times, the disease was first described in 1944, during a large outbreak of 200 cases of a severe hemorrhagic disease reported in the Crimean peninsula of the former Soviet Union [4]. Some years later, a virus causing a similar clinical picture was described in the Congo, Africa. Subsequently, it was found that both of these diseases are caused by the same virus, and it was named Crimean Congo hemorrhagic fever virus (CCHFV). CCHFV is the etiological agent of a human disease characterized by fever, prostration, severe hemorrhages, and death. Geographically, CCHFV is the second most widespread arbovirus of medical importance after dengue virus [5, 6]. CCHFV is known to be widely distributed throughout large areas of sub-Saharan Africa, the Balkans, northern Greece, European Russia, Pakistan, the Xinjiang province of northwest China, the Arabian Peninsula, Turkey, Iraq, and Iran [7–17].

CCHFV is classified within the Nairovirus genus of the family *Bunyaviridae*. CCHFV infects animals, but these remain asymptomatic as shown for cattle, sheep, goats, camels, and hares. The virus can be transmitted to humans through ticks of the genus *Hyalomma*, in particular *H. marginatum marginatum* [5, 6]. CCHFV has also been isolated from other *Hyalomma* species. Human infection also occurs by contact with blood or tissue material from infected animals or humans. In addition, person to person transmission can occur via bloody vomit, body fluids, or by aerosol from patients in advanced stages of disease [18–20]. Therefore, risk groups are found among professions with contact to infected animals (e.g., livestock breeders, abattoir workers) or infected humans (e.g., health care workers) [21].

BSL3 and BSL4 Agents: Epidemiology, Microbiology, and Practical Guidelines, First Edition.
Edited by Mandy C. Elschner, Sally J. Cutler, Manfred Weidmann, and Patrick Butaye.

14.1.2
Characteristics

The family *Bunyaviridae* is one of the largest virus groups comprising over 350 arthropod- and rodent-borne viruses [22]. The virus family is divided into five genera, Orthobunyavirus, Phlebovirus, Hantavirus, Nairovirus, and Tospovirus. Thirty-four described viruses are found within the Nairovirus genus and these viruses are further classified into seven serotypes. The only three members of this genus known to cause disease in humans are CCHFV, Dugbe virus, and Nairobi sheep disease virus. The virions in this family are enveloped and spherical with a diameter of approximately 100 nm. The genome of CCHFV is composed of three single-stranded RNA segments of negative sense, designated the small, medium, and large segments. The small segment encodes a nucleocapsid protein [23–27]. The M segment encodes a precursor for the two envelope glycoproteins Gn and Gc, and also a nonstructural protein (NSm) [1, 28–30]. The L segment of *CCHFV* codes for the RNA-dependent RNA polymerase. The terminal nucleotides of these RNA segments are partially complementary in sequence and predicted to form a stable pseudo-circular "panhandle" structure.

14.1.3
Epidemiology

CCHFV circulates unnoticed in nature in an enzootic tick–vertebrate–tick cycle. Enzootic foci of the virus occur mainly where one or several *Hyalomma* species represent the predominant tick parasitizing domestic and wild animals.

CCHFV is mainly transmitted by ixodid ticks, particularly from the genus *Hyalomma*. Virus has been isolated from both eggs and unfed immature stages of ticks, showing evidence of both transovarial transstadial transmission. In endemic areas, viremia and the presence of antibody are documented in a long list of domestic and wild vertebrates, including cattle, horses, sheep, goats, pigs, camels, donkey, mice, and dogs. But until today, there is no evidence that the virus causes disease in these animals. CCHFV is endemic in large parts of the world and has one of the most extensive geographic ranges of the tick-borne viruses causing disease. The disease has been reported from more than 28 countries in Africa, Asia, the Middle East, and Europe. In the twentyfirst century, CCHF outbreaks have become more frequent in Europe (cases or outbreaks have been recorded in Kosovo, Albania, Greece, and Bulgaria). A large outbreak in Turkey has been going on since 2002 with more than 4000 cases (up to 2010).

To date, a number of studies have addressed the genetic variability of CCHFV in different countries and from several outbreaks. Compiled data from all of these studies led to the subdivision of CCHFV into seven genetically distinct groups; Africa 1 (Senegal), Africa 2 (Democratic Republic of the Congo, South Africa), Africa 3 (southern and western Africa), Europe 1 (Russia, Turkey, Bulgaria, Kosovo, Albania), Europe 2 (Greece), Asia 1 (Middle East, Pakistan, Iran), and Asia 2 (China, Uzbekistan, Tajikistan, Kazakhstan).

14.1.4
Clinical and Pathological Findings

Humans are the only known hosts that develop disease after infection with CCHFV. The infection can results in severe hemorrhagic fever. The disease progression is rapid and can be subdivided into four different stages: incubation, pre-hemorrhagic, hemorrhagic, and the convalescence phase. The incubation period ranges from one day to up to one week, the length most probably depending on the transmission route, and the amount of inoculum. The prehemorrhagic phase usually begins with fever, myalgia, dizziness, headache, and vomiting, and it ends on average after three days. The hemorrhagic fever phase is short and characterized by epistaxis, bleeding from the gastrointestinal system, urinary and respiratory tracts, and also skin bleeding, ranging from petechiae to ecchymoses. Other symptoms include enlarged spleen and liver, in approximately 30% of patients. The average mortality rate is 30% but can be as high as 70% [3, 31, 32]. The severity of disease has been shown to correlate to the amount of virus in the blood (up to 10^9 genome equiv./ml blood) [33, 34]. Furthermore, It has been shown that antibody responses have a highly significant inverse correlation to viral loads.

Convalescence starts on average at 0–20 days post onset of illness. This phase is characterized by weakness, loss of hair, dizziness, nausea, loss of hearing, and loss of memory. It should be mentioned that the duration and symptoms in these different phases vary significantly between individuals [35].

14.1.5
Pathogenesis

The pathogenesis of CCHF is only poorly characterized due to several reasons such as: (i) infections occurring sporadically and in areas where facilities are limited for performing complete autopsies, (ii) virus handling requiring BSL-4 containment laboratories, and (iii) a lack of available animal models for the disease.

The limited knowledge about CCHF pathogenesis is mostly derived from blood analyses and liver biopsies of patients [36–40]. Swanepoel *et al.* [32] describe that cerebral hemorrhage, severe anemia, severe dehydration, and shock associated with prolonged diarrhea, lung edema, and pleural effusion are the factors causing a fatal outcome. Almost all patients who died in their study developed multiple organ failure.

In fatal cases, platelet counts can be extremely low, right from an early stage of illness. An increase in aspartate aminotransferase (AST) and alanine aminotransferase (ALT) levels in the serum, prolongation of prothrombin, and partial thromboplastin times have also been observed [32].

The existing knowledge concerning CCHF emanates from autopsies and clinical findings. The primary pathophysiological events appear to be leakage of erythrocytes and plasma through the vasculature into tissues [4]. Endothelial damage can contribute to coagulopathy by deregulated stimulation of platelet aggregation, which in turn activates the intrinsic coagulation cascade, ultimately leading to clotting factor deficiency causing hemorrhages. For CCHFV, vascular leakage may be caused either by the destruction of endothelial cells or by a disruption of the tight junctions which constitute the endothelial barrier between cells. Moreover, it is unclear whether these events are a direct consequence of infection or whether virus-induced host factors cause the endothelial dysfunction [41]. However, in an epithelial cell line model CCHFV caused neither disruption of tight junctions nor necrosis nor the apoptosis of cells [42]. This could suggest that the hemorrhages and coagulation disturbances may be caused indirectly, possibly by high levels of proinflammatory cytokines. Key players in disease progression are the cytokines interleukin (IL)-10, IL-1, IL-6, and tumor necrosis factor (TNF)-a [43]. Recently, Ergonul and coworkers demonstrated significantly higher levels of IL-6 and TNF-a in the CCHFV patients with fatal outcome compared to the nonfatal cases [37]. Another observation was a correlation between high levels of IL-6 and TNF-a to the onset of disseminated intravascular coagulation (DIC) in patients. However, elevated levels of IL-10 were not observed in these patients and the levels of IL-10 were negatively correlated to the DIC scores. In another study, Papa *et al.* confirmed that high TNF-a levels are correlated with the severity of CCHF disease, but they suggested that IL-6 could be found in both mild and severe cases [44]. The only fatal case in this study had high levels of both IL-6 and TNF-a compared to nonfatal cases. Another interesting observation is the elevated level of Neopterin in patients with Dengue fever or Ebola hemorrhagic fever [2, 45]. Neopterin derivates are produced by macrophages and dendritic cells upon stimulation by IFNs [46]. Neopterin is a useful tool to assess the intensity of cell-mediated immune response [46] and a recent study indeed noted a correlation between elevated levels of neopterin in CCHFV patients and disease severity [38].

All these results suggest that capillary fragility, a common feature of CCHF, is most probably due to multiple host-induced mechanisms in response to CCHFV infection. Endothelial damage would cause the characteristic rash and contribute to hemostatic failure. Interestingly, some authors have noticed similarities between various viral hemorrhagic fevers and the septic shock caused by severe bacterial infections [47].

14.1.6 **Diagnosis**

Early diagnosis is essential for the outcome of disease and for the prevention of transmission in the community. Suspected cases should be evaluated and their handling planned, including antiviral and supportive treatment.

Evaluation of a suspected case should be done by observation of the clinical symptoms [1] and evaluation of patient history [2], laboratory quantitative analysis of platelets and white blood cells, levels of aspartate aminotransferase, alanine aminotransferase, lactate dehydrogenase, and creatinine phosphokinase [3], and all the suspected cases should be verified by specific microbiological assays. To date there are several different tests available: virus isolation, molecular RNA detection, antigen detection, and serology.

14.1.6.1 **Virus Isolation**

Virus isolation studies should be done in high-containment BSL-4 laboratories. The virus can be isolated using several different cell lines such as Vero, SW13, and so on. Virus isolation can be achieved during the relatively high viremia encountered during the first five days of illness.

14.1.6.2 **Molecular Methods**

Reverse transcriptase PCR/real-time assays are the method of choice for rapid laboratory diagnosis of CCHF virus infection. The method is highly specific, sensitive, and rapid [34].

14.1.6.3 **Antigen Detection**

In spite of its relative lack of sensitivity, this method can detect the high viremia cases.

14.1.6.4 **Serology**

IgM and IgG antibodies are detectable by ELISA and immunofluorescence assays from about five days after the onset of disease. ELISA methods are quite specific and more sensitive than immunofluorescence assays.

References

1. Altamura, L.A., Bertolotti-Ciarlet, A., Teigler, J., Paragas, J., Schmaljohn, C.S., and Doms, R.W. (2007) Identification of a novel C-terminal cleavage of Crimean-Congo hemorrhagic fever virus PreGN that leads to generation of an NSM protein. *J. Virol.*, **81**, 6632–6642.
2. Baize, S., Leroy, E.M., Georges, A.J., Georges-Courbot, M.C., Capron, M., Bedjabaga, I., Lansoud-Soukate, J., and Mavoungou, E. (2002) Inflammatory responses in Ebola virus-infected patients. *Clin. Exp. Immunol.*, **128**, 163–168.
3. Baskerville, A., Satti, A., Murphy, F.A., and Simpson, D.I. (1981) Congo-Crimean haemorrhagic fever in Dubai: histopathological studies. *J. Clin. Pathol.*, **34**, 871–874.
4. Ergonul, O. (2006) Crimean-Congo haemorrhagic fever. *Lancet Infect. Dis.*, **6**, 203–214.
5. Hoogstraal, H. (1979) The epidemiology of tick-borne Crimean-Congo hemorrhagic fever in Asia, Europe, and Africa. *J. Med. Entomol.*, **15**, 307–417.
6. Watts, D.M., Ksiazek, T.G., Linthicum, K.J., and Hoogstraal, H. (1988) in *The Arboviruses: Epidemiology and Ecology*, vol. 2 (ed. T.P. Monath), CRC Press, Boca Raton, FL, pp. 177–260.
7. Burney, M.I., Ghafoor, A., Saleen, M., Webb, P.A., and Casals, J. (1980) Nosocomial outbreak of viral hemorrhagic fever caused by Crimean Hemorrhagic fever-Congo virus in Pakistan, January 1976. *Am. J. Trop. Med. Hyg.*, **29**, 941–947.
8. Drosten, C., Minnak, D., Emmerich, P., Schmitz, H., and Reinicke, T. (2002) Crimean-Congo hemorrhagic fever in Kosovo. *J. Clin. Microbiol.*, **40**, 1122–1123.
9. Dunster, L., Dunster, M., Ofula, V., Beti, D., Kazooba-Voskamp, F., Burt, F., Swanepoel, R., and DeCock, K.M. (2002) First documentation of human Crimean-Congo hemorrhagic fever, Kenya. *Emerg. Infect. Dis.*, **8**, 1005–1006.
10. el-Azazy, O.M. and Scrimgeour, E.M. (1997) Crimean-Congo haemorrhagic fever virus infection in the western

province of Saudi Arabia. *Trans. R. Soc. Trop. Med. Hyg.*, **91**, 275–278.

11. Nabeth, P., Cheikh, D.O., Lo, B., Faye, O., Vall, I.O., Niang, M., Wague, B., Diop, D., Diallo, M., Diallo, B., Diop, O.M., and Simon, F. (2004) Crimean-Congo hemorrhagic fever, Mauritania. *Emerg. Infect. Dis.*, **10**, 2143–2149.
12. Nabeth, P., Thior, M., Faye, O., and Simon, F. (2004) Human Crimean-Congo hemorrhagic fever, Senegal. *Emerg. Infect. Dis.*, **10**, 1881–1882.
13. Papa, A., Bino, S., Llagami, A., Brahimaj, B., Papadimitriou, E., Pavlidou, V., Velo, E., Cahani, G., Hajdini, M., Pilaca, A., Harxhi, A., and Antoniadis, A. (2002) Crimean-Congo hemorrhagic fever in Albania, 2001. *Eur. J. Clin. Microbiol. Infect. Dis.*, **21**, 603–606.
14. Papa, A., Christova, I., Papadimitriou, E., and Antoniadis, A. (2004) Crimean-Congo hemorrhagic fever in Bulgaria. *Emerg. Infect. Dis.*, **10**, 1465–1467.
15. Papa, A., Ma, B., Kouidou, S., Tang, Q., Hang, C., and Antoniadis, A. (2002) Genetic characterization of the M RNA segment of Crimean Congo hemorrhagic fever virus strains, China. *Emerg. Infect. Dis.*, **8**, 50–53.
16. Sheikh, A.S., Sheikh, A.A., Sheikh, N.S., Rafi, U.S., Asif, M., Afridi, F., and Malik, M.T. (2005) Bi-annual surge of Crimean-Congo haemorrhagic fever (CCHF): a five-year experience. *Int. J. Infect. Dis.*, **9**, 37–42.
17. Williams, R.J., Al-Busaidy, S., Mehta, F.R., Maupin, G.O., Wagoner, K.D., Al-Awaidy, S., Suleiman, A.J., Khan, A.S., Peters, C.J., and Ksiazek, T.G. (2000) Crimean-congo haemorrhagic fever: a seroepidemiological and tick survey in the Sultanate of Oman. *Trop. Med. Int. Health.*, **5**, 99–106.
18. Swanepoel, R., Shepherd, A.J., Leman, P.A., and Shepherd, S.P. (1985) Investigations following initial recognition of Crimean-Congo haemorrhagic fever in South Africa and the diagnosis of 2 further cases. *S. Afr. Med. J.*, **68**, 638–641.
19. van de Wal, B.W., Joubert, J.R., van Eeden, P.J., and King, J.B. (1985) A nosocomial outbreak of Crimean-Congo haemorrhagic fever at Tygerberg Hospital. Part IV. Preventive and prophylactic measures. *S. Afr. Med. J.*, **68**, 729–732.
20. van Eeden, P.J., van Eeden, S.F., Joubert, J.R., King, J.B., van de Wal, B.W., and Michell, W.L. (1985) A nosocomial outbreak of Crimean-Congo haemorrhagic fever at Tygerberg Hospital. Part II. Management of patients. *S. Afr. Med. J.*, **68**, 718–721.
21. Chinikar, S., Persson, S.M., Johansson, M., Bladh, L., Goya, M., Houshmand, B., Mirazimi, A., Plyusnin, A., Lundkvist, A., and Nilsson, M. (2004) Genetic analysis of Crimean-congo hemorrhagic fever virus in Iran. *J. Med. Virol.*, **73**, 404–411.
22. Nichol, S. (2001) in *Fields Virology*, 4th edn, vol. 1 (eds D. Knipe and P. Howley), Lippincott Williams and Wilkins, Philadelphia, PA, pp. 1603–1633.
23. Blakqori, G., Delhaye, S., Habjan, M., Blair, C.D., Sanchez-Vargas, I., Olson, K.E., Attarzadeh-Yazdi, G., Fragkoudis, R., Kohl, A., Kalinke, U., Weiss, S., Michiels, T., Staeheli, P., and Weber, F. (2007) La Crosse bunyavirus nonstructural protein NSs serves to suppress the type I interferon system of mammalian hosts. *J. Virol.*, **81**, 4991–4999.
24. Fuller, F., Bhown, A.S., and Bishop, D.H. (1983) Bunyavirus nucleoprotein, N, and a non-structural protein, NSS, are coded by overlapping reading frames in the S RNA. *J. Gen. Virol.*, **64** (Pt 8), 1705–1714.
25. Jaaskelainen, K.M., Kaukinen, P., Minskaya, E.S., Plyusnina, A., Vapalahti, O., Elliott, R.M., Weber, F., Vaheri, A., and Plyusnin, A. (2007) Tula and Puumala hantavirus NSs ORFs are functional and the products inhibit activation of the interferon-beta promoter. *J. Med. Virol.*, **79**, 1527–1536.
26. Kormelink, R., Kitajima, E.W., De Haan, P., Zuidema, D., Peters, D., and Goldbach, R. (1991) The nonstructural protein (NSs) encoded by the ambisense S RNA segment of tomato

spotted wilt virus is associated with fibrous structures in infected plant cells. *Virology*, **181**, 459–468.

27. Le May, N., Dubaele, S., Proietti De Santis, L., Billecocq, A., Bouloy, M., and Egly, J.M. (2004) TFIIH transcription factor, a target for the Rift Valley hemorrhagic fever virus. *Cell*, **116**, 541–550.

28. Bergeron, E., Vincent, M.J., and Nichol, S.T. (2007) Crimean-Congo hemorrhagic fever virus glycoprotein processing by the endoprotease SKI-1/S1P is critical for virus infectivity. *J. Virol.*, **81**, 13271–13276.

29. Fontana, J., Lopez-Montero, N., Elliott, R.M., Fernandez, J.J., and Risco, C. (2008) The unique architecture of Bunyamwera virus factories around the Golgi complex. *Cell Microbiol.*, **35**, 64–67

30. Pollitt, E., Zhao, J., Muscat, P., and Elliott, R.M. (2006) Characterization of Maguari orthobunyavirus mutants suggests the nonstructural protein NSm is not essential for growth in tissue culture. *Virology*, **348**, 224–232.

31. Schwarz, T.F., Nsanze, H., and Ameen, A.M. (1997) Clinical features of Crimean-Congo haemorrhagic fever in the United Arab Emirates. *Infection*, **25**, 364–367.

32. Swanepoel, R., Gill, D.E., Shepherd, A.J., Leman, P.A., Mynhardt, J.H., and Harvey, S. (1989) The clinical pathology of Crimean-Congo hemorrhagic fever. *Rev. Infect. Dis.*, **11** (Suppl. 4), S794–S800.

33. Duh, D., Saksida, A., Petrovec, M., Ahmeti, S., Dedushaj, I., Panning, M., Drosten, C., and Avsic-Zupanc, T. (2007) Viral load as predictor of Crimean-Congo hemorrhagic fever outcome. *Emerg. Infect. Dis.*, **13**, 1769–1772.

34. Wolfel, R., Paweska, J.T., Petersen, N., Grobbelaar, A.A., Leman, P.A., Hewson, R., Georges-Courbot, M.C., Papa, A., Gunther, S., and Drosten, C. (2007) Virus detection and monitoring of viral load in Crimean-Congo hemorrhagic fever virus patients. *Emerg. Infect. Dis.*, **13**, 1097–1100.

35. Weber, F. and Mirazimi, A. (2008) Interferon and cytokine responses to Crimean Congo hemorrhagic fever virus; an emerging and neglected viral zonoosis. *Cytokine Growth Factor Rev.*, **19**, 395–404.

36. Celikbas, A., Ergonul, O., Dokuzoguz, B., Eren, S., Baykam, N., and Polat-Duzgun, A. (2005) Crimean Congo hemorrhagic fever infection simulating acute appendicitis. *J. Infect.*, **50**, 363–365.

37. Ergonul, O., Tuncbilek, S., Baykam, N., Celikbas, A., and Dokuzoguz, B. (2006) Evaluation of serum levels of interleukin (IL)-6, IL-10, and tumor necrosis factor-alpha in patients with Crimean-Congo hemorrhagic fever. *J. Infect. Dis.*, **193**, 941–944.

38. Onguru, P., Akgul, E.O., Akinci, E., Yaman, H., Kurt, Y.G., Erbay, A., Bayazit, F.N., Bodur, H., Erbil, K., Acikel, C.H., and Cevik, M.A. (2008) High serum levels of neopterin in patients with Crimean-Congo hemorrhagic fever and its relation with mortality. *J. Infect.*, **56**, 366–370.

39. Ozkurt, Z., Kiki, I., Erol, S., Erdem, F., Yilmaz, N., Parlak, M., Gundogdu, M., Tasyaran, M.A. (2005) Crimean-Congo hemorrhagic fever in Eastern Turkey: clinical features, risk factors and efficacy of ribavirin therapy. *J. Infect.*, **53**, 354–357.

40. Yilmaz, M., Aydin, K., Akdogan, E., Sucu, N., Sonmez, M., Omay, S.B., and Koksal, I. (2008) Peripheral blood natural killer cells in Crimean-Congo hemorrhagic fever. *J. Clin. Virol.*, **42**, 415–417.

41. Schnittler, H.J. and Feldmann, H. (2003) Viral hemorrhagic fever–a vascular disease? *Thromb. Haemost.*, **89**, 967–972.

42. Connolly-Andersen, A.M., Magnusson, K.E., and Mirazimi, A. (2007) Basolateral entry and release of Crimean-Congo hemorrhagic fever virus in polarized MDCK-1 cells. *J. Virol.*, **81**, 2158–2164.

43. Chen, H.C., Hofman, F.M., Kung, J.T., Lin, Y.D., and Wu-Hsieh, B.A. (2007) Both virus and tumor necrosis factor alpha are critical for endothelium damage in a mouse model of dengue virus-induced hemorrhage. *J. Virol.*, **81**, 5518–5526.

44. Papa, A., Bino, S., Velo, E., Harxhi, A., Kota, M., and Antoniadis, A. (2006) Cytokine levels in Crimean-Congo hemorrhagic fever. *J. Clin. Virol.*, **36**, 272–276.
45. Chan, C.P., Choi, J.W., Cao, K.Y., Wang, M., Gao, Y., Zhou, D.H., Di, B., Xu, H.F., Leung, M.F., Bergmann, A., Lehmann, M., Nie, Y.M., Cautherley, G.W., Fuchs, D., Renneberg, R., and Zheng, B.J. (2006) Detection of serum neopterin for early assessment of dengue virus infection. *J. Infect.*, **53**, 152–158.
46. Wirleitner, B., Reider, D., Ebner, S., Bock, G., Widner, B., Jaeger, M., Schennach, H., Romani, N., and Fuchs, D. (2002) Monocyte-derived dendritic cells release neopterin. *J. Leukoc. Biol.*, **72**, 1148–1153.
47. Whitehouse, C.A. (2004) Crimean-Congo hemorrhagic fever. *Antiviral. Res.*, **64**, 145–160.

14.2
Rift Valley Fever Virus: a Promiscuous Vector Borne Virus

Manfred Weidmann, F. Xavier Abad, and Janusz T. Paweska

14.2.1
Introduction

Rift Valley fever virus (RVFV) was first isolated by inoculating lambs with the serum from moribund sheep in Kenya, in 1931 [1]. However this virus apparently had been circulating among lambs and sheep as early as 1911, when an outbreak causing lambs and sheep deaths was reported [2]. Several countries in southern and central Africa (e.g., South Africa, Zimbabwe, Zambia, Kenya) reported major outbreaks in the 1960s, 1970s, and 1990s and in the first decade of the new century. Molecular phylogeny confirms that the virus spread from East Africa to West Africa (Mauritania) and sub-Saharan countries (Egypt in 1977, 1993, 1997). In 2000 the virus even jumped the continental borders and spread to Saudi Arabia and Yemen (Asia). The key factors in this last spread might have been viremic animal trade or infected mosquitoes [1].

RVFV outbreaks have caused and cause great economical impact resulting from the death of livestock and restrictions or ban of trade and export, ranging from several months to years after the end of an epizootic [3, 4]. However most of the outbreaks in humans have been small in size, largely confined to Africa and quarantine succeeded as a countermeasure in controlling epidemics.

Since RVFV is promiscuous in its choice of transmission vectors it has the potential to escape its traditional boundaries to other parts of the world. The spread to Saudi Arabia and Yemen in 2000 led to heightened alertness in the neighboring countries of the region and in Europe. In Europe the recent spread of an African Reovirus (bluetongue virus) in livestock by endemic *Culicoides* species has shaken the veterinary scientific community and raised the question whether this would also be possible for RVFV. RVFV was included in the offensive bioweapons program of the United States until 1969 [5].

14.2.2
Characteristics of the Agent

RVFV is a member of the *Bunyaviridae* family, genus *Phlebovirus*. Most of the viruses of this genus are transmitted by phlebotomine sandflies, hence the genus

BSL3 and BSL4 Agents: Epidemiology, Microbiology, and Practical Guidelines, First Edition.
Edited by Mandy C. Elschner, Sally J. Cutler, Manfred Weidmann, and Patrick Butaye.

name. In this respect, RVFV is an exception, as it is transmitted by many different mosquitoes.

RVFV consists of a single serotype (no antigenic differences have been demonstrated) but differences in pathogenicity have been shown [6, 7]. It is an enveloped spherical virus, 80–120 nm in diameter, with a single-stranded RNA genome in three segments: large (L), medium (M), and small (S). The L and M segments show negative polarity and express, the RNA-dependent RNA polymerase, and the precursor of the glycoproteins G_N and G_C, respectively. From this glycoprotein precursor, a nonstructural protein (NSm) of unknown role also arises from post-translational cleavage. The S segment has a bi-directional ambisense coding organization and encodes for the nucleoprotein N in antisense and for the nonstructural protein NSs in sense orientation [8]. For more details see [9].

As shown for other segmented RNA virus such as influenza virus [10], other bunyaviruses or hantaviruses [11], RVFV has high capabilities to exchange genetic material (genetic shift) between different isolates from distant geographic areas [12, 13].

14.2.3
Epidemiology

RVFV transmission to man can occur by contact to infected tissues, for example, from the abortus of livestock [14, 15], and consequently nomadic tribes in Kenya [16, 17] or livestock workers and wildlife rangers in Nigeria [18, 19] show a very high seroprevalence toward RVFV. Raw milk has also been linked to RVFV transmission although very rarely [20, 21]. Rift Valley fever (RVF) can spread to man by the above modes of transmission but it mainly spreads by transmission via a variety of infected mosquitoes (from the genera *Aedes, Culex, Anopheles, Eretmapodites, Mansonia*) and occasionally from ticks (*Rhipicephalus*).

RVFV is maintained in nature via transovarial transmission in mosquitoes (*Aedes* in Kenya and Senegal) [22, 23] and no vertebrate reservoirs have as yet been described.

The analysis of rainfall records and vegetation index data coupled with satellite observations can predict RVFV outbreaks in East Africa [24, 25]. And indeed the outbreak of 2007 in Kenya/Tanzania was predicted after heavy rainfalls at the end of 2006.

Here massive mosquito proliferation following flooding events usually leads to establishment of a transmission cycle in relative resistant indigenous ruminants and zebu-type cattle (*Bos indicus*). The first indications of the outbreak are infections in exotic breeds of cattle, sheep, or goats and their crosses, and of course human beings. In other more semi-desert regions of Africa elevated abortions in up to 15% of pregnant livestock are the earliest indication of a beginning RVFV outbreak, but these may be overlooked.

Since between 15 and 35% of sheep, goats, and cattle have been found to be RVFV seropositive without any clear sign of acute clinical cases in animal or man there may be considerable undetected RVFV activity [26].

Although major advances have been made on RVFV molecular evolution, significant gaps remain to be filled to obtain a better understanding of its genetic variability and subsequently of the patterns and processes involved in its dispersal, distribution, and its spread. Indeed, at the continent level previous studies showed that exchanges between different subregions of Africa occurred most likely over the last century with east–central Africa playing a major role as a source of diversity generation. However, the recent spread to the Middle East as well the changing epidemiology in the different subregions implies that further detailed analysis is necessary to clearly understand the transmission dynamics for RVFV.

Although RVFV presents as one single serotype, phylogenetic analysis has shown the existence of three major lineages of the segmented RVFV genome linked to geographic variants from west Africa, Egypt, and central–east Africa. Reassortments between strains from different areas (or lineages) have been strongly suggested, pointing to a potential sylvatic cycle in the tropical rain forest [13]. These reassortment phenomena had been previously demonstrated in tissue cultures [27] and in dually infected mosquitoes [28].

14.2.4
Clinical and Pathological Findings in Humans

In humans, to which the disease mainly spreads via arthropods but also by aerosols from slaughtered animals or aborted animal carcasses, the symptoms are mainly asymptomatic and can be confused with a moderate to severe, nonfatal influenza-like illness [29, 30]. The symptoms can however range from mild fever to encephalitis, retinitis, and fatal hepatitis with hemorrhages. The more severe forms occur in less than 1–8% of patients of which up to 50% may die.

In humans, few fatal cases were reported before 1977. That year, in Egypt between 18 000 and 200 000 people were estimated as infected, with 623 recorded deaths from encephalitis and/or hemorrhagic fevers [31]. Ten years later, more than 200 deaths were attributed to RVFV infections in Mauritania [32, 33], and in 1997/1998 after the El Niño–Southern Oscillation (ENSO) floods at the horn of Africa, around 1100 people in Kenya, spreading rapidly to Somalia and Tanzania, were infected [34]. In the last reported outbreak, in Saudi Arabia and Yemen, in 2000, more than 1900 people required hospitalization; if we accept that only a small fraction (around 1%) of people affected develop clinical signs then, a strong indication of the magnitude of this outbreak arises. Similar figures were reported from the outbreak in Kenya in 2006–2007, with an estimated 180 000 people infected and about 400 severely ill [35].

14.2.5
Prophylaxis and Treatment

RVFV has caused serious human infection in laboratory workers; staff should either be vaccinated and work under containment level 3, or containment level 4 conditions, or wear respiratory protection; such care has to be increased when working with infected animals or when performing post mortem examinations (OIE terrestrial Manual 2008).

Because of the economic impact of the disease in sheep and cattle and the fatal illness in some patients several attempts were made to produce a veterinary but also a human vaccine. A first attenuated vaccine from the Entebbe virulent strain passaged in suckling mice and embryonated eggs [36] was shown to be neurotropic, led to abortion and stillbirth in ewes and cattle [37], tended to be teratogenic, and could induce abortions in up to 15% of pregnant animals in the susceptible breeds of sheep and goats [26]. However it is cheap, effective, highly immunogenic, confers lifelong immunity, and is therefore still extensively used in Africa.

Other attenuated strains were produced and assayed. As most promising appeared the mutated strain MP12, from a virulent isolate in Egypt, in 1977, containing mutations in each one of the three genome segments evoked by serial passage in the presence of 5-fluorouracil. It was reported as an efficient immunogenic in adult and young animals [38–41] but it also showed teratogenic features and induced abortions in pregnant ewes [42] and was reported neurovirulent for monkeys [43]. Another promising candidate, clone 13, was found to be avirulent in mice and hamsters and highly immunogenic in lambs, sheep, young, and adult goats [44], but was also non-abortogenic in pregnant sheep, conferring an high degree of protection [45]. This candidate is deeply interesting as its possible reversion toward the virulent phenotype is really quite unlikely as this virus possesses a large deletion in the S-RNA fragment coding for the nonstructural proteins, which is a major determinant for attenuation [46].

Additionally several attempts were performed to produce a safer inactivated vaccine [47–51]. A formalin-inactivated vaccine requiring three boosters has been in use for 20 years and administered to protect laboratory workers [9]. However, the ability to produce this vaccine has been lost [52]. Most recently an approach using virus-like particles was published. This vaccine is currently being tested in animals [53].

Concerning therapeutic treatments, the administration of antibodies, interferon, interferon inducer, or rivabirin as assayed in experimentally infected RVFV mice, rats, or monkeys were shown as efficient in protecting against the disease [33, 54], but these treatments have never been tested to treat RVFV-infected patients.

14.2.6 Pathogenesis

The incubation period for RVFV, as reported in several experimental infections with sheep [18, 55, 56] is from 18 h to 7 days, and the viremia may persist for 1–7 days in most cases. Pyrexia is the most clear clinical sign at the early stage, and lasts for 1–7 days, peaking between day 2 and day 4 post infection [57, 58]. The level and duration of viremia is probably associated with the virus strain but also with the genotype of the animal and its relative susceptibility. Titers from 3.4 to 7.5 $\log_{10}$ PFU/ml of blood have been reported [18, 38, 57, 58]. Some data pointed out that wool sheep exotic to Africa are in general highly susceptible, with viremias lasting for 4–7 days; indigenous East Africa hair sheep breeds are relatively insusceptible, with viremia periods of one or two days [26].

RVFV affects primarily liver, causing hepatocellular changes progressing to massive necrosis [59] (in domestic animals and humans). Hepatocyte destruction can be so marked that the normal architecture of the liver disappears. Rod shaped or oval eosinophilic intranuclear inclusions which contain NSs protein [60] are commonly found in RVFV-infected livers. Changes in hematological and serum chemistry such as profound leukopenia and thrombocytopenia [59] and elevated serum enzymes are linked to liver damage (alanine and aspartate aminotransferases [18], bilirubin [59]). Additionally low levels of coagulation proteins (decreased fibrinogen, elevated fibrin degradation products) [61], promoting the occurrence of disseminated intravascular coagulopathy [18] and impaired blood flow are observed.

Similar to other arboviruses, upon inoculation by a mosquito, RVFV viruses are transported to regional lymph nodes by lymphatic drainage where viral replication occurs. Spreading by the blood (viremia) allows RVFV to reach the target organs. The main replication sites are the liver (in the hepatocytes), the spleen, the kidney (adrenocortical cells and glomeruli), and the brain in the case of animals dying from encephalitis. High loads of virus are also detected in several organs of the fetus, from the fresh placenta and in the serosanguineous fluid found in the thoracic cavity of late gestation abortuses.

Young lambs and sheep are most susceptible to hemorrhages in the liver. A wide range of animals from laboratory animals to frog, tortoises, and rhesus monkeys are susceptible to be infected by RVFV showing various clinical signs [9].

Different field isolates of RVFV vary in their virulence for laboratory rodents, suggesting that viral and possibly host factors are important in the pathogenesis of the disease. RVFV replicates in various tissues, including skin, lymphatic organs, liver, and neuronal tissue [9]. However, the cell types involved and mechanisms of cell tropism and cell destruction are yet unknown. The interaction of the virus with the innate immunity revealed that RVFV counteracts the interferon system by blocking RNA polymerase II-mediated transcription, which leads to the suppression of type I interferon production [62] and by triggering the specific degradation of the antiviral kinase protein kinase R (PKR) [63]. The data of the interaction of RVFV with the adapted immune system are at present very limited and the factors that are involved in organ tropism and the impairment of endothelial- and liver cell function remain to be investigated.

14.2.7
Diagnosis and Surveillance

Heavy rain falls, followed by high mosquito densities can lead to RVF outbreaks with elevated abortus counts in livestock herds.

The clear demonstration of the occurrence of RVFV is virus isolation using adult or suckling mice and hamsters [64] as susceptible animals, continuous cell lines (such as Vero, BHK-21, baby hamster kidney cells), mosquito cell lines [32], or embryonated chicken eggs.

RVFV-specific IgM of IgG in animal or human sera can be detected, by using enzyme-linked immunoassays (ELISA) [65–67].

In West Africa a sentinel herd system using goats whose serum is tested for RVFV antibodies by neutralization test and ELISA is used to monitor RVFV prevalence and the emergence of enzootics [68]. The diagnostic window of RVFV offers a high viremia of RVFV (lasting for up to five days) and a swiftly rising IgM afterwards [69]. The presence of IgM indicates a recent infection. For serology an ELISA using the RVFV nucleocapsid to attract antiRVFV IgM/IgG has been developed and extensively evaluated with samples from domestic livestock and from game [70, 71]. The first microsensor-based system for the detection of IgG has recently been published [72].

Real-time PCR assays [73–75] and a first isothermal assay have been developed for laboratory use [76]. The viremic period in animals and man is short, but the usefulness of PCR in an outbreak situation has been shown [77].

14.2.8
Conclusions

RVFV is a segmented RNA virus that can re-assort. All isolates constitute one single serotype, but three major phylogenetic lineages have been described which correspond to distinct geographic areas: West Africa, Egypt, and Central–East Africa. The virus affects mainly livestock (sheep, goats, cattle, but also wild ruminants) but occasionally overspills to man when the vector population (mosquitoes, e.g., *Aedes, Culex*) increases, as other ways of transmission are negligible. The virus is maintained in nature via transovarial transmission in mosquitoes.

In humans, RVFV infections are mainly asymptomatic, the most common clinical manifestations are fever, headache, and in only about 1–8% hemorrhages, which can lead to severe complications such as retinitis, encephalitis, and hepatitis with fatal hemorrhagic fever. RVFV primarily affects the liver causing hepatocellular changes progressing to massive necrosis.

The ability to forecast regional RVFV activity, based on checking sea surface temperature oscillations and normalized vegetation index (as affected by the ENSO phenomenon), can permit the vaccination of domestic animals and pre-treatment of mosquito habitats [25]. The question is how to coordinate the measures and under which circumstances vaccination campaigns can be successful.

References

1. Daubney, R., Hudson, J.R., and Graham, P.C. (1931) Enzootic hepatitis of Rift Valley fever, an undescribed virus disease of sheep, cattle and man from East Africa. *J. Pathol. Bacteriol.*, **34**, 545–579.
2. Stordy, R. (1913) Mortality among lambs. *Annu. Rep. Dept Agric. Br. East Afr.*
3. Domenech, J., Lubroth, J., Eddi, C., Martin, V., and Roger, F. (2006) Regional and international approaches on prevention and control of animal transboundary and emerging diseases. *Impact Emerg. Zoonotic Dis. Anim. Health*, **1081**, 90–107.
4. Rich, K.M. and Wanyoike, F. (2008) An assessment of the regional and national

socio-economic impacts of the 2007 Rift Valley fever outbreak in Kenya. *Am. J. Trop. Med. Hyg.*, **83**, 52–57.

5. Anonymous (2000) *Chemical and Biological Weapons: Possession and Programs Past and Present*, Center for Nonproliferation Studies.
6. Bird, B.H., Khristova, M.L., Rollin, P.E., Ksiazek, T.G., and Nichol, S.T. (2007) Complete genome analysis of 33 ecologically and biologically diverse Rift Valley fever virus strains reveals widespread virus movement and low genetic diversity due to recent common ancestry. *J. Virol.*, **81**, 2805–2816.
7. Swanepoel, R., Struthers, J.K., Erasmus, M.J., Shepherd, S.P., McGillivray, G.M., Shepherd, A.J., Hummitzsch, D.E., Erasmus, B.J., and Barnard, B.J. (1986) Comparative pathogenicity and antigenic cross-reactivity of Rift Valley fever and other African phleboviruses in sheep. *J. Hyg. (Lond)*, **97**, 331–346.
8. Giorgi, C., Accardi, L., Nicoletti, L., Gro, M.C., Takehara, K., Hilditch, C., Morikawa, S., and Bishop, D.H. (1991) Sequences and coding strategies of the S RNAs of Toscana and Rift Valley fever viruses compared to those of Punta Toro, Sicilian Sandfly fever, and Uukuniemi viruses. *Virology*, **180**, 738–753.
9. Flick, R. and Bouloy, M. (2005) Rift valley fever virus. *Curr. Mol. Med.*, **5**, 827–834.
10. Webster, R.G., Bean, W.J., Gorman, O.T., Chambers, T.M., and Kawaoka, Y. (1992) Evolution and ecology of influenza A viruses. *Microbiol. Rev.*, **56**, 152–179.
11. Rodriguez, L.L., Owens, J.H., Peters, C.J., and Nichol, S.T. (1998) Genetic reassortment among viruses causing hantavirus pulmonary syndrome. *Virology*, **242**, 99–106.
12. Sall, A.A., de A Zanotto, P.M., Vialat, P., Sene, O.K., and Bouloy, M. (1998) Origin of 1997–98 Rift Valley fever outbreak in East Africa. *Lancet*, **352**, 1596–1597.
13. Sall, A.A., Zanotto, P.M., Sene, O.K., Zeller, H.G., Digoutte, J.P., Thiongane, Y., and Bouloy, M. (1999) Genetic reassortment of Rift Valley fever virus in nature. *J. Virol.*, **73**, 8196–8200.
14. Hoch, A.L., Gargan, T.P. II, and Bailey, C.L. (1985) Mechanical transmission of Rift Valley fever virus by hematophagous diptera. *Am. J. Trop. Med. Hyg.*, **34**, 188–193.
15. Logan, T.M., Davies, F.G., Linthicum, K.J., and Ksiazek, T.G. (1992) Rift Valley fever antibody in human sera collected after an outbreak in domestic animals in Kenya. *Trans. R. Soc. Trop. Med. Hyg.*, **86**, 202–203.
16. LaBeaud, A.D., Muchiri, E.M., Ndzovu, M., Mwanje, M.T., Muiruri, S., Peters, C.J., and King, C.H. (2008) Interepidemic Rift Valley fever virus seropositivity, northeastern Kenya. *Emerg. Infect. Dis.*, **14**, 1240–1246.
17. LaBeaud, A.D., Ochiai, Y., Peters, C.J., Muchiri, E.M., and King, C.H. (2007) Spectrum of Rift Valley fever virus transmission in Kenya: insights from three distinct regions. *Am. J. Trop. Med. Hyg.*, **76**, 795–800.
18. Olaleye, O.D., Tomori, O., Fajimi, J.L., and Schmitz, H. (1996) Experimental infection of three Nigerian breeds of sheep with the Zinga strain of the Rift Valley Fever virus. *Rev. Elev. Med. Vet. Pays. Trop.*, **49**, 6–16.
19. Olaleye, O.D., Tomori, O., Ladipo, M.A., and Schmitz, H. (1996) Rift Valley fever in Nigeria: infections in humans. *Rev. Sci. Tech.*, **15**, 923–935.
20. Gerdes, G.H. (2004) Rift valley fever. *Rev. Sci. Tech.*, **23**, 613–623.
21. Shimshony, A. and Barzilai, R. (1983) Rift Valley fever. *Adv. Vet. Sci. Comp. Med.*, **27**, 347–425.
22. Linthicum, K.J., Davies, F.G., Kairo, A., and Bailey, C.L. (1985) Rift Valley fever virus (family Bunyaviridae, genus Phlebovirus). Isolations from Diptera collected during an inter-epizootic period in Kenya. *J. Hyg. (Lond)*, **95**, 197–209.
23. Zeller, H.G., Fontenille, D., Traore-Lamizana, M., Thiongane, Y., and Digoutte, J.P. (1997) Enzootic activity of Rift Valley fever virus in Senegal. *Am. J. Trop. Med. Hyg.*, **56**, 265–272.
24. Harvell, C.D., Mitchell, C.E., Ward, J.R., Altizer, S., Dobson, A.P., Ostfeld, R.S.,

and Samuel, M.D. (2002) Climate warming and disease risks for terrestrial and marine biota. *Science*, **296**, 2158–2162.

25. Linthicum, K.J., Anyamba, A., Tucker, C.J., Kelley, P.W., Myers, M.F., and Peters, C.J. (1999) Climate and satellite indicators to forecast Rift Valley fever epidemics in Kenya. *Science*, **285**, 397–400.

26. Davies, F.G. (2006) Risk of a rift valley fever epidemic at the haj in Mecca, Saudi Arabia. *Rev. Sci. Tech.*, **25**, 137–147.

27. Saluzzo, J.F. and Smith, J.F. (1990) Use of reassortant viruses to map attenuating and temperature-sensitive mutations of the Rift-Valley fever virus Mp-12 vaccine. *Vaccine*, **8**, 369–375.

28. Turell, M.J., Saluzzo, J.F., Tammariello, R.F., and Smith, J.F. (1990) Generation and transmission of Rift Valley fever viral reassortants by the mosquito Culex pipiens. *J. Gen. Virol.*, **71** (Pt 10), 2307–2312.

29. McIntosh, B.M., Russell, D., dos Santos, I., and Gear, J.H. (1980) Rift Valley fever in humans in South Africa. *S. Afr. Med. J.*, **58**, 803–806.

30. Meegan, J. (1981) Rift valley fever in Egypt: an overview of the epizootics in 1977 and 1978. *Contrib. Epidemiol. Biostat.*, **3**, 100–103.

31. Meegan, J.M., Hoogstraal, H., and Moussa, M.I. (1979) An epizootic of Rift Valley fever in Egypt in 1977. *Vet. Rec.*, **105**, 124–125.

32. Digoutte, J.P., Jouan, A., Leguenno, B., Riou, O., Philippe, B., Meegan, J., Ksiazek, T.G., and Peters, C.J. (1989) Isolation of the rift-valley fever virus by Inoculation into Aedes-pseudoscutellaris cells – comparison with other diagnostic methods. *Res. Virol.*, **140**, 31–41.

33. Peters, C.J., Reynolds, J.A., Slone, T.W., Jones, D.E., and Stephen, E.L. (1986) Prophylaxis of Rift Valley fever with antiviral drugs, immune serum, an interferon inducer, and a macrophage activator. *Antiviral. Res.*, **6**, 285–297.

34. Bengis, R.G., Leighton, F.A., Fischer, J.R., Artois, M., Morner, T., and Tate, C.M. (2004) The role of wildlife in emerging and re-emerging zoonoses. *Rev. Sci. Tech.*, **23**, 497–511.

35. Nguku, P.M., Sharif, S.K., Mutonga, D., Amwayi, S., Omolo, J., Mohammed, O., Farnon, E.C., Gould, L.H., Lederman, E., Rao, C., Sang, R., Schnabel, D., Feikin, D.R., Hightower, A., Njenga, M.K., and Breiman, R.F. (2008) An investigation of a major outbreak of Rift Valley fever in Kenya: 2006–2007. *Am. J. Trop. Med. Hyg.*, **83**, 5–13.

36. Smithburn, K.C. (1949) Rift Valley fever; the neurotropic adaptation of the virus and the experimental use of this modified virus as a vaccine. *Br. J. Exp. Pathol.*, **30**, 1–16.

37. Botros, B., Omar, A., Elian, K., Mohamed, G., Soliman, A., Salib, A., Salman, D., Saad, M., and Earhart, K. (2006) Adverse response of non-indigenous cattle of European breeds to live attenuated Smithburn Rift Valley fever vaccine. *J. Med. Virol.*, **78**, 787–791.

38. Hubbard, K.A., Baskerville, A., and Stephenson, J.R. (1991) Ability of a mutagenized virus variant to protect young lambs from rift-valley fever. *Am. J. Vet. Res.*, **52**, 50–55.

39. Meadors, G.F. III, Gibbs, P.H., and Peters, C.J. (1986) Evaluation of a new Rift Valley fever vaccine: safety and immunogenicity trials. *Vaccine*, **4**, 179–184.

40. Morrill, J.C., Mebus, C.A., and Peters, C.J. (1997) Safety and efficacy of a mutagen-attenuated Rift Valley fever virus vaccine in cattle. *Am. J. Vet. Res.*, **58**, 1104–1109.

41. Morrill, J.C., Mebus, C.A., and Peters, C.J. (1997) Safety of a mutagen-attenuated Rift Valley fever virus vaccine in fetal and neonatal bovids. *Am. J. Vet. Res.*, **58**, 1110–1114.

42. Hunter, P., Erasmus, B.J., and Vorster, J.H. (2002) Teratogenicity of a mutagenised Rift Valley fever virus (MVP 12) in sheep. *Onderstepoort J. Vet. Res.*, **69**, 95–98.

43. Morrill, J.C. and Peters, C.J. (2003) Pathogenicity and neurovirulence of a mutagen-attenuated Rift Valley fever vaccine in rhesus monkeys. *Vaccine*, **21**, 2994–3002.

44. Muller, R., Saluzzo, J.F., Lopez, N., Dreier, T., Turell, M., Smith, J., and

Bouloy, M. (1995) Characterization of clone 13, a naturally attenuated avirulent isolate of Rift Valley fever virus, which is altered in the small segment. *Am. J. Trop. Med. Hyg.*, **53**, 405–411.

45. Hunter, P. and M. Bouloy (2001) Presented at the 5th International Sheep Veterinary Congress, South Africa, January 21–25 2001.

46. Vialat, P., Billecocq, A., Kohl, A., and Bouloy, M. (2000) The S segment of rift valley fever phlebovirus (Bunyaviridae) carries determinants for attenuation and virulence in mice. *J. Virol.*, **74**, 1538–1543.

47. Barnard, B.J. (1979) Rift Valley fever vaccine–antibody and immune response in cattle to a live and an inactivated vaccine. *J. S. Afr. Vet. Assoc.*, **50**, 155–157.

48. Barnard, B.J. and Botha, M.J. (1977) An inactivated rift valley fever vaccine. *J. S. Afr. Vet. Assoc.*, **48**, 45–48.

49. Randall, R., Gibbs, C.J., Aulisio, C.G., Binn, L.N., and Harrison, V.R. Jr. (1962) The development of a formalin-killed Rift Valley fever virus vaccine for use in man. *J. Immunol.*, **89**, 660–671.

50. Randall, R., Harrison, V.R., and Binn, L.N. (1964) Immunization against Rift Valley fever virus – studies on immunogenicity of lyophilized formalin-inactivated vaccine. *J. Immunol.*, **93**, 293–29&.

51. Yedloutschnig, R.J., Dardiri, A.H., Walker, J.S., Peters, C.J., and Eddy, G.A. (1979) Immune response of steers, goats and sheep to inactivated Rift Valley fever vaccine. *Proc. Annu. Meet. U.S. Anim. Health Assoc.*, **1979**, 253–260.

52. Pittman, P.R., Liu, C.T., Cannon, T.L., Makuch, R.S., Mangiafico, J.A., Gibbs, P.H., and Peters, C.J. (1999) Immunogenicity of an inactivated Rift Valley fever vaccine in humans: a 12-year experience. *Vaccine*, **18**, 181–189.

53. Weber, F., Pichlmair, A., Habjan, M., and Unger, H. (2010) Virus-like particles expressing the nucleocapsid gene as an efficient vaccine against Rift Valley fever virus. *Vector-Borne Zoonotic Dis.*, **10**, 701–703.

54. Huggins, J.W. (1989) Prospects for treatment of viral hemorrhagic fevers with ribavirin, a broad-spectrum antiviral drug. *Rev. Infect. Dis.*, **4** (Suppl. 11), S750–S761.

55. Busquets, N., F. Xavier, R. Martin-Folgar, G. Lorenzo, I. Galindo-Cardiel, B.P. del Val, R. Rivas, J. Iglesias, F. Rodriguez, D. Solanes, M. Domingo, and Brun, A. (2006) Experimental infection of young adult European breed sheep with Rift Valley fever virus field isolates. *Vector Borne Zoonotic Dis.*, **10**, 689–696.

56. Easterday, B.C., Murphy, L.C., and Bennett, D.G. (1962) Experimental rift valley fever in calves, goats, and pigs. *Am. J. Vet. Res.*, **23**, 1224–1122–.

57. Baskerville, A., Hubbard, K.A., and Stephenson, J.R. (1992) Comparison of the pathogenicity for pregnant sheep of Rift Valley fever virus and a live attenuated vaccine. *Res. Vet. Sci.*, **52**, 307–311.

58. Harrington, D.G., Lupton, H.W., Crabbs, C.L., Peters, C.J., Reynolds, J.A., and Slone, T.W. (1980) Evaluation of a formalin-Inactivated rift-valley fever vaccine in sheep. *Am. J. Vet. Res.*, **41**, 1559–1564.

59. Geisbert, T.W. and Jahrling, P.B. (2004) Exotic emerging viral diseases: progress and challenges. *Nat. Med.*, **10**, S110–S121.

60. Swanepoel, R. and Blackburn, N.K. (1977) Demonstration of nuclear immunofluorescence in Rift-Valley fever infected-cells. *J. Gen. Virol.*, **34**, 557–561.

61. Lacy, M.D. and Smego, R.A. (1996) Viral hemorrhagic fevers. *Adv. Pediatr. Infect. Dis.*, **12**, 21–53.

62. Le May, N., Dubaele, S., De Santis, L.P., Billecocq, A., Bouloy, M., and Egly, J.M. (2004) TFIIH transcription factor, a target for the Rift Valley hemorrhagic fever virus. *Cell*, **116**, 541–550.

63. Habjan, M., Pichlmair, A., Elliott, R.M., Overby, A.K., Glatter, T., Gstaiger, M., Superti-Furga, G., Unger, H., and Weber, F. (2009) NSs protein of rift valley fever virus induces the specific degradation of the double-stranded RNA-dependent protein kinase. *J. Virol.*, **83**, 4365–4375.

64. Anderson, G.W. Jr., Saluzzo, J.F., Ksiazek, T.G., Smith, J.F., Ennis, W.,

Thureen, D., Peters, C.J., and Digoutte, J.P. (1989) Comparison of in vitro and in vivo systems for propagation of Rift Valley fever virus from clinical specimens. *Res. Virol.*, **140**, 129–138.

65. Niklasson, B., Peters, C.J., Grandien, M., and Wood, O. (1984) Detection of human immunoglobulins G and M antibodies to Rift Valley fever virus by enzyme-linked immunosorbent assay. *J. Clin. Microbiol.*, **19**, 225–229.
66. Paweska, J.T., Burt, F.J., Anthony, F., Smith, S.J., Grobbelaar, A.A., Croft, J.E., Ksiazek, T.G., and Swanepoel, R. (2003) IgG-sandwich and IgM-capture enzyme-linked immunosorbent assay for the detection of antibody to Rift Valley fever virus in domestic ruminants. *J. Virol. Methods*, **113**, 103–112.
67. Paweska, J.T., Burt, F.J., and Swanepoel, R. (2005) Validation of IgG-sandwich and IgM-capture ELISA for the detection of antibody to Rift Valley fever virus in humans. *J. Virol. Methods*, **124**, 173–181.
68. Thiongane, Y., Zeller, H., Lo, M.M., Fati, N.A., Akakpo, J.A., and Gonzalez, J.P. (1994) Decrease in the natural immunity of Rift-Valley fever neutralizing antibodies from domestic ruminants in the senegal river basin 3 years after the 1987 epizootic. *Bull. Soci. Pathol. Exot.*, **87**, 5–6.
69. Sall, A.A., Thonnon, J., Sene, O.K., Fall, A., Ndiaye, M., Baudez, B., Mathiot, C., and Bouloy, M. (2001) Single-tube and nested reverse transcriptase-polymerase chain reaction for detection of Rift Valley fever virus in human and animal sera. *J. Virol. Methods*, **91**, 85–92.
70. Paweska, J.T., van Vuren, P.J., Kemp, A., Buss, P., Bengis, R.G., Gakuya, F., Breirnan, R.F., Njenga, M.K., and Swanepoel, R. (2008) Recombinant nucleocapsid-based ELISA for detection of IgG antibody to Rift Valley fever virus in African buffalo. *Vet. Microbiol.*, **127**, 21–28.
71. Paweska, J.T., van Vuren, P.J., and Swanepoel, R. (2007) Validation of an indirect ELISA based on a recombinant nucleocapsid protein of Rift Valley fever virus for the detection of IgG antibody in humans. *J. Virol. Methods*, **146**, 119–124.
72. Sobarzo, A., Paweska, J.T., Herrmann, S., Amir, T., Marks, R.S., and Lobel, L. (2007) Optical fiber immunosensor for the detection of IgG antibody to Rift Valley fever virus in humans. *J. Virol. Methods*, **146**, 327–334.
73. Drosten, C., Gottig, S., Schilling, S., Asper, M., Panning, M., Schmitz, H., and Gunther, S. (2002) Rapid detection and quantification of RNA of Ebola and Marburg viruses, Lassa virus, Crimean-Congo hemorrhagic fever virus, Rift Valley fever virus, dengue virus, and yellow fever virus by real-time reverse transcription-PCR. *J. Clin. Microbiol.*, **40**, 2323–2330.
74. Garcia, S., Crance, J.M., Billecocq, A., Peinnequin, A., Jouan, A., Bouloy, M., and Garin, D. (2001) Quantitative real-time PCR detection of Rift Valley fever virus and its application to evaluation of antiviral compounds. *J. Clin. Microbiol.*, **39**, 4456–4461.
75. Weidmann, M., Sanchez-Seco, M.P., Sall, A.A., Ly, P.O., Thiongane, Y., Lo, M.M., Schley, H., and Hufert, F.T. (2008) Rapid detection of important human pathogenic Phleboviruses. *J. Clin. Virol.*, **41**, 138–142.
76. Peyrefitte, C.N., Boubis, L., Coudrier, D., Bouloy, M., Grandadam, M., Tolou, H.J., and Plumet, S. (2008) Real-time reverse-transcription loop-mediated isothermal amplification for rapid detection of rift valley Fever virus. *J. Clin. Microbiol.*, **46**, 3653–3659.
77. Njenga, M.K., Paweska, J., Wanjala, R., Rao, C.Y., Weiner, M., Omballa, V., Luman, E.T., Mutonga, D., Sharif, S., Panning, M., Drosten, C., Feikin, D.R., and Breiman, R.F. (2009) Using a field quantitative real-time PCR test to rapidly identify highly viremic rift valley fever cases. *J. Clin. Microbiol.*, **47**, 1166–1171.

14.3 Hantaviruses: the Most Widely Distributed Zoonotic Viruses on Earth

Jonas Klingström

14.3.1 Introduction

Hantavirus is one of five genera within the *Bunyaviridae* family [1]. Infection of humans with *Orthobunyavirus, Phlebovirus, Nairovirus,* and *Tospovirus* occur through direct transmission by arthropods, while *Hantavirus* infection of humans occurs via inhalation of contaminated excreta from hantavirus-infected hosts [2]. Infection of humans can lead to hemorrhagic fever with renal syndrome (HFRS) in Eurasia, or hantavirus cardiopulmonary syndrome (HCPS) in the Americas, depending on the hantavirus species involved [3–5]. HFRS/HCPS causing hantaviruses are all carried by rodents, but there are also several other rodent-borne hantaviruses that do not cause disease in humans. Rodents are ubiquitous, and this reflects the fact that hantaviruses are the most widely distributed zoonotic viruses on earth [6]. Furthermore, recently hantaviruses carried by other natural hosts than rodents have been described. It is unclear if these viruses can cause HFRS, HCPS, or perhaps other diseases.

HFRS includes diseases formerly known as *Korean hemorrhagic fever, epidemic hemorrhagic fever,* and nephropathia epidemica (NE) [7], caused by Hantaan virus (HTNV), Dobrava virus (DOBV), Seoul virus (SEOUV), and Puumala virus (PUUV). There are Chinese medical records from the tenth century describing a disease resembling HFRS, and NE was described in Sweden 1934 [8, 9]. An important outbreak of HFRS occurred during the Korean conflict 1951–1954, with 3200 clinically diagnosed cases among the United Nations forces in Korea [10], and mortality rates of 5–10% [11]. This outbreak was the starting point for the research on hantaviruses and the diseases they cause.

HCPS is caused by Andes virus (ANDV), sin nombre virus (SNV), and related viruses [3, 4]. The mortality rate in HCPS is up to 50%. HCPS was discovered in 1993 when a cluster of fatal cases of an acute respiratory distress syndrome was detected in the southwest of the United States. Within two months the causative agent, later termed SNV, was isolated and the host, the deer mouse (*Peromyscus maniculatus*), was identified [12, 13]. Soon thereafter cases of HCPS were observed in South America, and ANDV and related viruses were shown to be the etiological agents [3, 14].

BSL3 and BSL4 Agents: Epidemiology, Microbiology, and Practical Guidelines, First Edition.
Edited by Mandy C. Elschner, Sally J. Cutler, Manfred Weidmann, and Patrick Butaye.

Hantaviruses that cause HFRS and HCPS are classified into biosafety level 3 (BSL3), except for PUUV, which is classified into BSL2. Experiments involving laboratory animals are classified into BSL3 for all hantaviruses, except for ANDV and SNV, which are classified into BSL4.

14.3.2
Characteristics of the Agent

Hantavirus particles, are 80–120 nm in diameter and have a lipid bilayer envelope with two surface glycoproteins, Gn (68–76 kDa) and Gc (52–58 kDa). Inside the virus there are three single-stranded RNA segments of negative polarity covered by nucleocapsid (N) protein (50–54 kDa). The RNA-dependent RNA polymerase (RdRp; approximately 240 kDa), is also associated with the genomic segments.

The three negative stranded RNA segments are of different sizes: the small (S), the medium (M), and the large (L) segment have a length of approximately 1.8, 3.7, and 6.5 kb. They encode the N protein, the Gn and Gc proteins, and the RdRp protein respectively. It has been suggested that certain, but not all, hantaviruses also encode a nonstructural protein, NSs, in their S segment [15].

Noncoding regions (NCRs) flank the genome encoding the viral proteins. The 14, 17, and 15 terminal nucleotides of the S-, M-, and L-segment are identical for all hantaviruses [16] and are distinct from those of the terminal nucleotides of the viruses of the other genera of *Bunyaviridae* [17, 18]. The panhandle structure thought to be formed by those structures is believed to be involved in the regulation of replication and genomic RNA structure [19–21]. The 3' and 5' NCR length varies from 40–50 nucleotides (nt) for the 5' NCRs on S, M, and L, to 300–700 nt for the 3' NCR of S [15].

The hantavirus N protein, approximately 430 amino acids long, is expressed at high levels, and inclusion bodies consisting of large granular to filamentous aggregates of N proteins are casually formed in infected cells. The N protein forms homodimers and homotrimers [22, 23]. Trimers recognize and bind to the panhandle structure of viral genomic RNA [21]. The N protein interacts with several cellular proteins, like actin [24], Daxx [25], and SUMO-1 [26, 27], but it is currently not known if this has an effect on viral replication or on pathogenesis, or if this is a unique interaction with human proteins. The N protein has RNA chaperone activity [28, 29], and binds with high affinity to the 5'cap of cellular mRNAs, thereby protecting the 5'cap from being degraded [30]. The structure of the N-terminal part of N (amino acids 1–74) has been shown to contain a coiled coil domain consisting of two long helices that intertwine [31, 32].

The type 1 transmembrane proteins Gn and Gc (earlier designated G1 and G2) are synthesized as a precursor glycoprotein of 1132–1148 amino acids [33, 34], that is subsequently cleaved immediately after the conserved WAASA amino acid motif [35]. Gn and Gc are both glycosylated. The two glycoproteins are the most variable among hantavirus proteins, but they have a high content of conserved cysteine residues, indicating similar tertiary structures for Gn and Gc from different hantavirus species [36]. The Gn cytoplasmic tail contains a functional immunoreceptor tyrosine activation motif (ITAM), and is a target for ubiquitation and subsequent degradation [37, 38].

The RdRp (also called the *L protein* because it is encoded by the L segment), 2151–2156 amino acids long, is the most conserved protein among the different hantaviruses [15]. It has five conserved regions, also found in other viral RNA-dependent polymerases [39, 40]. Exactly how hantavirus RdRp functions is not known, but it harbors replicase, transcriptase, endonuclease, and possibly RNA helicase activities.

Hantavirus replication is non-cytopathogenic [41–45]. There are several cellular receptors reported to be involved in binding of hantaviruses to cells. The first to be described was the $\beta 3$ integrin for pathogenic hantaviruses, and the $\beta 1$ integrin for nonpathogenic hantaviruses [46, 47]. A 30-kDa Vero E6 cell membrane protein has been suggested to function as a receptor [48]. A decay-accelerating factor (DAF) has been described as a critical co-factor for entry [49]. The receptor for the globular head domain of complement C1q (gC1qR/p32) was recently suggested to be of importance during infection with HTNV [50]. It remains to be shown if hantaviruses can use all these receptors, or if certain combinations of receptors are needed for hantavirus entry, and if hantaviruses use the same receptor(s) in humans and in their natural hosts. After attachment to the receptor(s), receptor-mediated endocytosis takes place via the chlatrin-coated pit pathway [51]. Hantaviruses are then targeted to early endosomes, which progress to lysosomes. Here, the nucleocapsids are released into the cytoplasm, and viral replication can start. As the hantavirus genome is of negative polarity the virus must carry its own RdRp with it. The first step of the replication is the production of viral full-length positive stranded RNA, which can also function as viral mRNA. Positive stranded full-length viral RNA is then transcribed into negative stranded full-length genomic RNA that can be incorporated into newly formed virus particles. The transcription of hantavirus RNA takes place via a prime and realign model [52], and the process also involves cap snatching, which involves the acquisition of capped 5' oligonucleotides from cellular mRNA as originally described for influenza A virus [53]. Unexpectedly, it was recently shown that the hantavirus N protein can replace the entire cellular eIF4F complex [54], in a viral mRNA specific manner, thus mediating efficient translation of viral mRNA in competition with cellular mRNAs. Translation of the S and L mRNAs takes place on free ribosomes, while translation of the M mRNA takes place on membrane-bound ribosomes. Nucleocapsids are formed, and complex with RdRp. These complexes then interact with Gn and Gc in the Golgi compartment to form new virus particles. Maturation and budding takes place in the Golgi complex and mature viruses are then transported and released from the cell via vesicular secretory pathways, at least for the Old World hantaviruses [55, 56]. The plasma membrane has been suggested as the budding site for New World hantaviruses [24, 57, 58].

14.3.3 Epidemiology

Of all hemorrhagic fever viruses, only hantaviruses have a worldwide distribution. Confirmed clinical cases are documented from Europe, Asia, and the Americas. As the reservoirs for HFRS/HCPS-causing hantaviruses are rodents, the distribution of the specific hantaviruses are dependent on the geographical distribution of the corresponding reservoirs. However, in certain areas potential rodent reservoir species are not infected by hantavirus, showing that the presence of a reservoir host is not always connected to the presence of the corresponding hantavirus. This is for instance the case in southern Sweden, where bank voles for unknown reasons are not carrying PUUV, and where PUUV does not circulate in nature [59].

HTNV, SEOV, PUUV, and DOBV cause HFRS in Eurasia, while ANDV, SNV, and ANDV/SNV-like hantaviruses cause HCPS in the Americas [3–5]. HTNV is prevalent in the Far East, SEOV worldwide, PUUV in Russia and western Europe, DOBV in central and eastern Europe, ANDV in Central and South America, and SNV in North America.

The first hantavirus to be isolated was Thottapalayam virus (TPMV), isolated from a musk shrew in India [60]. This was later followed by the discovery of several rodent-borne hantaviruses; the HFRS-causing hantaviruses HTNV in the striped field mouse [61], SEOV in the black and the brown rat [62], DOBV in the yellow-necked field mouse [63], and PUUV in the bank vole [64]. Several nonpathogenic hantaviruses have also been discovered, for example, Prospect Hill virus (PHV), isolated from meadow voles [65]. The first recognized outbreak of HCPS lead to the discoveries of SNV in the deer mouse [12], and of ANDV in long-tailed pygmy rice rats [14]. Infection of nonnatural hosts occurs frequently [66], but these infections are most likely transient. For a long time it was believed that hantaviruses were strictly rodent-borne and that observations in other animals were spill-over infections. As of today, the natural hosts for all known HFRS/HPS causing hantaviruses are rodents. However, the first hantavirus to be isolated, TPMV, is shrew-borne, and recent findings of several other hantaviruses in shrews [67–72] and moles [73–75] show that there is an expanding list of animals that can act as natural hosts for hantaviruses. If these nonrodent borne hantaviruses can cause disease in humans is currently unknown.

Humans are normally infected by aerosolized hantavirus-contaminated rodent excreta [4, 5], and with the exception of ANDV, human to human transmission of hantaviruses do not occur. Hantaviruses are highly stable in the environment [76, 77]. ANDV is the only hantavirus so far shown to be able to spread between humans [78–80]. ANDV antigen has been detected in the secretory cells of the salivary glands of HCPS patients [81], and compared to other contacts, the risk of being infected with ANDV is higher among people having sex and/or people who are involved in deep kissing with the index case [82]. Together, this suggests that ANDV needs close person to person contact to be transmitted between humans. ANDV is clearly more resistant to the antiviral effects of human saliva than HTNV and PUUV [83], suggesting that saliva might be the route of transmission between humans. However, although human to human transmission of ANDV has been documented, almost all of the ANDV infections are caused by rodent to human transmission. Transmission of hantavirus via rodent bite has been described [84], but seems to be extremely uncommon.

Hantavirus infections are more common in rural than urban areas, however, as SEOV is carried by rats, SEOV infection also occur in urban areas. There is an increased risk to be infected with hantaviruses for individuals involved in heavy farm work, military exercises, outdoor sleeping, and for individuals living in homes infested with rodents during for instance cold weather, or during cleaning of summer cottages and other buildings where infected rodents have been living.

Currently it is not exactly known why some infected individuals develop asymptomatic infection while others develop severe disease and even succumb to the disease. There are some genetic factors that seem to play a role for the risk of developing severe forms of HFRS. Certain human leukocyte antigen (HLA) haplotypes have been associated to outcome of PUUV infection [85–87], and it was recently suggested that the distributions of genotype and allele frequencies of human platelet alloantigen (HPA)-3 differ between HFRS patients and controls [88].

It is believed that patients recovering from HFRS or HCPS mounts immune responses that protects against subsequent infections, and there are no records of any individual that has been diagnosed twice with HFRS or HCPS.

14.3.4 Clinical Findings

Hantavirus infection can result in different clinical syndromes depending on the specific hantavirus types, but generally hantaviruses cause HFRS in Eurasia and HCPS in the Americas. Infection can be asymptomatic, cause disease, or even be lethal. Mortality is clearly higher for HCPS (up to 50% lethality) than for HFRS (up to 10% lethality), and within HFRS the risk of mortal outcome is clearly higher after infection with HTNV or DOBV, than after infection with PUUV (0.1–1.0% lethality).

Common for both diseases is the increased capillary leakage, leading to hypotension, vasodilation, and hemoconcentration in patients. Furthermore, acute thrombocytopenia, CD8+ T cell activation, and an increased leukocyte count with an increased number of immature leukocytes (left shift) is often found [4, 5, 89, 90]. The differences between HFRS and HCPS are explained by the different vascular beads that are affected: in HCPS mainly pulmonary capillaries, and in HFRS mainly renal medulla capillaries. However, although renal symptoms are a hallmark for HFRS, and cardiopulmonary symptoms for HCPS, there are HFRS patients with cardiopulmonary symptoms, and HCPS patients with renal symptoms [91–95].

14.3.4.1 HFRS

The clinical presentations of HFRS range from a mild febrile disease to fulminant hemorrhagic shock and death [89]. The incubation time is normally two to three weeks for HFRS, but can be as long as six weeks [95]. HFRS starts with an abrupt onset of flu-like symptoms including fever, myalgia, nausea, vomiting, and headache. Flushing of the face, conjuntival injections, and blurred vision is often observed. There are often, but not always, five distinct disease stages in HFRS: febrile phase (three to five days), hypotensive phase (a few hours up to two days), oliguric phase (a few days up to two weeks), polyuric phase, and convalescent phase [4, 5, 11, 96].

Early during the disease, the platelet count drops abruptly, reaching the lowest count at the end of the febrile phase. The platelet function can be disturbed; this in turn can cause petechia, ecchymoses, conjunctival suffusion, hematemesis, epistaxis, hematuria, melena, and fatal intracranial hemorrhage. Disseminated intravascular coagulation is often observed in HFRS patients. During the hypotensive phase patients show falling blood pressure due to the vascular leakage which can result in fatal shock syndrome. In the oliguric phase the patients often have severe abdominal or back pain, and the kidney function is decreased. The kidney is the main target, and HFRS patients often show signs of interstitial infiltrates with immune cells and interstitial hemorrhage in renal tissue. Acute tubulointerstitial nephritis is the most common histopathological lesion. The interstitium is broadened by edema, and intertubular capillaries are congested, indicating capillary damage. Glomerular pathology is occasionally observed, for example, hypercellularity and expansion of the mesangium, and this might explain the proteinuria observed in patients [97]. Urinary sediment contains hantavirus antigen positive tubular cells with enlarged nucleoli [98, 99]. Patients with severe symptoms during the oliguric phase might need to be treated by hemodialysis. When the kidney starts to function again, the patient goes into the polyuric phase, and then rapidly into the convalescent phase, which can last for several months.

In NE the five phases are not easy to distinguish, and only one-third of patients show clear signs of hemorrhage. Instead, pulmonary involvement is observed in 10–20% of NE patients [94, 95].

14.3.4.2 **HCPS**

The clinical symptoms during HCPS vary from mild hypoxemia to respiratory failure and cardiogenic shock [90]. The initial phase of HCPS lasts for three to six days and includes fever, chills, myalgia, nausea, headache, and gastroinstestinal symptoms such as vomiting, abdominal pain, and diarrhea. The next phase, the cardiopulmonary phase, often has an abrupt start with progressive cough, tachypnea, tachycardia, and hypotension. Pulmonary edema and respiratory failure can develop rapidly. Capillary leak syndrome in the lungs causes adult respiratory distress syndrome. The cardiopulmonary phase last for two to four days and can be complicated by cardiogenic shock [100, 101], lactic acidosis, and massive hemoconcentration. Patients that survive this phase enter the polyuric phase and then the convalescent phase. Convalescence proceeds slowly and it can take months before patients fully recover.

14.3.5 Pathogenesis

HFRS and HCPS are prime examples of acute viral infections and normally patients that survive the acute infection recover completely. Viremia in humans is transient and seems to be cleared within 14 days after the onset of disease; however complete recovery from disease can take months. The main characteristic for HFRS and HCPS, like for all hemorrhagic fevers, are vascular dysfunction, characterized by vasodilation, and increased permeability. The degree of increased permeability of infected endothelium is believed to be the main factor determining the course and the severity of HFRS/HCPS. It is poorly understood how hantaviruses induce the capillary leakage. Infected endothelial cells shows no histological signs of damage and no visible cytopathogenic effect [41–43, 102], it therefore seems likely that the pathogenesis is indirectly caused by the infection, and it has been suggested that HFRS/HCPS may be immunomediated diseases, caused by an over-active immune response against infected cells [4, 96, 100, 103]. It remains to be shown if this is the sole explanation to these diseases, and if so how and why this over-active immune response develops in humans and not the natural hosts.

14.3.6 Diagnosis

HFRS/HCPS have a high impact on the public health systems. Hantavirus infections have high case-fatality rates, and treatment of acutely infected patients causes considerable burden on health systems. There are also indications that hantavirus infections can have long-term consequences [104–108], further emphasizing the need for correct diagnosis and follow-up of patients.

During the first well-known outbreak of HFRS, during the Korean conflict, diagnoses were made solely based on symptoms [10]. Kidney biopsies were used for diagnostic purposes; however, the findings were not specific, and later studies showed that viral markers are only rarely found in the affected kidneys at the time when biopsies are taken [109]. Later serum/plasma samples were used for diagnoses based on serology, and today hantavirus N protein-specific IgM capture enzyme-linked immunosorbent assay (ELISA) is often used. Viral RNA has been detected in peripheral blood mononuclear cells (PBMCs), blood clots, serum/plasma, urinary sediments, and saliva. Isolation of virus *in vitro* from patient samples is extremely cumbersome and not used for diagnostic purposes.

14.3.6.1 Serology

In almost all HFRS/HCPS-patients, specific IgM and/or IgG antibody responses are detected at hospitalization, and therefore detection of hantavirus-specific IgM responses, or increased IgG responses, are sensitive markers for infection [109–113].

With the discovery of hantaviruses as the etiological agent that caused HFRS [61] came the first diagnoses based on specific antibody responses. However, diagnosis of disease was only feasible by detection of a greater than fourfold increase in patient serum IgG levels by indirect immunefluorescence assay (IFA) based on fixed lung section of hantavirus-infected rodents [61, 64]. Although this method is of high specificity, there is an obvious risk of false negative results, as the method relays on paired samples; if the first sample is taken to late during the course of infection, a fourfold increase in titers might not develop at the time of the second sample. When hantaviruses were adapted to grow in cell lines [18, 114], diagnosis became more available, although still performed on acetone-fixed hantavirus-infected cells. The next step was the single-sample assay for low avidity IgG antibodies [115], which is a characteristic for recent infections. IFA based on IgM can also be used for diagnosis of HFRS/HCPS. The first strip immunoblot assay based on recombinant SNV and SEOV N protein was introduced next [116]. Today, an IgM-capture ELISA, initially based on antigens from infected Vero E6 cells [111], but soon replaced with recombinant N protein [117–121] is frequently in use. The latest antibody-based invention in diagnoses of hantavirus infections is a rapid point of care (POC) IgM antibody test [122, 123].

In order to determine which hantavirus species previously infected a nonviremic person, a focus reduction neutralization test (FRNT) can be used [124–126]. This is not performed during routine diagnosis, but can be of interest, for example, in studies addressing identification of different hantavirus species in areas where more than one hantavirus circulate. A fourfold difference in neutralization titer against different hantaviruses is usually used to identify the specific hantavirus which previously infected the individual.

14.3.6.2 **Virus Detection**

Detection of viral antigen in neutrophils and peripheral blood mononuclear cells with hantavirus-specific monoclonal or polyclonal antibodies has been used to diagnose hantavirus infections [102] and can also be used for post-mortem diagnoses of fatal cases [127].

Hantavirus RNA can generally by detected in hantavirus-infected patients, and nested reversed transcription (RT)-PCR can be used to detect hantavirus RNA in samples [12, 128–133]. Because of the highly sensitive and specific antibody-based diagnostic tools available, detection of viral RNA is not a standard method for diagnoses of HFRS or HCPS. Furthermore, although hantaviruses are rather conserved, they, as all RNA viruses, have a high mutation rate [134], and there is a large variation in sequences between different hantaviruses. Therefore protocols for the detection of a specific hantavirus strain rarely can detect other hantaviruses, and sometimes not even the same hantavirus species from another geographical region. However, recent discoveries indicate that PCR-based diagnoses might be feasible, at least during local outbreaks where the exact sequence of the circulating hantavirus is known [135]. For scientific purposes, quantitative RT-PCR on patient samples can be of interest, as this will give a chance to study possible associations between levels of viral replication/viremia and other markers of disease in patients. Recent findings indicate that there is a correlation between virus RNA load, the outcome of HCPS [136], and the severity of HFRS [137]; therefore determination of viral RNA concentrations might be of beneficial for the patients, as it could point toward an oncoming severe disease.

References

1. Elliott, R.M., Bouloy, M., Calisher, C.H., Goldbach, R., Moyer, J.T., Nichol, S.T., Petterson, R., Plyusnin, A., and Schmaljohn, C. (2000) in *Virus Taxonomy: the Classification and Nomenclature of Viruses. The Seventh Report of the International Committee on Taxonomy of Viruses* (eds M.H.V. van Regenmortl, C.M. Fauquet, D.H.L. Bishop, E.B. Castens, M.K. Estes, S.M. Lemon, J. Maniloff, M.A. Mayo, D.J. McGeoch, C.R. Pringle, and R.B. Wickner), Academic Press, San Diego, pp. 599–621.
2. Lee, H.W. and Johnson, K.M. (1982) Laboratory-acquired infections with Hantaan virus, the etiological agent of Korean hemorrhagic fever. *J. Infect. Dis.*, **146**, 645–651.
3. Schmaljohn, C.S. and Hjelle, B. (1997) Hantaviruses: a global disease problem. *Emerg. Inf. Dis.*, **3**, 95–104.
4. Schönrich, G., Rang, A., Lütteke, N., Raftery, M.J., Charbonnel, N., and Ulrich, R.G. (2008) Hantavirus-induced immunity in rodent reservoirs and humans. *Immunol. Rev.*, **225**, 163–189.
5. Vapalahti, O., Mustonen, J., Lundkvist, Å., Henttonen, H., Plyusnin, A., and Vaheri, A. (2003) Hantavirus infections in Europe. *Lancet Infect. Dis.*, **3**, 653–661.
6. Johnson, K.M. (2001) Hantaviruses: history and overview. *Curr. Top. Microbiol. Immunol.*, **256**, 1–14.
7. Gajdusek, G.C. (1962) Viral hemorrhagic fevers – special references to hemorrhagic fever with renal syndrome (epidemic hemorrhagiv fever). *J. Pediatr.*, **60**, 841–857.
8. Myrman, G. (1934) En njursjukdom med egenartad symptombild [in Swedish]. *Nord. Med. Tidskr.*, **7**, 793–794.
9. Zetterholm, S.G. (1934) Akuta nefriter simulerande akuta bukfall [in Swedish]. *Sven. Läkartidn.*, **31**, 425–429.
10. Earle, D.P. (1954) Symposium on epidemic hemorrhagic fever. *Am. J. Med.*, **16**, 617–709.
11. Sheedy, J.A., Froeb, H.F., Batson, H.A., Conley, C.C., Murphy, J.P., Hunter, R.B., Cugell, D.W., Giles, R.B., Bershadsky, S.C., Vester, J.W., and Yoe, R.H. (1954) The clinical course of epidemic hemorrhagic fever. *Am. J. Med.*, **16**, 619–628.
12. Nichol, S.T., Spiropoulou, C.F., Morzunov, S., Rollin, P.E., Ksiazek, T.G., Feldmann, H., Sanchez, A., Childs, J., Zaki, S., and Peters, C.J. (1993) Genetic identification of hantavirus associated with an outbreak of acute respiratory illness. *Science*, **262**, 914–917.
13. Elliott, L.H., Ksiazek, T.G., Rollin, P.E., Spiropoulou, C.F., Morzunov, S., Monroe, M., Goldsmith, C.S., Humphrey, C.D., Zaki, S.R., Krebs, J.W., Maupin, G., Gage, K., Childs, J.E., Nichol, S.T., and Peters, C.J. (1994) Isolation of the causative agent of hantavirus pulmonary syndrome. *Am. J. Trop. Med. Hyg.*, **51**, 102–108.
14. Lopez, N., Padula, P., Rossi, C., Lazaro, M.E., and Franze-Fernandez, M.T. (1996) Genetic identification of a new hantavirus causing severe pulmonary syndrome in Argentina. *Virology*, **220**, 223–226.
15. Plyusnin, A. (2002) Genetics of hantavirus: implications to taxonomy. *Arch. Virol.*, **147**, 665–682.
16. Plyusnin, A., Vapalahti, O., and Vaheri, A. (1996) Hantaviruses: genome structure, expression and evolution. *J. Gen. Virol.*, **77**, 2677–2687.
17. Schmaljohn, C.S. and Dalrymple, J.M. (1983) Analysis of Hantaan virus RNA: evidence for a new genus of Bunyaviridae. *Virology*, **131**, 482–491.
18. Schmaljohn, C.S., Hasty, S.E., Dalrymple, J.M., LeDuc, J.W., Lee, H.W., von Bonsdorff, C.H.,

Brummer-Korvenkontio, M., Vaheri, A., Tsai, T.F., Regnerv, H.L., Goldgaber, D., and Lee, P.W. (1985) Antigenic and genetic properties of viruses linked to hemorrhagic fever with renal syndrome. *Science*, **227**, 1041–1044.

19. Raju, R. and Kolakofsky, D. (1989) The ends of La Crosse virus genome and antigenomic RNAs within nucleocapsids are based paired. *J. Virol.*, **63**, 122–128.
20. Flick, R., Elgh, F., and Pettersson, R.F. (2002) Mutational analysis of the Uukuniemi virus (*Bunuyaviridae* family) promotor reveals two elements of functional importance. *J. Virol.*, **76**, 10849–10860.
21. Mir, M.A. and Panganiban, A.T. (2004) Trimeric hantavirus nucleocapsid protein binds specifically to the viral RNA panhandle. *J. Virol.*, **78**, 8281–8288.
22. Alfadhli, A., Love, Z., Arvidson, B., Seds, J., Willey, J., and Barklis, E. (2001) Hantavirus nucleocapsid protein oligomerization. *J. Virol.*, **75**, 2019–2023.
23. Kaukinen, P., Koistinen, V., Vapalahti, O., Vaheri, A., and Plyusnin, A. (2001) Interaction between molecules of hantavirus nucleocapsid protein. *J. Gen. Virol.*, **82**, 1845–1853.
24. Ravkov, E.V., Nichol, S.T., Peters, C.J., and Compans, R.W. (1998) Role of actin microfilaments in Black Creek Canal virus morphogenesis. *J. Virol.*, **72**, 2865–2870.
25. Li, X.D., Mäkelä, T.P., Guo, D., Soliymani, R., Koistinen, V., Vapalahti, O., Vaheri, A., and Lankinen, H. (2002) Hantavirus nucleocapsid protein interacts with the Fas-mediated apoptosis enhancer Daxx. *J. Gen. Virol.*, **83**, 759–766.
26. Kaukinen, P., Vaheri, A., and Plyusnin, A. (2003) Non-covalent interaction between nucleocapsid protein of Tula hantavirus and small ubiquitin-related modifier-1, SUMO-1. *Virus Res.*, **92**, 37–45.
27. Lee, B.H., Yoshimatsu, K., Maede, A., Ochial, K., Morimatsu, M., Araki, K., Ogino, M., Morikawa, S., and Arikawa, J. (2003) Association of the nucleocapsid protein of the Seoul and Hantaan hantavirus with small ubiquitin-like modifier-1 related molecules. *Virus Res.*, **98**, 83–91.
28. Mir, M.A. and Panganiban, A.T. (2006) Characterization of the RNA chaperone activity of hantavirus nucleocapsid protein. *J. Virol.*, **80**, 6276–6285.
29. Mir, M.A. and Panganiban, A.T. (2006) The bunyavirus nucleocapsid protein is an RNA chaperone: possible roles in viral RNA panhandle formation and genome replication. *RNA*, **12**, 272–282.
30. Mir, M.A., Duran, W.A., Hjelle, B.L., Ye, C., and Panganiban, A.T. (2008) Storage of cellular 5' mRNA caps in P bodies for viral cap-snatching. *Proc. Natl. Acad. Sci. U.S.A.*, **105**, 19294–19299.
31. Wang, Y., Boudreaux, D.M., Estrada, D.F., Egan, C.W., St Jeor, S.C., and De Guzman, R.N. (2008) NMR structure of the N-terminal coiled coil domain of the Andes hantavirus nucleocapsid protein. *J. Biol. Chem.*, **283**, 28297–28304.
32. Alfadhli, A., Steel, E., Finlay, L., Bächinger, H.P., and Barklis, E. (2002) Hantavirus nucleocapsid protein Coiled-coil domains. *J. Biol. Chem.*, **277**, 27103–27108.
33. Schmaljohn, C.S., Schmaljohn, Al., and Dalrymple, J.M. (1987) Hantaan virus M RNA: coding strategy, nucleotide sequence, and gene order. *Virology*, **157**, 31–39.
34. Elliott, R.M. (1990) Molecular biology of *Bunyaviridae*. *J. Gen. Virol.*, **71**, 501–522.
35. Löber, C., Anheier, B., Lindow, S., Klenk, H.D., and Feldmann, H. (2001) The Hantaan virus glycoprotein precursor is cleaved at the conserved pentapeptid WAASA. *Virology*, **289**, 224–229.
36. Giebel, L.B., Stohwasser, R., Zoller, L., Bautz, E.K., and Darai, G. (1989) Determination of the coding capacity of the M genome segment of nephropathia epidemica virus strain Hällnäs B1 by molecular cloning and nucleotide sequence analysis. *Virology*, **172**, 498–505.

37. Geimonen, E., Fernandez, I., Gavrilovskaya, I.N., and Mackow, E.R. (2003) Tyrosine residues direct the ubiquitination and degradation of the NY-1 hantavirus G1 cytoplasmic tail. *J. Virol.*, **77**, 10760–10868.
38. Geimonen, E., LaMonica, R., Springer, K., Farooqui, Y., Gavrilovskaya, I.N., and Mackow, E.R. (2003) Hantavirus pulmonary syndrome-associated hantaviruses contain conserved and functional ITAM signalling elements. *J. Virol.*, **77**, 1638–1643.
39. Poch, O., Sauvaget, I., Delarue, M., and Tordo, N. (1989) Identification of four conserved motifs among the RNA-dependent polymerase encoding elements. *EMBO J.*, **8**, 3867–3874.
40. Müller, R., Poch, O., Delarue, M., Bishop, D.H., and Bouloy, M. (1994) Rift Valley fever virus L segment: correction of the sequence and possible functional role of newly identified regions conserved in RNA-dependent polymerases. *J. Gen. Virol.*, **75**, 1345–1352.
41. Yanagihara, R. and Silverman, D.J. (1990) Experimental infection of human vascular endothelial cells by pathogenic and non-pathogenic hantaviruses. *Arch. Virol.*, **111**, 281–286.
42. Pensiero, M.N., Sharefkin, J.B., Dieffenbach, C.W., and Hay, J. (1992) Hantaan virus infection of human endothelial cells. *J. Virol.*, **66**, 5929–5936.
43. Temonen, M., Vapalahti, O., Holthöfer, H., Brummer-Korvenkontio, M., Vaheri, A., and Lankinen, H. (1993) Susceptibility of human cells to Puumala virus infection. *J. Gen. Virol.*, **74**, 515–518.
44. Raftery, M.J., Kraus, A.A., Ulrich, R., Krüger, D.H., and Schönrich, G. (2002) Hantavirus infection of dendritic cells. *J. Virol.*, **76**, 10724–10733.
45. Hardestam, J., Klingström, J., Mattsson, K., and Lundkvist, Å. (2005) HFRS causing hantaviruses do not induce apoptosis in confluent Vero E6 and A-549 cells. *J. Med. Virol.*, **76**, 234–240.
46. Gavrilovskaya, I.N., Shepley, M., Shaw, R., Ginsberg, M.H., and Mackow E.R. (1998) b3 integrins mediate the cellular entry of hantaviruses that cause respiratory failure. *Proc. Natl. Acad. Sci. U.S.A.*, **95**, 7074–7079.
47. Gavrilovskaya, I.N., Brown, E.J., Ginsberg, M.H., and Mackow, E.R. (1999) Cellular entry of hantaviruses which cause hemorrhagic fever with renal syndrome is mediated by beta3 integrins. *J. Virol.*, **73**, 3951–3959.
48. Kim, T.Y., Choi, Y., Cheong, Y., H.S., and Choe, J. (2002) Identification of a cell surface 30 kDa protein as a candidate receptor for Hantaan virus. *J. Gen. Virol.*, **83**, 767–773.
49. Krautkrämer, E. and Zeier, M. (2008) Hantavirus causing hemorrhagic fever with renal syndrome enters from the apical surface and requires decay-accelerating factor (DAF/CD55). *J. Virol.*, **82**, 4257–4264.
50. Choi, Y., Kwon, Y.C., Kim, S.I., Park, J.M., Lee, K.H., and Ahn, B.Y. (2008) A hantavirus causing hemorrhagic fever with renal syndrome requires gC1qR/p32 for efficient cell binding and infection. *Virology*, **381**, 178–183.
51. Jin, M., Park, J., Lee, S., Park, B., Shin, J., Song, K.J., Ahn, T.I., Hwang, S.Y., Ahn, B.Y., and Ahn, K. (2002) Hantaan virus enters cells by clathrin-dependent receptor-mediated endocytosis. *Virology*, **294**, 60–69.
52. Garzin, D., Lezzi, M., Dobbs, M., Elliot, R.M., Scmaljohn, C., Kang, C.Y., and Kolakofsky, D. (1995) The 5' ends of Hantaan virus (*Bunyaviridae*) RNAs suggest a prime-and-realign mechanisms for the initiation of RNA synthesis. *J. Virol.*, **69**, 5754–5762.
53. Krug, R.M. (1981) Priming of influenza viral RNA transcription by capped heterologous RNAs. *Curr. Top. Microbiol. Immunol.*, **93**, 125–149.
54. Mir, M.A. and Panganiban, A.T. (2008) A protein that replaces the entire cellular eIF4F complex. *EMBO J.*, **27**, 3129–3139.
55. Jonsson, C.B. and Schmaljohn, C.S. (2001) Replication of Hantaviruses. *Curr. Top. Microbiol. Immunol.*, **256**, 15–32.
56. Spiropoulou, C.F. (2001) Hantavirus maturation. *Curr. Top. Microbiol. Immunol.*, **256**, 33–46.

57. Goldsmith, C.S., Elliot, L.H., Peters, C.J., and Zaki, S.R. (1995) Ultrastructural characteristics of Sin Nombre virus, causative agent of hantavirus pulmonary syndrome. *Arch. Virol.*, **140**, 2107–2122.
58. Ravkov, E.V., Nichol, S.T., and Compans, R.W. (1997) Polarized entry and release in epithelial cells of Black Creek Canal virus its association with human disease and *Sigmodon hispidus* infection. *Virology*, **210**, 482–489.
59. Niklasson, B. and LeDuc, J.W. (1987) Epidemiology of nephropathia epidemica in Sweden. *J. Infect. Dis.*, **155**, 269–276.
60. Carey, D.E., Reuben, R., Panicker, K.N., Shope, R.E., and Myers, R.M. (1971) Thottapalayam virus: a presumptive arbovirus isolated from a shrew in India. *Indian J. Med. Res.*, **59**, 1758--1760.
61. Lee, H.W., Lee, P.W., and Johnson, K.M. (1978) Isolation of the etiologic agent of Korean Hemorrhagic fever. *J. Infect. Dis.*, **137**, 298–308.
62. Lee, H.W., Baek, L.J., and Johnson, K.M. (1982) Ioslation of Hantaan virus, the etiological agent of Korean hemorrhagic fever, from wild urban rats. *J. Infect. Dis.*, **146**, 638–644.
63. Avsic-Zupanc, T., Xiao, S.Y., Stojanovic, R., Glicic, A., van der Groen, G., and LeDuc, J.W. (1992) Characterization of Dobrava virus: a hantavirus from Slovenia, Yugoslavia. *J. Med. Virol.*, **38**, 132–137.
64. Brummer-Korvenkontio, M., Vaheri, A., Hovi, T., von Bonsdorff, C.H., Vuorimies, J., Manni, T., Penttinen, K., Oker-Blom, N., and Lähdevirta, J. (1980) Nephropathia epidemica: detection of antigen in bank voles and serologic diagnosis of human infection. *J. Infect. Dis.*, **141**, 131–134.
65. Lee, P.W., Amyx, H.L., Gajdusek, D.C., Yanagihara, R.T., Goldgaber, D., and Gibbs, C.J. Jr. (1982) New hemorrhagic fever with renal syndrome-related virus in rodents in the United States. *Lancet*, **2**, 1405.
66. Klingström, J., Heyman, P., Escutenaire, S., Sjölander, K.B., De Jaegere, F., Henttonen, H., and Lundkvist, Å. (2002) Rodent host specificity of European hantaviruses: evidence of Puumala virus interspecific spillover. *J. Med. Virol.*, **68**, 581–588.
67. Klempa, B., Fichet-Calvet, E., Lecompte, E., Auste, B., Aniskin, V., Meisel, H., Barrière, P., Koivogui, L., ter Meulen, J., and Krüger, D.H. (2007) Novel hantavirus sequences in Shrew, Guinea. *Emerg. Infect. Dis.*, **13**, 520–522.
68. Song, J.W., Kang, H.J., Song, K.J., Truong, T.T., Bennett, S.N., Arai, S., Truong, N.U., and Yanagihara, R. (2007) Newfound hantavirus in Chinese mole shrew, Vietnam. *Emerg. Infect. Dis.*, **13**, 1784–1787.
69. Song, J.W., Gu, S.H., Bennett, S.N., Arai, S., Puorger, M., Hilbe, M., and Yanagihara, R. (2007) Seewis virus, a genetically distinct hantavirus in the Eurasian common shrew (*Sorex araneus*). *Virol. J.*, **4**, 114.
70. Song, J.W., Kang, H.J., Gu, S.H., Moon, S.S., Bennett, S.N., Song, K.J., Baek, L.J., Kim, H.C., O'Guinn, M.L., Chong, S.T., Klein, T.A., and Yanagihara, R. (2009) Characterization of Imjin virus, a newly isolated hantavirus from the Ussuri white-toothed shrew (*Crocidura lasiura*). *J. Virol.*, **83**, 6184–6191.
71. Arai, S., Song, J.W., Sumibcay, L., Bennett, S.N., Nerurkar, V.R., Parmenter, C., Cook, J.A., Yates, T.L., and Yanagihara, R. (2007) Hantavirus in northern short-tailed shrew, United States. *Emerg. Infect. Dis.*, **13**, 1420–1423.
72. Arai, S., Bennett, S.N., Sumibcay, L., Cook, J.A., Song, J.W., Hope, A., Parmenter, C., Nerurkar, V.R., Yates, T.L., and Yanagihara, R. (2008) Phylogenetically distinct hantaviruses in the masked shrew (*Sorex cinereus*) and dusky shrew (*Sorex monticolus*) in the United States. *Am. J. Trop. Med. Hyg.*, **78**, 348–351.
73. Arai, S., Ohdachi, S.D., Asakawa, M., Kang, H.J., Mocz, G., Arikawa, J., Okabe, N., and Yanagihara, R. (2008) Molecular phylogeny of a newfound hantavirus in the Japanese shrew mole

(*Urotrichus talpoides*). *Proc. Natl. Acad. Sci. U.S.A.*, **105**, 16296–16301.

74. Kang, H.J., Bennett, S.N., Sumibcay, L., Arai, S., Hope, A.G., Mocz, G., Song, J.W., Cook, J.A., and Yanagihara, R. (2009) Evolutionary insights from a genetically divergent hantavirus harbored by the European common mole (*Talpa europaea*). *PLoS ONE*, **4**, e6149.

75. Kang, H.J., Bennett, S.N., Dizney, L., Sumibcay, L., Arai, S., Ruedas, L.A., Song, J.W., and Yanagihara, R. (2009) Host switch during evolution of a genetically distinct hantavirus in the American shrew mole (*Neurotrichus gibbsii*). *Virology*, **388**, 8–14.

76. Hardestam, J., Simon, M., Hedlund, K.O., Vaheri, A., Klingström, J., and Lundkvist, Å. (2007) Ex vivo stability of the rodent-borne Hantaan virus in comparison to that of arthropod-borne members of the Bunyaviridae family. *Appl. Environ. Microbiol.*, **73**, 2547–2551.

77. Kallio, E.R., Klingström, J., Gustafsson, E., Manni, T., Vaheri, A., Henttonen, H., Vapalahti, O., and Lundkvist, Å. (2006) Prolonged survival of Puumala hantavirus outside the host: evidence for indirect transmission via the environment. *J. Gen. Virol.*, **87**, 2127–2134.

78. Martinez, V.P., Bellomo, C., San Juan, J., Pinna, D., Forlenza, R., Elder, M., and Padula, P.J. (2005) Person-to-person transmission of Andes virus. *Emerg. Infect. Dis.*, **11**, 1848–1853.

79. Padula, P.J., Edelstein, A., Miguel, S.D., Lopez, N.M., Rossi, C.M., and Rabinovich, R.D. (1998) Hantavirus pulmonary syndrome outbreak in Argentina: molecular evidence for person-to-person transmission of Andes virus. *Virology*, **241**, 323–330.

80. Wells, R.M., Sosa Estani, S., Yadon, Z.E., Enria, D., Padula, P., Pini, N., Mills, J.N., Peters, C.J., and Segura, E.L. (1997) An unusual hantavirus outbreak in southern Argentina: person-to-person transmission? *Emerg. Infect. Dis.*, **3**, 171–174.

81. Navarrete, M., Pizarro, E., Méndez, C., Salazar, P., Padula, P., Zaror, L. *et al.* (2007) Hantavirus ANDV distribution in human lungs and salivary glands. Abstracts of the VII International Conference on HFRS, HPS and Hantaviruses, Buenos Aires, June 13–15, 2007, Abstract 103.

82. Ferres, M., Vial, P., Marco, C., Yanez, L., Godoy, P., Castillo, C., Hjelle, B., Delgado, I., Lee, S.J., and Mertz, G.J. (2007) Prospective evaluation of household contacts of persons with hantavirus cardiopulmonary syndrome in Chile. *J. Infect. Dis.*, **195**, 1563–1571.

83. Hardestam, J., Lundkvist, Å., and Klingström, J. (2009) Sensitivity of Andes hantavirus to antiviral effect of human saliva. *Emerg. Infect. Dis.*, **15**, 1140–1142.

84. Dournon, E., Moriniere, B., Matheron, S., Girard, P., Gonzalez, J., Hirsch, F., and McCormick, J.B. (1984) HFRS after a wild rodent bite in the hautesavoie – and risk of exposure to Hantaan-like virus in Paris laboratory. *Lancet*, **1**, 676–677.

85. Mäkelä, S., Mustonen, J., Ala-Houhala, I., Hurme, M., Partanen, J., Vapalahti, O., Vaheri, A., and Pasternack, A. (2002) Human leukocyte antigen-B8-DR3 is a more important risk factor for severe Puumala hantavirus infection than the tumor necrosis factor-alpha(-308) G/A polymorphism. *J. Infect. Dis.*, **186**, 843–846.

86. Mustonen, J., Huttunen, N.P., Partanen, J., Baer, M., Paakkala, A., Vapalahti, O., and Uhari, M. (2004) Human leukocyte antigens B8-DRB1*03 in pediatric patients with nephropathia epidemica caused by Puumala hantavirus. *Pediatr. Infect. Dis. J.*, **23**, 959–961.

87. Mustonen, J., Partanen, J., Kanerva, M., Pietilä, K., Vapalahti, O., Pasternack, A., and Vaheri, A. (1996) Genetic susceptibility to severe course of nephropathia epidemica caused by Puumala hantavirus. *Kidney Int.*, **49**, 217–221.

88. Liu, Z., Gao, M., Han, Q., Lou, S., and Fang, J. (2009) Platelet glycoprotein

IIb/IIIa (HPA-1 and HPA-3) polymorphisms in patients with hemorrhagic fever with renal syndrome. *Hum. Immunol.*, **70**, 452–456.

89. Linderholm, M. and Elgh, F. (2001) Clinical characteristics of hantavirus infections on the Eurasian continent. *Curr. Top. Microbiol. Immunol.*, **256**, 135–151.

90. Enria, D.A., Briggiler, A.M., Pini, N., and Lewis, S. (2001) Clinical manifestations of New World hantaviruses. *Curr. Top. Microbiol. Immunol.*, **256**, 117–134.

91. Mäkelä, S., Kokkonen, L., Ala-Houhala, I., Groundstroem, K., Harmoinen, A., Huhtala, H., Hurme, M., Paakkala, A., Porsti, I., Virtanen, V., Vaheri, A., and Mustonen, J. (2009) More than half of the patients with acute Puumala hantavirus infection have abnormal cardiac findings. *Scand J. Infect. Dis.*, **41**, 57–62.

92. Dara, S.I., Albright, R.C., and Peters, S.G. (2005) Acute sin nombre hantavirus infection complicated by renal failure requiring hemodialysis. *Mayo Clin. Proc.*, **80**, 703–704.

93. Passaro, D.J., Shieh, W.J., Hacker, J.K., Fritz, C.L., Hogan, S.R., Fischer, M., Hendry, R.M., and Vugia, D.J. (2001) Predominant kidney involvement in a fatal case of hantavirus pulmonary syndrome caused by Sin Nombre virus. *Clin. Infect. Dis.*, **33**, 263–264.

94. Linderholm, M., Billström, A., Settergren, B., and Tärnvik, A. (1992) Pulmonary involvement in nephropathia epidemica as demonstrated by computed tomography. *Infection*, **20**, 263–266.

95. Kanerva, M., Paakkala, A., Mustonen, J., Paakkala, T., Lahtela, J., and Pasternack, A. (1996) Pulmonary involvement in nephropathia epidemica: radiological findings and their clinical correlations. *Clin. Nephrol.*, **46**, 369–378.

96. Khaiboullina, S.F. and St Jeor, S.C. (2002) Hantavirus immunology. *Viral. Immunol.*, **15**, 609–625.

97. Muranyi, W., Bahr, U., Zeier, M., and van der Woude, F.J. (2005) Hantavirus Infection. *J. Am. Soc. Nephrol.*, **16**, 3669–3679.

98. Mustonen, J., Helin, H., Pietilä, K., Brummer-Korvenkontio, M., Hedman, K., Vaheri, A., and Pasternack, A. (1994) Renal biopsy findings and clinicopathologic correlations in nephropathia epidemica. *Clin. Nephrol.*, **41**, 121–126.

99. Groen, J., Bruijn, J.A., Gerding, M.N., Jordans, J.G., Moll van Charante, A.W., and Osterhaus, A.D. (1996) Hantavirus antigen detection in kidney biopsies from patients with nephropathia epidemica. *Clin. Nephrol.*, **46**, 379–383.

100. Borges, A., and Figueiredo, L.T. (2008) Mechanisms of shock in hantavirus pulmonary syndrome. *Curr. Opin. Infect. Dis.*, **21**, 293–297.

101. Hallin, G.W., Simpson, S.Q., Crowell, R.E., James, D.S., Koster, F.T., Mertz, G.J., and Levy, H. (1996) Cardiopulmonary manifestations of hantavirus pulmonary syndrome. *Crit. Care Med.*, **24**, 252–258.

102. Yao, Z.Q., Wang, W.S., Zhang, W.B., and Bai, X.F. (1989) The distribution and duration of Hantaan virus in the body fluids of patients with hemorrhagic fever with renal syndrome. *J. Infect. Dis.*, **160**, 218–224.

103. Terajima, M., Vapalahti, O., Van Epps, H.L., Vaheri, A., and Ennis, F.A. (2004) Immune responses to Puumala virus infection and the pathogenesis of nephropathia epidemica. *Microbes Infect.*, **6**, 238–245.

104. Miettinen, M.H., Mäkelä, S.M., Ala-Houhala, I.O., Huhtala, H.S., Koobi, T., Vaheri, A.I., Pasternack, A.I., Porsti, I.H., and Mustonen, J.T. (2009) Tubular proteinuria and glomerular filtration 6 years after puumala hantavirus-induced acute interstitial nephritis. *Nephron Clin. Pract.*, **112**, c115–c120.

105. Mäkelä, S., Ala-Houhala, I., Mustonen, J., Koivisto, A.M., Kouri, T., Turjanmaa, V., Vapalahti, O., Vaheri, A., and Pasternack, A. (2000) Renal function and blood pressure five years after puumala virus-induced nephropathy. *Kidney Int.*, **58**, 1711–1718.

106. Pergam, S.A., Schmidt, D.W., Nofchissey, R.A., Hunt, W.C., Harford, A.H., and Goade, D.E. (2009) Potential renal sequelae in survivors of hantavirus cardiopulmonary syndrome. *Am. J. Trop. Med. Hyg.*, **80**, 279–285.
107. Rubini, M.E., Jablon, S., and Mc, D.M. (1960) Renal residuals of acute epidemic hemorrhagic fever. *Arch. Int. Med.*, **106**, 378–387.
108. Glass, G.E., Watson, A.J., LeDuc, J.W., Kelen, G.D., Quinn, T.C., and Childs, J.E. (1993) Infection with a ratborne hantavirus in US residents is consistently associated with hypertensive renal disease. *J. Infect. Dis.*, **167**, 614–620.
109. Vaheri, A., Vapalahti, O., and Plyusnin, A. (2008) How to diagnose hantavirus infections and detect them in rodents and insectivores. *Rev. Med. Virol.*, **18**, 277–288.
110. Bostik, P., Winter, J., Ksiazek, T.G., Rollin, P.E., Villinger, F., Zaki, S.R., Peters, C.J., and Ansari, A.A. (2000) Sin nombre virus (SNV) Ig isotype antibody response during acute and convalescent phases of hantavirus pulmonary syndrome. *Emerg. Infect. Dis.*, **6**, 184–187.
111. Niklasson, B. and Kjelsson, T. (1988) Detection of nephropathia epidemica (Puumala virus)-specific immunoglobulin M by enzyme-linked immunosorbent assay. *J. Clin. Microbiol.*, **26**, 1519–1523.
112. Lundkvist, Å., Hörling, J., and Niklasson, B. (1993) The humoral response to Puumala virus infection (nephropathia epidemica) investigated by viral protein specific immunoassays. *Arch. Virol.*, **130**, 121–130.
113. Jenison, S., Yamada, T., Morris, C., Anderson, B., Torrez-Martinez, N., Keller, N., and Hjelle, B. (1994) Characterization of human antibody responses to Four Corners hantavirus infections among patients with hantavirus pulmonary syndrome. *J. Virol.*, **68**, 3000–3006.
114. French, G.R., Foulker, R.S., Brand, O.M., Eddy, G.A., Lee, H.W., and Lee, P.W. (1981) Korean hemorrhagic fever: propagation of the etiologic agent in a cell line of human origin. *Science*, **211**, 1046–1048.
115. Hedman, K., Vaheri, A., and Brummer-Korvenkontio, M. (1991) Rapid diagnosis of hantavirus disease with an IgG-avidity assay. *Lancet*, **338**, 1353–1356.
116. Hjelle, B., Jenison, S., Torrez-Martinez, N., Herring, B., Quan, S., Polito, A., Pichuantes, S., Yamada, T., Morris, C., Elgh, F., Lee, H.W., Artsob, H., and Dinello, R. (1997) Rapid and specific detection of Sin Nombre virus antibodies in patients with hantavirus pulmonary syndrome by a strip immunoblot assay suitable for field diagnosis. *J. Clin. Microbiol.*, **35**, 600–608.
117. Zöller, L.G., Yang, S., Gött, P., Bautz, E.K., and Darai, G. (1993) A novel mu-capture enzyme-linked immunosorbent assay based on recombinant proteins for sensitive and specific diagnosis of hemorrhagic fever with renal syndrome. *J. Clin. Microbiol.*, **31**, 1194–1199.
118. Wang, M., Rossi, C., and Schmaljohn, C.S. (1993) Expression of non-conserved regions of the S genome segments of three hantaviruses: evaluation of the expressed polypeptides for diagnosis of haemorrhagic fever with renal syndrome. *J. Gen. Virol.*, **74**, 1115–1124.
119. Vapalahti, O., Lundkvist, A., Kallio-Kokko, H., Paukku, K., Julkunen, I., Lankinen, H., and Vaheri, A. (1996) Antigenic properties and diagnostic potential of puumala virus nucleocapsid protein expressed in insect cells. *J. Clin. Microbiol.*, **34**, 119–125.
120. Elgh, F., Lundkvist, A., Alexeyev, O.A., Stenlund, H., Avsic-Zupanc, T., Hjelle, B., Lee, H.W., Smith, K.J., Vainionpää, R., Wiger, D., Wadell, G., and Juto, P. (1997) Serological diagnosis of hantavirus infections by an enzyme-linked immunosorbent assay based on detection of immunoglobulin G and M responses to recombinant nucleocapsid proteins of five viral serotypes. *J. Clin. Microbiol.*, **35**, 1122–1130.
121. Padula, P.J., Rossi, C.M., Della Valle, M.O., Martinez, P.V., Colavecchia,

S.B., Edelstein, A., Miguel, S.D.L., Rabinovich, R.D., and Segura, E.L. (2000) Development and evaluation of a solid-phase enzyme immunoassay based on Andes hantavirus recombinant nucleoprotein. *J. Med. Microbiol.*, **49**, 149–155.

122. Hujakka, H., Koistinen, V., Eerikäinen, P., Kuronen, I., Laatikainen, A., Kauppinen, J., Vaheri, A., Vapalahti, O., and Närvänen, A. (2001) Comparison of a new immunochromatographic rapid test with a commercial EIA for the detection of Puumala virus specific IgM antibodies. *J. Clin. Virol.*, **23**, 79–85.

123. Hujakka, H., Koistinen, V., Kuronen, I., Eerikäinen, P., Parviainen, M., Lundkvist, Å., Vaheri, A., Vapalahti, O., and Närvänen, A. (2003) Diagnostic rapid tests for acute hantavirus infections: specific tests for Hantaan, Dobrava and Puumala viruses versus a hantavirus combination test. *J. Virol. Methods*, **108**, 117–122.

124. Lundkvist, Å., Hukic, M., Hörling, J., Gilljam, M., Nichol, S., and Niklasson, B. (1997) Puumala and Dobrava viruses cause hemorrhagic fever with renal syndrome in Bosnia-Herzegovina: evidence of highly cross-neutralizing antibody responses in early patient sera. *J. Med. Virol.*, **53**, 51–59.

125. Klempa, B., Tkachenko, E.A., Dzagurova, T.K., Yunicheva, Y.V., Morozov, V.G., Okulova, N.M., Slyusareva, G.P., Smirnov, A., and Kruger, D.H. (2008) Hemorrhagic fever with renal syndrome caused by 2 lineages of Dobrava hantavirus, Russia. *Emerg. Infect. Dis.*, **14**, 617–625.

126. Golovljova, I., Vasilenko, V., Mittzenkov, V., Prükk, T., Seppet, E., Vene, S., Settergren, B., Plyusnin, A., and Lundkvist, Å. (2007) Characterization of hemorrhagic fever with renal syndrome caused by hantaviruses, Estonia. *Emerg. Infect. Dis.*, **13**, 1773–1776.

127. Zaki, S.R., Greer, P.W., Coffield, L.M., Goldsmith, C.S., Nolte, K.B., Foucar, K., Feddersen, R.M., Zumwalt, R.E., Miller, G.L., Khan, A.S., Rollin, P.E., Ksiazek, T.G., Nichol, S.T., Mahy, B.W.J., and Peters, C.J. (1995) Hantavirus pulmonary syndrome. Pathogenesis of an emerging infectious disease. *Am. J. Pathol.*, **146**, 552–579.

128. Ahn, C., Cho, J.T., Lee, J.G., Lim, C.S., Kim, Y.Y., Han, J.S., Kim, S., and Lee, J.S. (2000) Detection of Hantaan and Seoul viruses by reverse transcriptase-polymerase chain reaction (RT-PCR) and restriction fragment length polymorphism (RFLP) in renal syndrome patients with hemorrhagic fever. *Clin. Nephrol.*, **53**, 79–89.

129. Grankvist, O., Juto, P., Settergren, B., Ahlm, C., Bjermer, L., Linderholm, M., Tärnvik, A., and Wadell, G. (1992) Detection of nephropathia epidemica virus RNA in patient samples using a nested primer-based polymerase chain reaction. *J. Infect. Dis.*, **165**, 934–937.

130. Antoniadis, A., Stylianakis, A., Papa, A., Alexiou-Daniel, S., Lampropoulos, A., Nichol, S.T., Peters, C.J., and Spiropoulou, C.F. (1996) Direct genetic detection of Dobrava virus in Greek and Albanian patients with hemorrhagic fever with renal syndrome. *J. Infect. Dis.*, **174**, 407–410.

131. Khan, A.S., Spiropoulou, C.F., Morzunov, S., Zaki, S.R., Kohn, M.A., Nawas, S.R., McFarland, L., and Nichol, S.T. (1995) Fatal illness associated with a new hantavirus in Louisiana. *J. Med. Virol.*, **46**, 281–286.

132. Yashina, L.N., Patrushev, N.A., Ivanov, L.I., Slonova, R.A., Mishin, V.P., Kompanez, G.G., Zdanovskaya, N.I., Kuzina, I.I., Safronov, P.F., Chizhikov, V.E., Schmaljohn, C., and Netesov, S.V. (2000) Genetic diversity of hantaviruses associated with hemorrhagic fever with renal syndrome in the far east of Russia. *Virus Res.*, **70**, 31–44.

133. Hjelle, B., Spiropoulou, C.F., Torrez-Martinez, N., Morzunov, S., Peters, C.J., and Nichol, S.T. (1994) Detection of Muerto Canyon virus RNA in peripheral blood mononuclear cells from patients with hantavirus pulmonary syndrome. *J. Infect. Dis.*, **78**, 47–55.

134. Ramsden, C., Melo, F.L., Figueiredo, L.M., Holmes, E.C., and Zanotto, P.M.

(2008) High rates of molecular evolution in hantaviruses. *Mol. Biol. Evol.*, **25**, 1488–1492.

135. Evander, M., Eriksson, I., Pettersson, L., Juto, P., Ahlm, C., Olsson, G.E., Bucht, G., and Allard, A. (2007) Puumala hantavirus viremia diagnosed by real-time reverse transcriptase PCR using samples from patients with hemorrhagic fever and renal syndrome. *J. Clin. Microbiol.*, **45**, 2491–2497.

136. Xiao, R., Yang, S., Koster, F., Ye, C., Stidley, C., and Hjelle, B. (2006) Sin Nombre viral RNA load in patients with hantavirus cardiopulmonary syndrome. *J. Infect. Dis.*, **194**, 1403–1409.

137. Saksida, A., Duh, D., Korva, M., and Avsic-Zupanc, T. (2008) Dobrava virus RNA load in patients who have hemorrhagic fever with renal syndrome. *J. Infect. Dis.*, **197**, 681–685.

Part B
Practical Guidelines

BSL3 and BSL4 Agents: Epidemiology, Microbiology, and Practical Guidelines, First Edition.
Edited by Mandy C. Elschner, Sally J. Cutler, Manfred Weidmann, and Patrick Butaye.

Part I
Bacteria

BSL3 and BSL4 Agents: Epidemiology, Microbiology, and Practical Guidelines, First Edition.
Edited by Mandy C. Elschner, Sally J. Cutler, Manfred Weidmann, and Patrick Butaye.

1
Bacillus anthracis

Markus Antwerpen, Paola Pilo, Pierre Wattiau, Patrick Butaye, Joachim Frey, and Dimitrios Frangoulidis

Organism	Laboratory category	Transmissible by aerosol
Bacillus anthracis	BSL-3	Yes

Recommended Respiratory Protection

In awareness of aerosolized spores or a bacteria suspension filtering facepiece (FFP)3 masks or positive ventilated masks attached to FFP3 filters are recommended to avoid an infection via the respiratory route. There are several suppliers "on the market" (3M, Dräger).

Recommended Personal Protective Equipment

Gloves, fully protected single-use body suits, face shield for protection against splattering liquids as well as respiratory protection in awareness of aerosols (see above).

Best Disinfection

Spores are high resistant to environmental stress!

Surface and Equipment

- 10% formaldehyde or 4% glutaraldehyde (pH 8.0–8.5);
- 3% H_2O_2 or 1% peracetic acid – incubation of 2 h;
- Sodium hypochlorite bleach – fivefold diluted commercial bleach; incubation of 2 h;

BSL3 and BSL4 Agents: Epidemiology, Microbiology, and Practical Guidelines, First Edition.
Edited by Mandy C. Elschner, Sally J. Cutler, Manfred Weidmann, and Patrick Butaye.

Skin/Wound Disinfection

Clean the wound with lots of water and soap (mechanical purification) followed by a 2-min wash using 0.2% peracetic acid.

Best Decontamination

See Best Disinfection, additionally fumigation of room using formaldehyde.

Prevention

The United States and United Kingdom have developed nonliving human vaccines based on strain V770 and $34F_2$ (in the USA the vaccine is FDA licensed and is produced and sold by Emergent Biosolutions, USA).

A good overview with self-explaining tables can be found here:

Anthrax in Humans and Animals, 4th edition, WHO, Geneva, WHO Library Cataloguing in Publication Data, 2008, *www.who.int/entity/csr/resources/publications/anthrax_web.pdf*.

Case Reports, Ongoing Clinical Trials

See the newest ACIP recommendations regarding prophylaxis and therapy:

Use of Anthrax Vaccine in the United States, Recommendations of the Advisory Committee on Immunization Practices (ACIP), 2009, *MMWR*, July 23, 2010/ vol. **59**/No. RR-6.

Post Exposure Prophylaxis

Antibiotics: ciprofloxacin, doxycycline, and chloramphenicol (in the case of meningitis) alternative antibiotics: clindamycin, vancomycin, and rifampin.

Known Laboratory Accidents

- (2005) Inadvertent laboratory exposure to *Bacillus anthracis* – California, 2004. *MMWR Morb. Mortal. Wkly Rep.*, **54** (12), 301–304.
- Centers for Disease Control and Prevention (CDC) (2002) Update: cutaneous anthrax in a laboratory worker – Texas, 2002. *MMWR Morb. Mortal. Wkly Rep.*, **51** (22), 482.

Procedure Recommended in the Case of Laboratory Spill or Other Type of Accident

Disinfection and decontamination procedures depend on the situation and can include fumigation of the room using formaldehyde.

Treatment of Disease

A specific treatment of the disease is available (specific, see above). For details see national agency recommendations: CDC (USA), RKI (Germany), HPA (UK), and Institute Pasteur (France).

Clinical Guidelines

- Bossi, P., Tegnell, A., Baka, A. *et al.* (2004) Bichat guidelines for the clinical management of anthrax and bioterrorism-related anthrax. *Eur. Surveil.*, **9** (12), E3–E4.
- Chitlaru, T., Altboum, Z., Reuveny, S., and Shafferman, A. (2011) Progress and novel strategies in vaccine development and treatment of anthrax. *Immunol. Rev.*, **239** (1), 221–236. doi: 10.1111/j.1600-065X.2010.00969.x

2
Brucella Species

Sally J. Cutler, Michel S. Zygmunt, and Bruno Garin-Bastuji

Organism	Laboratory category	Transmissible by aerosol
Brucella spp.	BSL-3/4	Yes

Recommended Respiratory Protection

A specific respiratory protection (self-contained breathing apparatus) is usually not necessary, except in the case of potential aerosol formation. However, if potentially infectious materials or animals have to be handled outside a biosafety cabinet, a filtering facepiece (FFP)2 mask and an eye shield are highly recommended.

Recommended Personal Protective Equipment

Manipulators should wear at least splash protection suits (one-piece coveralls or multi-piece combinations), FFP2 masks, disposable gloves (two pairs during manipulations), and boots.

Best Disinfection

Some possible examples are:

- **Walls and floors:**
  - 1,3-Diamine – CAS No. 2372-82-9 (51 mg/g), didecyldimethylammonium chloride – CAS No. 7173-51-5 (25 mg/g).
  - Didecyldimethylammonium chloride – CAS No. 7173-51-5 (65 mg/g), biguanide polyhexamethylene hydrochloride – CAS No. 27083-27-8 (12 mg/g – active against *M. tuberculosis*).

BSL3 and BSL4 Agents: Epidemiology, Microbiology, and Practical Guidelines, First Edition.
Edited by Mandy C. Elschner, Sally J. Cutler, Manfred Weidmann, and Patrick Butaye.

- **Devices (spray):**
  - Denatured ethanol (41% v/v), polyhexamethylene biguanide hydrochloride, N-(3-aminopropyl)-N-dodecylpropane-1,3-diamine, didecyldimethylammonium chloride (active against *M. tuberculosis*).
  - Didecyldimethylammonium chloride, polyhexamethylene biguanide, ethanol, glycolic acid.

Best Decontamination

- Formaldehyde, N-3-aminopropyl N-dodecylpropane 1,3-diamine, ethanol;
- Hydrogen peroxide (50 mg/g);
- Formaldehyde, glutaraldehyde, ethanol.

Prevention

A human vaccine was used in France in the 1980s, but it proved inefficient and was abandoned. Animal vaccines are available for cattle (S19, RB51) and for small ruminants (Rev.1). While S19 and Rev.1 have been proved efficient in reducing the spread of the disease within and between herds, it could not prevent infection. The efficacy of the RB51 vaccine remains controversial. No vaccine is available for pigs.

Post Exposure Prophylaxis

Tetracyclines (doxycycline) plus eventually rifampicin, gentamicin, or streptomycin are advised in adults and children at least eight years old. In children less than eight years old, rifampicin plus potentiated sulfonamides are advised.

Known Laboratory Accidents

Brucellosis is the most frequently reported laboratory-associated bacterial infection. Several reports have been published as well as concise reviews.

- Fiori, P.L., Mastrandrea, S., Rappelli, P., and Cappuccinelli, P. (2000) *Brucella abortus* infection acquired in microbiology laboratories. *J. Clin. Microbiol.*, **38**, 2005–2006.
- Yagupsky, P., Peled, N., Riesenberg, K., and Banai, M. (2000) Exposure of hospital personnel to *Brucella melitensis* and occurrence of laboratory-acquired disease in an endemic area. *Scand. J. Infect. Dis.*, **32**, 31–35.
- Memish, Z.A. and Mah, M.W. (2001) Brucellosis in laboratory workers at a Saudi Arabian hospital. *Am. J. Infect. Control*, **29**, 48–52.
- Robichaud, S., Libman, M., Behr, M., and Rubin, E. (2004) Prevention of laboratory-acquired brucellosis. *Clin. Infect. Dis.*, **38**, e119–e122.

- Noviello, S., Gallo, R., Kelly, M. *et al.* (2004) Laboratory-acquired brucellosis. *Emerg. Infect. Dis.*, **10**, 1848–1850.
- Wallach, J.C., Ferrero, M.C., Delpino, M.V., Fossati, C.A., and Baldi, P.C. (2008) Occupational infection due to *Brucella abortus* S19 among workers involved in vaccine production in Argentina. *Clin. Microbiol. Infect.*, **14**, 805–807.
- Singh, K. (2009) Laboratory-acquired infections. *Clin. Infect. Dis.*, **49**, 142–147.

Procedure Recommended in Case of Laboratory Spill or Other Type of Accident

Disinfection and decontamination procedures depend on the situation and can include fumigation of the room using formaldehyde.

Treatment of Disease

A similar treatment as described for post exposure prophylaxis can be used.

Clinical Guidelines

- WHO/FAO/OIE (2006) *Brucellosis in Humans and Animals*, WHO, Geneva, *http://www.who.int/csr/resources/publications/deliberate/WHO_CDS_EPR_2006_7/en/*.
- Franco, M.P., Mulder, M., Gilman, R.H., and Smits, H.L. (2007) Human brucellosis. *Lancet Infect. Dis.*, **7**, 775–786.
- Pappas, G. (2008) Treatment of brucellosis. *Br. Med. J.*, **336**, 678–679.
- *http://www.bmj.com/content/336/7646/678.long*.
- Ariza, J., Bosilkovski, M., Cascio, A., Colmenero, J.D., Corbel, M.J., Falagas, M.E., Memish, Z.A., Roushan, M.R., Rubinstein, E., Sipsas, N.V., Solera, J., Young, E.J., and Pappas, G. (2007) *PLoS Med.*, **4**, e317.
- *http://www.plosmedicine.org/article/info%3Adoi%2F10.1371%2Fjournal.pmed.0040317*.

3
Burkholderia mallei: Glanders

Lisa D. Sprague and Mandy C. Elschner

Organism	Laboratory category	Transmissible by aerosol
Burkholderia mallei	BSL-3	Yes

Recommended Respiratory Protection

Filtering facepiece (FFP)3 respiratory protection masks.

Recommended Personal Protective Equipment

Gloves, safety goggles, and protective clothing.

Best Disinfection

The following disinfectants can be used depending on the surfaces and areas to be disinfected: 70% ethanol, 2% formaldehyde solution, 1% sodium hydroxide, 1% sodium hypochlorite, 5% calcium hypochlorite, 2% glutaraldehyde, and 1% potassium permanganate.

Best Decontamination

See Best Disinfection plus additional formaldehyde fumigation.

Prevention

Not available.

BSL3 and BSL4 Agents: Epidemiology, Microbiology, and Practical Guidelines, First Edition.
Edited by Mandy C. Elschner, Sally J. Cutler, Manfred Weidmann, and Patrick Butaye.

Case Reports, Ongoing Clinical Trials

Not available.

Post Exposure Prophylaxis

The utility of post exposure prophylaxis in humans is still being discussed. A prophylaxis with trimethoprim-sulfamethoxazole (co-trimoxazole) is recommended, although this is based on experimental data [1].

Known Laboratory Accidents

- Srinivasan, A., Kraus, C.N., and De Shazer, D. (2001) Glanders in a military microbiologist. *N. Engl. J. Med.*, **345**, 256–258.

Procedure Recommended in Case of Laboratory Spill or Other Type of Accident

Cover the spill with tissue and pour disinfectant on the tissue. Leave the laboratory and collect all possibly contaminated clothes for decontamination. After having changed into new clothes and safety equipment – collect all materials for decontamination. Disinfection of putative contaminated areas. A fumigation of the laboratory using formaldehyde is recommended. Contaminated and allegedly infected persons must be kept under medical supervision.

Treatment of Disease

Initial treatment: imipenem (i.v.) or meropenem (i.v.) or ceftazidime (i.v.) until improvement, then oral doses of doxycycline + co-trimoxazole or amoxicillin + clavulanate up to 20 weeks. For details see Bichat guidelines [1].

Clinical Guidelines

Bichat guidelines for the clinical management of glanders and melioidosis and bioterrorism-related glanders and melioidosis [1].

References

1. Bossi, P., Tegnell, A., Baka, A., Van Loock, F., Hendriks, J., Werner, A., Maidhof, H., and Gouvras, G. (2004) Bichat guidelines for the clinical management of glanders and melioidosis and bioterrorism-related glanders and melioidosis. *Eur. Surveil.*, **9** (12), E17–E18.

4
Burkholderia pseudomallei: Melioidosis

Lisa D. Sprague and Mandy C. Elschner

Organism	Laboratory category	Transmissible by aerosol
Burkholderia pseudomallei	BSL-3	Yes

Recommended Respiratory Protection

filtering facepiece (FFP)3 SL respiratory protection masks.

Recommended Personal Protective Equipment

Additional gloves, safety goggles, and protective clothing.

Best Disinfection

The following disinfectants can be used depending on the surfaces and places to be disinfected: 70% ethanol, 2% formaldehyde solution, 1% sodium hydroxide, 1% sodium hypochlorite, 5% calcium hypochlorite, 2% glutaraldehyde, and 1% potassium permanganate.

Best Decontamination

See Best Disinfection, plus additional formaldehyde fumigation.

Prevention

Not available.

BSL3 and BSL4 Agents: Epidemiology, Microbiology, and Practical Guidelines, First Edition.
Edited by Mandy C. Elschner, Sally J. Cutler, Manfred Weidmann, and Patrick Butaye.

Post Exposure Prophylaxis

Trimethoprim-sulfamethoxazole (co-trimoxazole) is recommended in case of a biological attack, although this is based on experimental data. The utility of post exposure prophylaxis in humans is still being discussed [1].

Known Laboratory Accidents

- Currie, B.J., Inglis, T.J., Vannier, A.M., Novak-Weekley, S.M., Ruskin, J., Mascola, L., Bancroft, E., Borenstein, L., Harvey, S., Rosenstein, N., Clark, T.A., and Nguyen, D.M. (2004) Laboratory exposure to Burkholderia pseudomallei – Los Angeles, California 2003. *Morbid. Mort. Wkly Rep.*, **53** (42), 988–990.
- Schlech, W.F. III, Turchik, J.B., Westlake, R.E. Jr., Klein, G.C., Band, J.D., and Weaver, R.E. (1981) Laboratory-acquired infection with Pseudomonas pseudomallei (melioidosis). *N. Engl. J. Med.*, **305** (19), 1133–1135.
- Green, R.N. and Tuffnell, P.G. (1968) Laboratory-acquired melioidosis. *Am. J. Med.*, **44**, 599–605.

Procedure Recommended in Case of Laboratory Spill or Other Type of Accident

In the case of spills, the liquid needs to be covered with a tissue and an appropriate disinfectant poured onto the tissue. The contaminated tissues are then collected for further decontamination by autoclave treatment. A fumigation of the contaminated area with formaldehyde is recommended. Persons involved in the accident must be placed under medical supervision.

Treatment of Disease

Initially with antibiotics (ceftazidime, imipenem, or meropenem – all i.v.), while a 20-week schedule should be completed with oral antibiotics such as doxycyline + co-trimoxazole, or amoxicillin + clavulanate, or ciprofloxacin [1].

In the case of a pulmonary disease, the treatment should be prolonged for up to 6–12 months. For the septicemic form, the duration of treatment is two weeks i.v. treatment followed by oral therapy for six months [2, 3]. Infected persons need a continuous follow up during the course of the disease.

Clinical Guidelines

Bichat guidelines for the clinical management of glanders and melioidosis and bioterrorism-related glanders and melioidosis [1].

References

1. Bossi, P., Tegnell, A., Baka, A., Van Loock, F., Hendriks, J., Werner, A., Maidhof, H., and Gouvras, G. (2004) Bichat guidelines for the clinical management of glanders and melioidosis and bioterrorism-related glanders and melioidosis. *Eurosurveillance*, **30** (9) *http://www.eurosurveillance.org* (accessed November 23 2010).
2. HPA-Collindale (2003) Glanders and Melioidosis. Interim Guidelines for Action in the Event of a Deliberate Release. HPA-Collindale, Ver2.2, issue date: 14 August 2003. *http://www.hpa.org.uk/infections/topics_az/deliberate_release/menu.htm* (accessed November 23 2010).
3. The European Agency for the Evaluation of Medicinal Products (2002) *EMEA/CPMP Guidance Document on Use of Medicinal Products for Treatment and Prophylaxis of Biological Agents that Might be Used as Weapons of Bioterrorism*, The European Agency for the Evaluation of Medicinal Products, Strasbourg, *http://www.emea.eu.int* (accessed November 23 2010).

5
Coxiella burnetii: Q Fever

Matthias Hanczaruk, Sally J. Cutler, Rudolf Toman, and Dimitrios Frangoulidis

Organism	Laboratory category	Transmissible by aerosol
Coxiella burnetii	BSL-3	Yes

Recommended Respiratory Protection

Due to its natural spread as an aerosol, it is definitely recommended to wear at least filtering facepiece (FFP)3 masks or positive ventilated masks attached to FFP3 filters to avoid infection via the respiratory route.

Recommended Personal Protective Equipment

It is advised to wear at least gloves. Based on the scenario, additional full-protection single-use body suits, face shield for protection against splattering liquids as well as respiratory protection in awareness of aerosols can be necessary.

Best Disinfection

Under normal hygiene situations (e.g., hospital): standard disinfectants (e.g., alcohol-based). When dried material and/or large amounts are faced which could contain the very resistant small cell variant, spore-inactivating methods are recommended.

Surface and Equipment

10% formaldehyde or 4% glutaraldehyde (pH 8.0–8.5) or 3% H_2O_2 or 1% peracetic acid (e.g., =2.5% Wofasteril®) – incubation time of 30 min.

BSL3 and BSL4 Agents: Epidemiology, Microbiology, and Practical Guidelines, First Edition.
Edited by Mandy C. Elschner, Sally J. Cutler, Manfred Weidmann, and Patrick Butaye.

Skin Disinfection

0.2% peracetic-acid (e.g., 0.5% Wofasteril,®).

Best Decontamination

See Best Disinfection. Additionally fumigation of room using formaldehyde might be advisable. For veterinary use: 2% NaOH or 20% Chlor chalk solution.

Prevention

There is one licensed human vaccine in Australia (CSL), using an inactivated Phase I Henzerling strain. A check of pre-existing antibodies is mandatory, because severe side effects have been documented. Therefore only one shot is possible. The duration of protection is uncertain (maybe five years). However, meta-analysis of (in general) poorly designed studies failed to demonstrate a significant protection.

An experimental Phase I, Q fever vaccine (IND) is available on a limited basis from the Special Immunizations Program (301-619-4653) of the USAMRIID (Fort Detrick, Md) for at-risk personnel under a cooperative agreement with the individual's requesting institution. Like the CSL vaccine it can be reactogenic in those with prior immunity, and thus it requires skin testing before administration.

An animal vaccine for ruminants, with inactivated Phase I bacteria (Coxevac®; CEVA, France) is available. Studies in France and very recently in the Netherlands demonstrated a reduction in shedding of the organism.

- Gidding, H.F., Wallace, C., Lawrence, G.L., and McIntyre, P.B. (2009) Australia's national Q fever vaccination program. *Vaccine*, **27** (14), 2037–2041.
- Waag, D.M. (2007) *Coxiella burnetii*: host and bacterial responses to infection. *Vaccine*, **25** (42), 7288–7295.

Post Exposure Prophylaxis

Antibiotics such as fluoroquinolone or tetracyclines can be used for post exposure prophylaxis.

Known Laboratory Accidents

- Hall, C.J., Richmond, S.J., Caul, E.O., Pearce, N.H., and Silver, I.A. (1982) Laboratory outbreak of Q fever acquired from sheep. *Lancet*, **1** (8279), 1004–1006.
- Johnson, J.E. and Kadull, P.J. (1966) Laboratory-acquired Q fever. A report of fifty cases. *Am. J. Med.*, **41** (3), 391–403.

- Rusnak, J.M., Kortepeter, M.G., Hawley, R.J., Anderson, A.O., Boudreau, E., and Eitzen, E. (2004) Risk of occupationally acquired illnesses from biological threat agents in unvaccinated laboratory workers. *Biosecur. Bioterror.*, **2** (4), 281–293.

Procedure Recommended in Case of Laboratory Spill or Other Type of Accident

Disinfection and decontamination procedures depend on the situation and can include fumigation of the room using formaldehyde.

Treatment of Disease

See also national agency recommendations, for example, CDC (USA), RKI (Germany), HPA (UK), Institute Pasteur (France).

Clinical Guidelines

- Bossi, P., Tegnell, A., Baka, A. *et al.* (2004) Bichat guidelines for the clinical management of Q fever and bioterrorism-related Q fever. *Eur. Surveil.*, **9** (12), E19–E20.
- Gikas, A., Kokkini, S., and Tsioutis, C. (2010) Q fever: clinical manifestations and treatment. *Expert Rev. Anti Infect. Ther.*, **8** (5), 529–539.

6
Francisella tularensis: Tularemia

Anders Johansson, Herbert Tomaso, Plamen Padeshki, Anders Sjostedt, Nigel Silman, and Paola Pilo

Organism	Laboratory category	Transmissible by aerosol
Francisella tularensis	BSL-3	Yes

Recommended Respiratory Protection

Given the proven track record of respiratory transmission, it is recommended to always handle cultures of the live agent inside a safety cabinet in a BSL-3 containment suite according to BSL-3 safe working practices. Person to person spread has not reported to date.

Recommended Personal Protective Equipment

Gloves, full-protection single-use body suits, face shield for protection against splattering liquids as well as respiratory protection in awareness of aerosols (see above).

Best Disinfection/Decontamination

Water/70% ethanol will decontaminate surfaces, while disinfection soap is used for hands.

F. tularensis do not show any particular resistance to decontaminant agents, thus products based on peracetic acid, potassium peroxymonosulfate, chlorine, or isopropanol (among others) can be used.

BSL3 and BSL4 Agents: Epidemiology, Microbiology, and Practical Guidelines, First Edition.
Edited by Mandy C. Elschner, Sally J. Cutler, Manfred Weidmann, and Patrick Butaye.

Prevention

Live vaccine strain (LVS) has been developed; however, this vaccine is not commercially available [1].

Case Reports, Ongoing Clinical Trials

A PubMed search will give an enormous amount of tularemia case reports. Some references to recent case reports are listed below:

- Edouard, S., Gonin, K., Turc, Y., Angelakis, E. *et al.* (2011) Eschar and neck lymphadenopathy caused by *Francisella tularensis* after a tick bite: a case report. *J. Med. Case Rep.*, **5**, 108.
- Kim, D.Y., Reilly, T.J., Schommer, S.K., and Spagnoli, S.T. (2010) Rabbit tularemia and hepatic coccidiosis in wild rabbit. *Emerg. Infect. Dis.*, **12**, 2016–2017.
- Evans, M.E., Gregory, D.W., Schaffner, W., and McGee, Z.A. (1985) Tularemia: a 30-year experience with 88 cases. *Medicine*, **64** (4), 251–269.

Post Exposure Prophylaxis

Recommendations are available on the European Medicines Agency (EMA) web site: *http://www.ema.europa.eu/docs/en_GB/document_library/Other/2010/08/WC500095414.pdf.*

Known Laboratory Accidents

- **Review of laboratory disease and vaccine protection**:
  - Burke, D.S. (1977) Immunization against tularemia: analysis of the effectiveness of live *Francisella tularensis* vaccine in prevention of laboratory-acquired tularemia. *J. Infect. Dis.*, **135**, 55–60.
- **Boston accident**:
  - Lawler, A. (2005) Biodefense labs. Boston university under fire for pathogen mishap. *Science*, **307** (5709), 501.

Procedure Recommended in Case of Laboratory Spill or Other Type of Accident

Although the standard operating procedures (SOPs) defined in each laboratory or institute have to be followed, these SOPs should include the following points:

- Immediately remove contaminated clothing and place them on the floor. In the case of aerosols, all persons must leave the laboratory until the aerosol has settled (at least 30 min).
- Call the contact person by use of the internal alarm list. Use disinfectants or spraying with Nu-Cidex, if necessary. Take a shower.

- Put up a sign "Decontamination in progress" at the entrance to the laboratory.
- Anyone who enters to clean up should bear disposable overalls, double gloves, and appropriate mask if necessary. Disinfectant is sprinkled over the spill. After 10 min, wipe up spills. Small spills can be covered with cloths that are drowned with appropriate disinfectant. Contaminated clothing and wipes are thrown in waste bins and autoclaved. Use forceps to pick up broken glassware and put in the sharps disposal container. Wipe the area once more with ethanol and allow to air-dry.
- Remember to avoid the creation of new aerosol formation by pouring disinfectant too quickly from a height.
- Report the incident immediately to the department head responsible for further processing.
- Consultation with infectious diseases physicians regarding further treatment of exposed individuals should be sought.

Treatment of Disease

Recommendations are available on the EMA web site: *http://www.ema.europa.eu/docs/en_GB/document_library/Other/2010/08/WC500095414.pdf* and at Bichat guidelines for the clinical management of tularemia and bioterrorism-related tularemia [2].

Clinical Guidelines

Bichat guidelines for the clinical management of tularemia and bioterrorism-related tularemia [2].

References

1. Conlan, J.W. (2011) Tularemia vaccines: recent developments and remaining hurdles. *Fut. Microbiol.*, **6** (4), 391–405.
2. Bossi, P., Tegnell, A., Baka, A., Van Loock, F. *et al.* (2004) Bichat guidelines for the clinical management of tularaemia and bioterrorism-related tularaemia. *Eur. Surveil.*, **15** (12), E9–10, *http://www.eurosurveillance.org/images/dynamic/em/v09n12/0912-234.pdf* (accessed July 21 2011).

7
Yersinia pestis: Plague

Anne Laudisoit, Werner Ruppitsch, Anna Stoeger, and Ariane Pietzka

Organism	Laboratory category	Transmissible by aerosol
Yersinia pestis	BSL-3	Yes

Recommended Respiratory Protection

Respiratory protection is highly advisable to provide respiratory protection against droplet transmission. A filtering facepiece (FFP3) mask or a positively ventilated mask attached to a FFP3 filter is recommended to avoid an infection via respiratory route. Similarly, eye protection is advisable.

Recommended Personal Protective Equipment

As for other BSL-3 pathogens, a laboratory coat, surgical gown (open-backed), with high collar, cuffed sleeves, closing on high-sleeved gloves together with double laboratory gloves should be used. The laboratory gown and gloves should always be changed when moving between laboratory areas.

Best Disinfection

Surface and Equipment

Y. pestis is fragile and does not survive for long periods outside a host. It is readily destroyed by desiccation or by the action of common disinfectants, namely sodium hypochlorite 1%, ethanol 70%, glutaraldehyde 2%, iodine-based agents, and germicidal solutions that contain phenol, formaldehyde, and quaternary ammonium compounds [1]. *Y. pestis* is susceptible to inactivation by sunlight, by moist heat (121 °C for at least 15 min), or dry heat (160–170 °C for at least 1 h).

BSL3 and BSL4 Agents: Epidemiology, Microbiology, and Practical Guidelines, First Edition.
Edited by Mandy C. Elschner, Sally J. Cutler, Manfred Weidmann, and Patrick Butaye.

Best Decontamination

A household bleach solution, with a contact time of 30 min, may be used effectively for decontamination prior to normal cleaning. Organic material will quickly denature a bleach solution; therefore, if organic material is present, prior cleaning should precede decontamination. A 0.5% solution of sodium hypochlorite (1 : 9 parts dilution of commercial bleach in water; 10% bleach) is the recommended material to use for wiping counters and decontaminating since it kills both bacteria and viruses and degrades DNA [2].

In most cases, a wet spray of 0.5% hypochlorite solution over the spill, or best over a cloth dropped over the spill, and left in contact for 10 min followed by a 70% alcohol wash is adequate for decontamination (not 100% since it only dehydrates but does not lyse bacterial cell walls, and it evaporates quickly).

Skin/Wound Disinfection

A study evaluating the survival of *Y. pestis* in formulated tap water with chlorine, revealed that common chlorination of tap water is effective at killing *Y. pestis* [3, 4]. In consequence, rigorous cleaning of either skin or wound with lots of water and soap (mechanical purification) followed by 2 min wash using skin antiseptic or 0.2% peracetic acid should be effective.

Prevention

A first formaldehyde-killed whole-cell "army vaccine" [5] was widely used during both World War II and the Vietnam War, but it is known to be reactogenic in humans, resulting in malaise, headaches, local erythema, and induration or mild lymphadenopathy in approximately 10% of recipients [6]. The current formaldehyde-killed *Y. pestis* vaccine (Sampar; produced in Indonesia) is also reactogenic; it provides protection against flea-transmitted plague, but not protection against pneumonic plague [7]. As a killed vaccine it requires multiple doses. The vaccine could be useful for high-risk personnel (e.g., laboratory workers) but does not seem appropriate, or practical, for mass vaccination.

Case Reports, Ongoing Clinical Trials

The human immune response to a new recombinant plague vaccines, comprising recombinant F1 (rF1) and rV antigens, have been assessed through Phase 1 safety and immunogenicity trials in healthy volunteers. All but one of the subjects produced specific immunoglobulin G in serum that peaked following a booster dose (day 21). However, no significant vaccination-related change in activation of peripheral blood mononuclear cells was detected. Investigation of potential serological immune correlates of protection has failed to demonstrate cellular marker trends specific to vaccination [6].

Post Exposure Prophylaxis

The Working Group on Civilian Biodefense recommends post-exposure prophylaxis using oral antibiotics: (i) Doxycycline, 100 mg orally twice daily, (ii) Ciprofloxacin, 500 mg orally twice daily, (iii) alternative choice: chloramphenicol, 25 mg/kg orally four times daily; see reference [8] for restrictions and use in children and pregnant women; for seven days for asymptomatic individuals who have been in close contact with patients with pneumonic plague [8].

Doxycycline has been proposed as the drug of choice for post exposure prophylaxis in the case of a community contact with pneumonic plague [9].

Recently, Levofloxacin (fluoroquinolone) has been demonstrated to be curative for established pneumonic plague in nonhuman primates when treatment is initiated after the onset of fever and can be considered for treatment of other forms of plague [10, 11]. On a cautionary note, reports are emerging of multidrug resistant *Y. pestis* [12–14]. One of these strains was resistant to streptomycin, the other one to all antibiotics that are normally recommended for plague control.

Known Laboratory Accidents

A scientist became the first American researcher to contract the plague in 50 years and died of the disease. The professor (60) was working with an attenuated vaccine strain of the plague bacterium when he died unexpectedly in September 2009 [15].

Subsequent investigations revealed that he suffered from a hereditary condition called "*hemochromatosis*," which causes an excessive build-up of iron, potentially making people more susceptible to bacterial infection. This accident highlighted the fact that, even when working with attenuated strains, researchers should always adhere to the recommended use of personal protective equipment.

Procedure Recommended in Case of Laboratory Spill or Other Type of Accident

In the event of a laboratory spill the room should be vacated for at least 30 min to let the aerosols settle. The area should then be covered with a cloth and sprayed with appropriate disinfectant such as sodium hypochlorite (see section above). A minimum of 30 min contact time should be allowed prior to removal of the cloth and washing of the surface area. Room fumigation using formaldehyde can be considered, depending upon the extent of the spill.

Treatment of Disease

Early antimicrobial therapy is of utmost importance in the treatment of plague and should be started immediately after taking clinical samples, without waiting for culture. Plague is treatable with different antibiotics, namely aminoglycosides, chloramphenicol, and tetracyclines [16–19]. Streptomycin is highly effective against plague and is the drug of choice especially in the case of pneumonic plague [20, 21]. Nowadays availability may be problematic, in which case gentamicin has proven to be a good alternative, alone or in combination with tetracyclines [22]. Both streptomycin and gentamicin must be injected, which may increase the workload and demand for trained medical staff under epidemic conditions.

New antibiotics are under development, such as cethromycin which is a once daily orally bioavailable ketolide considered for the treatment of bacterial pneumonia including anthrax, plague, and tularemia [23]. Treatment guidelines are available from WHO [24, 25] for further reading.

GIDEON (*http://www.gideononline.com/infectiousdiseases/drugs/*) proposes the following treatments for adults:

- Gentamicin 2 mg/kg i.v. loading dose, then 1.7 mg/kg q8 h;
- Streptomycin 15 mg/kg q12 h × 10 days;
- Doxycycline 100 mg p.o. BID × 10 days;
- Chloramphenicol 20 mg/kg p.o. QID.

Also as a typical pediatric therapy:

- Gentamicin 2 mg/kg i.v. loading dose, then 1.7 mg/kg q8 h;
- Streptomycin 10 mg/kg q8 h × 10 days;
- Chloramphenicol 15 mg/kg p.o. QID × 10 days.

The WHO Plague Manual [24] also mentions tetracyclines at 2 g/day for an adult. In vitro and animal tests suggest that ciprofloxacin is a valuable additional possibility, but the experience with human plague cases is limited to a single case. Nevertheless, ciprofloxacin was listed as an alternative treatment (at 400 mg/day i.v. BID) for a contained casualty setting in the "US Working Group recommendations for treatment of patients with pneumonic plague." In a mass casualty setting, it was proposed as an orally taken antibiotic at 500 mg/day p.o. BID [26].

Prophylactic Therapy

True prophylaxis – the administration of an antibiotic prior to exposure – may be indicated when persons must be present for short periods in plague-active areas under circumstances in which exposure to plague sources (fleas, pneumonic cases) is difficult or impossible to prevent.

Clinical Guidelines

- WHO operational guidelines on plague surveillance, diagnosis, prevention, and control [27].
- Bichat guidelines for the clinical management of plague and bioterrorism-related glanders and plague [28].

References

1. Rutala, W.A., Gergen, M.F., and Weber, D.J. (2008) Impact of an oil-based lubricant on the effectiveness of the sterilization processes. *Infect. Control Hosp. Epidemiol.*, **29**, 69–72.
2. Chu, M. (2000) Laboratory Manual of Plague Diagnostic Tests, US Department of Health and Human Services, Public Health Service, Division of Vector-Borne Infectious Diseases National Center for Infectious Diseases, Centers for Disease Control and Prevention and WHO, Division of Emerging and Other Communicable Diseases, Surveillance and Control, p. 129.
3. Shams, A.M., O'Connell, H., Arduino, M.J., and Rose, L.J. (2011) Chlorine dioxide inactivation of bacterial threat agents. *Lett. Appl. Microbiol.*, **53**, 225–230.
4. Wade, M., Biggs, T., Chambers, A., Zulich, A., and Insalaco, J. (2010) Fate of Bacterial and Viral Bio-Warfare Agents in Disinfected Waters, Science Applications International Corporation Gunpowder, MD 21010–0068, Chemical Biological Center US Army Research, Development, and Engineering Command ECBC-TR-810, 23 p.
5. Cornelius, C., Quenee, L., Anderson, D., and Schneewind, O. (2007) Protective immunity against plague. *Adv. Exp. Med. Biol.*, **603**, 415–424.
6. Williamson, E., Flick-Smith, H., LeButt, C., Rowland, C., Jones, S., Waters, E., Gwyther, R., Miller, J., Packer, P., and Irving, M. (2005) Human immune response to a plague vaccine comprising recombinant F1 and V antigens. *Infect. Immun.*, **73**, 3598–3608.
7. Titball, R.W. and Williamson, E.D. (2001) Vaccination against bubonic and pneumonic plague. *Vaccine*, **19**, 4175–4184.
8. Hassani, M., Patel, M., and Pirofskia, L. (2004) Short analytical review. Vaccines for the prevention of diseases caused by potential bioweapons. *Clin. Immunol.*, **111**, 1–15.
9. Wendte, J.M., Ponnusamy, D., Reiber, D., Blair, J.L., and Clinkenbeard, K.D. (2011) In vitro efficacy of antibiotics commonly used to treat human plague against intracellular Yersinia pestis. *Antimicrob. Agents Chemother.*, **55**, 3752–3757.
10. Layton, R., Mega, W., McDonald, J., Brasel, T., Barr, E., Gigliotti, A., and Koster, F. (2011) Levofloxacin cures experimental pneumonic plague in African green monkeys. *PloS Negl. Trop. Dis.*, **5**, e959.
11. Layton, R.C., Brasel, T., Gigliotti, A., Barr, E., Storch, S., Myers, L., Hobbs, C., and Koster, F. (2011) Primary pneumonic plague in the African Green monkey as a model for treatment efficacy evaluation. *J. Med. Primatol.*, **40**, 6–17.
12. Galimand, M., Guiyoule, A., Gerbaud, G., Rasoamanana, B., Chanteau, S., Carniel, E., Courvalin, P., and Engl, J. (1997) Multi-drug resistance in *Yersinia pestis* mediated by a transferable plasmid. *N. Engl. J. Med.*, **337**, 677–680.
13. Guiyoule, A., Gerbaud, G., Buchrieser, C., Galimand, M., Rahalison, L., Chanteau, S., Courvalin, P., and Carniel, E. (2001) Transferable plasmid-mediated resistance to streptomycin in a clinical isolate of *Yersinia pestis. Emerg. Infect. Dis.*, **7**, 43–48.
14. McCormick, J. (1998) Epidemiology of emerging/re-emerging antimicrobial-resistant bacterial pathogens. *Curr. Opin. Microbiol.*, **1**, 125–129.
15. Anonymous (2009) U. of C. researcher dies after exposure to plague bacteria. *Chicago Tribune* (Sep 19).
16. Frean, J., Arntzen, L., Capper, T., Bryskier, A., and Klugman, K. (1996) In vitro activities of 14 antibiotics against 100 human isolates of Yersinia pestis from a southern African plague focus. *Antimicrob. Agents Chemother.*, **40**, 2646–2647.
17. McCrumb, F. Jr., Mercier, S., Robic, J., Bouillat, M., Smadel, J., Woodward, T., and Goodner, K. (1953) Chloramphenicol and terramycin in the treatment of pneumonic plague. *Am. J. Med.*, **14**, 284–293.

18. Russell, P., Eley, S., Green, M., Stagg, A., Taylor, R., Nelson, M., Beedham, R., Bell, D., Rogers, D., Whittington, D., and Titball, R. (1998) Efficacy of doxycycline and ciprofloxacin against experimental *Yersinia pestis* infection. *Antimicrob. Agents Chemother.*, **41**, 301–305.

19. Welty, T. (1984) in *Current Therapy* (ed. H.F. Conn), Saunders, Philadelphia, pp. 44–45.

20. Byrne, W., Welkos, S., Pitt, M., Davis, K., Brueckner, R., Ezzell, J., Nelson, G., Vaccaro, J., Battersby, L., and Friedlander, A. (1998) Antibiotic treatment of experimental pneumonic plague in mice. *Antimicrob. Agents Chemother.*, **42**, 675–681.

21. Meyer, K.F. (1950) Modern therapy of plague. *J. Am. Med. Assoc.*, **144**, 982–985.

22. Boulanger, L.L., Ettestad, P., Fogarty, J.D., Dennis, D.T., Romig, D., and Mertz, G. (2004) Gentamicin and tetracyclines for the treatment of human plague: review of 75 cases in new Mexico, 1985–1999. *Clin. Infect. Dis.*, **38**, 663–669.

23. Rafie, S., MacDougall, C., and James, C.L. (2010) Cethromycin: a promising new ketolide antibiotic for respiratory infections. *Pharmacotherapy*, **30**, 290–303.

24. WHO (1999) Plague Manual Epidemiology, Distribution, Surveillance and Control, WHO/CDS/CSR/EDC/99.2, *http://www.who.int/csr/resources/publications/plague/WHO_CDS_CSR_EDC_99_2_EN/*.

25. WHO (2008) Interregional Meeting on Prevention and Control of Plague – Antananarivo, Madagascar, April 7–11 2006, *http://www.who.int/csr/resources/publications/WHO_HSE_EPR_2008_3w.pdf* (accessed July 21 2009).

26. Inglesby, T., Dennis, D., Henderson, D., Bartlett, J., Ascher, M., Eitzen, E., Fine, A., Friedlander, A., Hauer, J., Koerner, J., Layton, M., McDade, J., Osterholm, M., O'Toole, T., Parker, G., Perl, T., Russell, P., Schoch-Spana, M., and Tonat, K., Working Group on Civilian Biodefense (2000) Plague as a biological weapon: medical and public health management. *J. Am. Med. Assoc.*, **283**, 2281–2290.

27. WHO (2009) Operational Guidelines on Plague Surveillance, Diagnosis, Prevention and Control, Regional Office for South-East Asia, New Delhi, WHO, 89 pp. *http://www.searo.who.int/LinkFiles/Publication_op_guidelines_plague.pdf* (accessed July 21 2009).

28. Bossi, P., Tegnell, A., Baka, A., Van Loock, F., Hendriks, J., Werner, A., Maidhof, H., and Gouvras, G. (2004) Bichat guidelines for the clinical management of plague and bioterrorism-related plague. *Euro Surveill.*, **9** (12), E17–E18.

8
Rickettsia Species: Rickettsioses

Alice N. Maina, Stephanie Speck, Eva Spitalska, Rudolf Toman, Gerhard Dobler, and Sally J. Cutler

Organism	Laboratory category	Transmissible by aerosol
Rickettsia conorii	BSL-3	Yes
Rickettsia rickettsii	BSL-3	Yes
Spotted-fever group *Rickettsia*	BSL-3	No data available
Rickettsia prowazekii		
Rickettsia typhi	BSL-3	Yes
	BSL-3	Yes

The spotted-fever group *Rickettsia* consists of a wide variety of species of different laboratory categories, though most are recommended to be handled in a BSL3 environment (* indicates classification dependent on country): *R. aeschlimannii* (BSL-3), *R. africae* (BSL-3), *R. akari* (BSL-3), *R. australis* (BSL-3), *R. felis* (BSL-2/BSL-3*), *R. heilongjiangensis* (BSL-3), *R. helvetica* (BSL-2/BSL-3*), *R. honei* (BSL-2/BSL-3*), *R. japonica* (BSL-3), *R. massiliae* (BSL-2/BSL-3*), *R. monacensis* (BSL-2), *R. mongolotimonae* (BSL-3), *R. parkeri* (BSL-2/BSL-3*), *R. sibirica* (BSL-3), and *R. slovaca* (BSL-2/BSL-3*).

Recommended Respiratory Protection

In case of contacts with aerosols of *Rickettsia* valved FFP3 disposable respirators or respiratory protection using powered air purifying respirators (PAPRs) are recommended.

Recommend Personal Protective Equipment

Handling this organism should be conducted in a biological safety cabinet. Other personal equipment includes gloves and gown. Personnel entering the laboratory should remove street clothing and jewelry, and change into dedicated laboratory

BSL3 and BSL4 Agents: Epidemiology, Microbiology, and Practical Guidelines, First Edition.
Edited by Mandy C. Elschner, Sally J. Cutler, Manfred Weidmann, and Patrick Butaye.

clothing and shoes, or wear full coverage protective clothing (i.e., completely covering all street clothing). Additional protection may be worn over laboratory clothing when infectious materials are directly handled, such as solid-front gowns with tight-fitting wrists, gloves. Eye protection must be used where there is a known or potential risk of exposure to splashes.

Best Disinfection

70% ethanol.

Best Decontamination

Gram-negative bacteria are susceptible to 1% sodium hypochlorite, 4% formaldehyde, 2% glutaraldehyde, 70% ethanol, 2% peracetic acid, 3–6% hydrogen peroxide, and 0.16% iodine.

Prevention

For most rickettsial infections, there is no vaccine available currently.

The first vaccine against *R. prowazekii* was developed by Weigl in Poland in lice [1]. Later, the Madrid E nonpathogenic strain and other inactivated typhus vaccines were evaluated, including chicken egg (Cox) and rat lung (Durand) types [2]. No commercial vaccines have been licensed, but experimental vaccines are produced by military sources in the United States and may be available for high-risk situations. At present, the research and development of a new generation vaccine is not a priority as an antibiotic treatment is very efficient.

A residual insecticide treatment of the clothing and hair is recommended for people who may have been exposed to infected lice. Prevention of an infection caused by *R. typhi* can be achieved through rodent and flea control.

Post Exposure Prophylaxis

Rickettsia infections can be treated with broad spectrum antibiotics, such as tetracyclines, which are the antibiotics of choice. There is little or no resistance reported against antibiotics.

For *R. prowazekii* one may also use chloramphenicol (available for human use in developing countries only).

Known Laboratory Accidents

Laboratory acquired infections by inhalation, mucous membrane contamination, needle-stick, and cuts have been reported [3]. Demonstrating potential infectivity of a rickettsial aerosol are the numerous laboratory-associated cases of infection. Although direct transcutaneous inoculation of the pathogen in the laboratory milieu occurs, many cases appear to be related to infectious aerosols of *R. prowazekii* [4, 5].

A 34-year-old laboratory worker developed murine typhus after an accidental splashing of *R. typhi* over her right eye and lips. An indirect immunoperoxidase test showed a four-fold increase in titer to *R. typhi*. She responded well to doxycycline [6].

Procedure Recommended in Case of Laboratory Spill or Other Type of Accident

All procedures must be carried out wearing personal protective clothing. In the case of spills allow aerosols to settle, cover spill with paper towels and apply appropriate disinfectant, starting at the periphery, and working toward the center. Allow sufficient contact time before cleaning up (30 min). Disinfection and decontamination procedures depend on situation and can include fumigation of the room using formaldehyde.

Treatment of Disease

As for the post exposure prevention, tetracyclines are the treatment of choice.

Tetracycline and chloramphenicol are usually used in treatment of epidemic typhus. A single dose of 200 mg of doxycycline was extremely efficient and ciprofloxacin should be avoided [7]. Reports on the use of fluoroquinolones show conflicting results [8].

Clinical Guidelines

- Raoult, D. and Parola, P. (eds) (2007) *Rickettsial Diseases*, Informa Healthcarem, New York, p. 379.
- Mohammad, M. and Lutwick, L.I. (2009) Epidemic typhus fever, in *Beyond Anthrax* (eds M. Mooty and L.I. Lutwick), Springer Science+Business Media, LLC 2008, pp. 159–180.
- Raoult, D. and Parola, P. (1997) Rickettsioses as paradigms of new or emerging infectious diseases. *Clin. Microbiol. Rev.*, **10**, 694–719.

References

1. Boese, J.L., Wisseman, C.L.J., Walsh, W.T., and Fiset, P. (1973) Antibody and antibiotic action on *Rickettsia prowazekii* in body lice across the host-vector interface, with observations on strain virulence and retrieval mechanisms. *Am. J. Epidemiol.*, **98**, 262–282.
2. Woodward, T.E. (1986) Rickettsial vaccines with emphasis on epidemic typhus: initial report of an old vaccine trial. *S. Afr. Med. J.*, **11** (Suppl.), 73–76.
3. Sewell, D.L. (1995) Laboratory-associated infections and biosafety. *Clin. Microbiol. Rev.*, **8**, 389–405.

4. Pike, R.M. (1976) Laboratory-associated infections: summary and analysis of 3921 cases. *Health Lab. Sci.*, **13**, 105–114.
5. Oster, C.N., Burke, D.S., Kenyon, R.H. *et al.* (1977) Laboratory-acquired Rocky Mountain potted fever. The hazard of aerosol transmission. *N. Engl. J. Med.*, **297**, 859–863.
6. Norazah, A., Mazlah, A., Cheong, Y.M., and Kamel, A.G. (1995) Laboratory acquired murine typhus–a case report. *Med. J. Malays.*, **50**, 177–179.
7. Raoult, D. and Roux, V. (1999) The body louse as a vector of reemerging human diseases. *Clin. Infect. Dis.*, **29**, 888–911.
8. Gikas, A., Doukakis, S., Pediaditis, J., Kastanakis, S. *et al.* (2004) Comparison of the effectiveness of five different antibiotic regimens on infection with *Rickettsia typhi*: therapeutic data from 87 cases. *Am. J. Trop. Med. Hyg.*, **70**, 576–579.

9
Mycobacterium tuberculosis: Tuberculosis

Stefan Panaiotov, Massimo Amicosante, Marc Govaerts, Patrick Butaye, Elizabeta Bachiyska, Nadia Brankova, and Victoria Levterova

Organism	Laboratory category	Transmissible by aerosol
Mycobacterium tuberculosis	BSL-3	Yes

Recommended Respiratory Protection

The main route of transmission of tuberculosis (TB) infection (*Mycobacterium tuberculosis*) is by aerosols; consequently it is highly advisable to wear respiratory protection. A susceptible person can acquire an infection with *M. tuberculosis* by inhaling fewer than 10 bacilli. The use of respirators in the health care setting and laboratories is an important step in the efforts to prevent the transmission of TB. Level 3 respiratory masks are recommended, for example, positive ventilated masks attached to filtering facepiece (FFP)3 filters (e.g., 3M, Dräger). A number of studies have shown that simple surgical masks do not provide adequate protection; furthermore, surgical masks are not respirators and thus do not satisfy requirements for respiratory protection. Comprehensive reviews on respiratory protection against bioaerosols and in particular protection against *M. tuberculosis* aerosols are given by Hodous and Coffey [1] and Rengasamy *et al.* [2].

Recommended Personal Protective Equipment

Latex disposable gloves are widely recommended. Gloves are to be worn when any possible contact with potentially infectious material exists. Samples should be handled in an appropriate BSL-3 cabinet. Respiratory protection should be used where this is not possible.

BSL3 and BSL4 Agents: Epidemiology, Microbiology, and Practical Guidelines, First Edition.
Edited by Mandy C. Elschner, Sally J. Cutler, Manfred Weidmann, and Patrick Butaye.

Best Disinfection

Surface and Equipment

Sodium hypochlorite (bleach) requires a higher concentration of available chlorine to achieve an effective level of disinfection, with 5% solution being generally recommended. Phenol (5%) is highly effective against *M. tuberculosis*, thus phenol-containing disinfectants are effective. A solution of 2% glutaraldehyde is also effective against *M. tuberculosis*. Some of these disinfectants are prohibited in certain countries, thus local regulations should be consulted. Ethanol (70%) is effective against *M. tuberculosis* only in suspension in the absence of sputum. Povidone–iodine is not as efficacious when the organism is dried on a surface, compared with in suspension.

Best Decontamination

See Disinfection and additionally fumigation of equipment or room using formaldehyde.

Prevention

TB is one of the major causes of death worldwide with about 3 million new cases of active TB per year. One-third of the world population is estimated to be infected by *M. tuberculosis*. *M. bovis* BCG vaccine has been extensively used worldwide to prevent infection by *M. tuberculosis* (TB) since 1920. Efficacy varies under different epidemiological conditions. It can prevent the typical TB manifestations and complicating forms in children and newborns (such as TB meningitis).

Ongoing Clinical Trials

No new therapeutic drugs are currently ready to launch onto the market. Some new drugs targeting *Mycobacteria* specific sigma factors are in preclinical evaluation. Nevertheless, the use of different standard regimens have been largely evaluated over recent years to maximize their efficacy and compliance over the long anti-tuberculous therapy (four to nine months according to the different guidelines and therapy regimens).

Different new vaccines are undergoing phase I evaluations. None have yet completed phase II assessment of immunogenicity/efficacy. Among the 15 vaccines presently under evaluation, a range of different strategies have been employed, including: modified BCG, subunit vaccines, viral vectors carrying *M. tuberculosis* proteins, live attenuated *M. tuberculosis*, and cross-reactive *Mycobacteria* [3].

Post Exposure Prophylaxis

Following exposure with active cases and in the presence of a documented infection (conversion to or presence of positive tuberculin skin test or IFN gamma assays) a six to nine month prophylactic course with isoniazid (commercially available as Lanizid or Nydrazid) is recommended. Use of a short prophylactic treatment with once weekly treatment for three months with isoniazid and mycobutin is currently under evaluation, but still awaits efficacy data.

Known Laboratory Accidents

Previously TB was the fifth most frequently reported infection among laboratory workers in the United States [4]. Recent figures show similar levels of accidental laboratory-acquired TB [5].

Procedure Recommended in Case of Laboratory Spill or Other Type of Accident

Disinfection and decontamination procedures depend on the situation (see sections above) and can include fumigation of the room using formaldehyde.

Treatment of Disease

There are four recommended regimens for treating patients with TB caused by drug-susceptible organisms. Each regimen has an initial phase of two months. In most circumstances, the treatment regime for all adults with previously untreated TB should consist of a two-month initial phase of isoniazid, rifampin, pyrazinamide, and ethambutol. This initial phase is followed by a choice of several options for the continuation phase of four or seven months. In recent years the treatment of drug resistant and multidrug resistant TB has significantly improved.

WHO web reference offers detailed information regarding treatment regimens and editions in various languages [6].

Clinical Guidelines

The most complete guidelines and international respiratory society statements for diagnosis and treatment are available at:

- **American Thoracic Society**: *http://www.thoracic.org/statements/*;
- **CDC**: *http://www.cdc.gov/tb/publications/guidelines/List_date.htm*;
- **WHO**: *http://www.who.int/tb/publications/cds_tb_2003_313/en/*.

References

1. Hodous, T.K. and Coffey, C.C. (1994) The role of respiratory protective devices in the control of tuberculosis. *Occup. Med.*, **9**, 631–657.
2. Rengasamy, A., Zhuang, Z., and Berryann, R. (2004) Respiratory protection against bioaerosols: literature review and research needs. *Am. J. Infect. Control*, **32**, 345–354.
3. Parida, S.K. and Kaufmann, S.H. (2010) Novel tuberculosis vaccines on the horizon. *Curr. Opin. Immunol.*, **22**, 374–384.
4. Pike, R.M. (1976) Laboratory-associated infections: summary and analysis of 3921 cases. *Health Lab. Sci.*, **13**, 105–114.
5. Singh, K. (2009) Laboratory-acquired infections. *Clin. Infect. Dis.*, **49**, 142–147.
6. WHO (2009) Treatment of Tuberculosis: Guidelines for National Programmes, 4th edn, WHO, WHO/HTM/TB/2009.420 *http://www.who.int/tb/publications/tb_treatmentguidelines/en/index.html* (accessed August 13 2011).

Part II
Viruses

BSL3 and BSL4 Agents: Epidemiology, Microbiology, and Practical Guidelines, First Edition.
Edited by Mandy C. Elschner, Sally J. Cutler, Manfred Weidmann, and Patrick Butaye.

10 Influenza Virus: Highly Pathogenic Avian Influenza

Chantal J. Snoeck and Claude P. Muller

Organism	Laboratory category	Transmissible by aerosol
Influenza virus	BSL-2	Yes
Highly pathogenic influenza virus	BSL-3	Yes

Recommended Respiratory Protection

Filtering facepiece (FFP)2 masks are recommended for manipulations such as aliquoting clinical specimens, nucleic acid extraction, and diagnostic testing (for example, 3M® 9322 foldable respirator (valved)).

For manipulations involving virus propagation (virus cultures, egg inoculation, virus neutralization assay, etc.), or handling concentrated viruses, FFP3 masks (such as Moldex 3405 FFP3 face masks) or face shield powered respirators are recommended.

Recommended Personal Protective Equipment

For manipulations not involving virus propagation (see above): wrap-around laboratory coat, inner and outer pair of gloves, sleeve protectors that entirely cover the forearms, head cover, eye protection, and shoe covers [1–3].

For manipulations involving virus propagation (see above): dedicated trousers, inner and outer wrap-around laboratory coat, three pairs of gloves, sleeve protectors that entirely cover the forearms, head cover, eye protection, dedicated shoes, and shoe covers [1–3]. Discard outer pair of gloves inside the biosafety cabinet before taking hands out of the biosafety cabinet. When staff leave the BSL-3 their outer laboratory coat is discarded and autoclaved.

BSL3 and BSL4 Agents: Epidemiology, Microbiology, and Practical Guidelines, First Edition.
Edited by Mandy C. Elschner, Sally J. Cutler, Manfred Weidmann, and Patrick Butaye.

Best Disinfection

Fumigation using hydrogen peroxide (H_2O_2) or formaldehyde is suitable to decontaminate BSL2 or BSL3 facilities or biosafety cabinets used to handle influenza viruses. These compounds are irritant, toxic, and carcinogenic and must be handled by trained personnel. Waste leaving the BSL2 or BSL3 area where this pathogen was handled must be disinfected by dry or moist heat or incineration. Autoclaving (moist heat under pressure) is fast and efficient and is widely used [4].

Best Decontamination

Hypochlorite solutions, such as sodium hypochlorite (bleach), can be used as decontaminant and at a concentration of 1 g/l available chlorine. A higher chlorine concentration (5 g/l) is recommended in the case of a spill or when dealing with a large amount of organic matter. Contact time of at least 10 min should be allowed [3]. Although it is a cheap decontaminant, sodium hypochlorite has some disadvantages, such as corrosiveness. It is a strong irritant. It is inactivated by organic matter and it is very dangerous when mixed with acids as this can lead to the release of very toxic chlorine gas [4]. Virkon® (potassium peroxomonosulfate, sodium dodecylbenzenesulfonate, sulfamic acid, and potassium persulfate) has a broad activity spectrum. As powder, it is irritating but has a low toxicity in solution, which makes Virkon® interesting compared to other decontaminating agents. A solution of 1% Virkon® in water can be used as a general decontaminant and a contact time of 15 min should be allowed. A 1% Virkon® solution can also be used in a dunk bath to soak small laboratory objects such as racks, centrifuge buckets. In case of a biohazardous spill, a 2% Virkon® solution or Virkon® powder directly sprinkled on the spill are recommended.

Prevention

Laboratory staff should be vaccinated every year with the seasonal influenza vaccine. Vaccination will not prevent infection in the case of an accident but it will prevent the possibility of co-infections that could result in the emergence of a reassortant virus.

Since the emergence of the HPAI H5N1 virus in Hong Kong, growing concern of a potential pandemic caused by a highly pathogenic influenza virus has led to a burst of research to develop a vaccine against H5N1. Several approaches have been developed to conform with the requirements for vaccine production in the preparation of a pandemic, such as the necessity of developing a broad long-lasting immunity using only low antigen doses. Most strategies include the production of split- or whole-virus, inactivated vaccines in embryonated chicken eggs. However, this approach is limited by the safe supply of eggs in the case of a pandemic and the time required for the production of these vaccines [5]. Cell-based vaccines are now encouraged to overcome this problem. The use of adjuvants in vaccine preparations has shown promising results in eliciting strong immune response despite lower antigen doses [6]. Live-attenuated strains, already used in seasonal vaccines, offer the advantages of eliciting a rapid and broader immune response and are easier to produce and administrate. They are now being investigated as alternatives for the preparation of pandemic vaccines [7–9].

In pre-clinical studies vaccines have been used that were based on new approaches, such as recombinant proteins (HA, NA, and M1), virus-like particles produced by baculovirus or lentivirus systems, DNA vaccines or vaccines based on viral vectors such as adenoviruses, alphaviruses, newcastle disease virus, poxviruses, or vesicular stomatitis virus as egg-independent alternatives.

Vaccines with Marketing Authorization from the European Medicines Agency

- **Prepandrix (GlaxoSmithKline Biologicals)**: split-virion, inactivated, adjuvanted (AS03) [10];
- **Daronrix (GlaxoSmithKline Biologicals)**: whole-virion, inactivated, adjuvanted (aluminum phosphate and aluminum hydroxide) [11];
- **Pumarix (GlaxoSmithKline Biologicals)**: split-virion, inactivated, adjuvanted (AS03) [12];
- **Aflunov (Novartis)**: surface antigen (HA, NA), inactivated, adjuvanted (MF56) [13];
- **Pandemic influenza vaccine H5N1 (Baxter)**: whole-virion, inactivated, grown on vero cells [14];
- **Foclivia (Novartis)**: surface antigens (HA, NA), inactivated, adjuvanted (MF59) [15].

Post Exposure Prophylaxis

People at high to moderate risk of exposure (having close contacts with an H5N1 infected patient in health care facilities or at home, handling sick or dead poultry with suboptimal protective clothing) can be prescribed with an antiviral chemoprophylaxis made of neuraminidase inhibitors, namely oseltamivir (Tamiflu, from Roche) and zanamivir (Relenza, from GlaxoSmithKline), or M2 inhibitors, namely amantadine (Symmetrel, from Endo Pharmaceuticals; generic available) or rimantadine (Flumadine, from Forest Pharmaceuticals; generic available) hydrochlorides. As a general rule, neuraminidase inhibitors should be preferred to M2 inhibitors because of the increased propensity of influenza viruses to develop resistance against the latter and the lower tolerability in humans. All drugs should be administered for 7–10 days after the last exposure [16].

Procedure Recommended in Case of Laboratory Spill or Other Type of Accident

Influenza Virus: Highly Pathogenic Avian InfluenzaReferencesIn case of a laboratory spill, all manipulations must be interrupted immediately and all open tubes must be closed. Decontaminate the spill with a 2% Virkon® solution or by applying Virkon® powder on the spill. After a contact time of 15 min, the liquid or the soaked powder is absorbed or collected and disposed into a biohazard bag. Decontaminate the surface a second time with Virkon 2%. Discard all laboratory material into a biohazard bag.

Decontaminate the whole area (biosafety cabinet and BSL2/3 room) with Virkon® and autoclave the waste. Biosafety cabinets are best decontaminated by formalin fumigation and the BSL2/3 laboratory is decontaminated by hydrogen peroxide fumigation.

In the case of contamination of personnel, all protective and personal clothes are removed and autoclaved, and the laboratory worker must take a shower before leaving the BSL2/3 laboratory. In case of skin, eye, or mucosal surface exposure, the exposed body parts are disinfected with appropriate solutions present inside the BSL2/3 rooms. In the case of bleeding, the wound is encouraged to bleed before disinfection. All accidents involving contact with pathogens must be regarded as medical emergencies and a medical doctor must be seen and reported.

Treatment of Disease

As a specific treatment, oseltamivir is the recommended drug treatment in the case of human H5N1 influenza virus infections [16]. This seems to increase survival rates, but the optimal dose and treatment duration is not clear [17]. Although the global mortality rate of highly pathogenic avian influenza (HPAI) H5N1 infected patient reaches 60% [18], recent studies have demonstrated that early hospitalization is crucial for survival rates [19–24]. Symptomatic treatment and supportive care at the extensive care unit including mechanical ventilation, antibiotics, and so on, are critical for the management of severe cases [16].

References

1. Bischoff, W.E., Reid, T., Russell, G.B., and Peters, T.R. (2011) Transocular entry of seasonal influenza-attenuated virus aerosols and the efficacy of n95 respirators, surgical masks, and eye protection in humans. *J. Infect. Dis.*, **204**, 193–199.
2. OSHA (2007) Avian Flu: General Precautions, *http://www.osha.gov/Publications/3306-10-06-english-06-27-2007.html* (accessed 22 June 2011).
3. WHO (2005) WHO Laboratory Biosafety Guidelines for Handling Specimens Suspected of Containing Avian Influenza A Virus, *http://www.who.int/csr/disease/avian_influenza/guidelines/handlingspecimens/en/* (accessed 17 June 2011).
4. WHO (2004) Laboratory Biosafety Manual, *http://www.who.int/entity/csr/resources/publications/biosafety/Biosafety7.pdf* (accessed 17 June 2011).

5. Gerdil, C. (2003) The annual production cycle for influenza vaccine. *Vaccine*, **21**, 1776–1779.
6. Stephenson, I., Hayden, F., Osterhaus, A., Howard, W. *et al.* (2010) Report of the fourth meeting on 'Influenza vaccines that induce broad spectrum and long-lasting immune responses', World Health Organization and Wellcome Trust, London, United Kingdom, 9-10 November 2009. *Vaccine*, **28**, 3875–3882.
7. Rudenko, L., Desheva, J., Korovkin, S., Mironov, A. *et al.* (2008) Safety and immunogenicity of live attenuated influenza reassortant H5 vaccine (phase I-II clinical trials). *Influenza Other Respi. Viruses*, **2**, 203–209.
8. Steel, J., Lowen, A.C., Pena, L., Angel, M. *et al.* (2009) Live attenuated influenza viruses containing NS1 truncations as vaccine candidates against H5N1 highly pathogenic avian influenza. *J. Virol.*, **83**, 1742–1753.
9. Suguitan, A.L., McAuliffe, J., Mills, K.L. *et al.* (2006) Live, attenuated influenza A H5N1 candidate vaccines provide broad cross-protection in mice and ferrets. *PLoS Med.*, **3**, e360.
10. EMEA (2009) Prepandrix, http://www.ema.europa.eu/ema/index.jsp?curl=pages/medicines/human/medicines/000822/human_med_000986.jsp&murl=menus/medicines/medicines.jsp&mid=WC0b01ac058001d125 (accessed 16 June 2011).
11. EMEA (2007) Daronrix, http://www.ema.europa.eu/ema/index.jsp?curl=pages/medicines/human/medicines/000706/human_med_000738.jsp&murl=menus/medicines/medicines.jsp&mid=WC0b01ac058001d125 (accessed 16 June 2011).
12. EMEA (2011) Pumarix, http://www.ema.europa.eu/ema/index.jsp?curl=pages/medicines/human/medicines/001212/human_med_001412.jsp&murl=menus/medicines/medicines.jsp&mid=WC0b01ac058001d125 (accessed 16 June 2011).
13. EMEA (2011) Aflunov, http://www.ema.europa.eu/ema/index.jsp?curl=pages/medicines/human/medicines/002094/human_med_001396.jsp&murl=menus/medicines/medicines.jsp&mid=WC0b01ac058001d125 (accessed 16 June 2011).
14. EMEA (2009) Pandemic Influenza Vaccine H5N1 Baxter, http://www.ema.europa.eu/ema/index.jsp?curl=pages/medicines/human/medicines/001200/human_med_001215.jsp&murl=menus/medicines/medicines.jsp&mid=WC0b01ac058001d125 (accessed 16 June 2011).
15. EMEA (2009) Foclivia, http://www.ema.europa.eu/ema/index.jsp?curl=pages/medicines/human/medicines/001208/human_med_001216.jsp&murl=menus/medicines/medicines.jsp&mid=WC0b01ac058001d125 (accessed 17 June 2011).
16. WHO (2006) Rapid Advice Guidelines on Pharmacological Management of Humans Infected with Avian Influenza (H5n1) Virus. http://www.who.int/entity/medicines/publications/WHO_PSM_PAR_2006.6.pdf (accessed 16 June 2011).
17. Abdel-Ghafar, A.N., Chotpitayasunondh, T., Gao, Z., Hayden, F.G. *et al.* (2008) Update on avian influenza A (H5N1) virus infection in humans. *N. Engl. J. Med.*, **358**, 261–273.
18. WHO (2011) Cumulative Number of Confirmed Human Cases of Avian Influenza A/(H5N1) Reported to WHO, http://www.who.int/csr/disease/avian_influenza/country/cases_table_2011_06_10/en/index.html (accessed 16 June 2011).
19. Kayali, G., Webby, R.J., Ducatez, M.F., El Shesheny, R.A. *et al.* (2011) The epidemiological and molecular aspects of influenza H5N1 viruses at the human-animal interface in Egypt. *PLoS ONE*, **6**, e17730.
20. Chotpitayasunondh, T., Ungchusak, K., Hanshaoworakul, W., Chunsuthiwat, S. *et al.* (2005) Human disease from influenza A (H5N1), Thailand, 2004. *Emerg. Infect. Dis.*, **11**, 201–209.
21. Kandun, I.N., Tresnaningsih, E., Purba, W.H., Lee, V. *et al.* (2008) Factors associated with case fatality of human H5N1 virus infections in Indonesia: a case series. *Lancet*, **372**, 744–749.

22. Sedyaningsih, E.R., Isfandari, S., Setiawaty, V., Rifati, L. *et al.* (2007) Epidemiology of cases of H5N1 virus infection in Indonesia, July 2005-June 2006. *J. Infect. Dis.*, **196**, 522–527.

23. Thanh, T.T., van Doorn, H.R., and de Jong, M.D. (2008) Human H5N1 influenza: current insight into pathogenesis. *Int. J. Biochem. Cell Biol.*, **40**, 2671–2674.

24. Gambotto, A., Barratt-Boyes, S.M., de Jong, M.D., Neumann, G. *et al.* (2008) Human infection with highly pathogenic H5N1 influenza virus. *Lancet*, **371**, 1464–1475.

11
Variola: Smallpox

Andreas Nitsche and Hermann Meyer

Organism	Laboratory category	Transmissible by aerosol
Variola virus (smallpox)	BSL-4	Yes

Recommended Respiratory Protection

Current international regulations state that work with Variola virus (VARV) is to be done using biosafety level 4 laboratories sanctioned by WHO. Therefore, the personal protective equipment consists of a positive pressure suit.

Recommended Personal Protective Equipment

When seeing potential smallpox patients Centers for Disease Control and Prevention (CDC) recommends wearing a fit-tested 3M particulate respirator NIOSH N-95 mask and appropriate eye protection, gloves, and gowns. In addition, an N-95 mask should be placed on the patient when the patient is not in a negative pressure respiratory isolation room. In Europe, equivalent respirator masks are designated as filtering facepiece-3 (FFP-3). In addition, it is recommended that personnel should be protected by current vaccination.

Best Disinfection

Whenever possible articles potentially contaminated should be destroyed by burning. If this cannot be done, articles should be autoclaved instead. In the later stages of the disease smallpox is still infectious because of virus particles dried on the skin, on bedclothes, and possibly within smallpox crusts, or dried pus, or blood. Smallpox crusts can be sterilized by exposure to dry heat at 160 °C for 60 min or to steam

BSL3 and BSL4 Agents: Epidemiology, Microbiology, and Practical Guidelines, First Edition.
Edited by Mandy C. Elschner, Sally J. Cutler, Manfred Weidmann, and Patrick Butaye.

under pressure at 121 °C for 15 min, or 126 °C for 10 min, or 134 °C for 3 min, as in regular autoclaving. If burning or autoclaving are both impracticable, chemical disinfection should be considered. Valuables may be disinfected by exposure to formaldehyde vapor or to ethylene oxide in a special chamber.

Since the envelope surrounding the pox particle contains lipids, this property renders poxviruses sensitive to disinfection with a number of chemicals, such as ethyl alcohol (40%), isopropyl alcohol (30%), sodium hypochlorite (200 ppm), or iodophor (75 ppm; the values in brackets refer to inactivation of vaccinia virus after 10 min contact time at room temperature).

Best Decontamination

If the patient has been removed from the premises, formaldehyde fumigation should be performed. The most practical method for formaldehyde fumigation is the vaporization of formaldehyde solution by boiling; this ensures the necessary high relative humidity. Formaldehyde fumigation is also the method of choice to decontaminate laboratories for subsequent technical inspections and to decontaminate equipment in material locks.

Prevention

WHO has built a strategic stock of smallpox vaccines of 30.5 million doses. It consists of ACAM2000™ smallpox vaccine, a cell culture-produced second-generation vaccine, which has been licensed in the United States since 2007.

A third-generation candidate vaccine is Imvamune® which consists of Vaccinia virus MVA, a virus which does not replicate fully in human and most other mammalian cells. However, a higher dose or multiple doses of MVA are required to achieve the immune protection produced using a single dose of replicating VACV.

Case Reports, Ongoing Clinical Trials

Phase I and II clinical trials with MVA have been completed, and phase III clinical trials started in 2011.

Post Exposure

Vaccination within three days of exposure may significantly modify smallpox, vaccination four to seven days after exposure likely offers some protection from disease. Vaccination after exposure to smallpox virus infection may be effective in minimizing casualties from smallpox. Several studies have evaluated this possibility in animal models. Most have concluded that the vaccine has to be administered no later than one to two days after exposure to a virulent poxvirus in order to protect against death [1–3]. Interestingly, MVA elicited a more rapid protective response than the NYCBH strain in a monkeypox challenge model, probably due to the high dose used in the trial [4].

Laboratory Accidents

The last known fatal smallpox infection occurred in 1978 by a laboratory accident. A photographer working in the anatomy department at the University of Birmingham Medical School became infected by VARV which was being handled in a laboratory one floor below. Transmission most probably occurred through a service shaft connecting both floors.

- Anonymous (1978) Smallpox in Birmingham. *Br. Med. J.*, **16**, 837–838.
- Shooter, R. (1980) Report of the Investigation into the Cause of the 1978 Birmingham Smallpox Occurrence, London.

Procedure Recommended in Case of Laboratory Spill or Other Type of Accident

Events (breach in the integrity of the space suit used in the high containment laboratory) that may result in exposure must be reported. In the case of potential percutaneous exposure the wound must be immediately treated with disinfectant and the person will exit with deliberate speed. Additional application of disinfectant and thorough washing of the infected area with soap, water, and brush will be carried out.

In both cases an evaluation will be made and further management may range from no action to close monitoring of the person for fever, to isolation and possibly treatment with prophylactic drugs or other modalities.

Spills that may cause damage to the space suit must be dealt with promptly and in a manner to eliminate the hazard. Areas in which spills of potentially infectious material occur will be thoroughly treated with a 3% (or greater) solution of Microchem or undiluted household bleach (Clorox) or any appropriate disinfectant, wiped up, and disposed of in autoclave pans or biohazard bags.

Treatment of Disease

It is the aim of WHO to obtain two approved oral antiviral compounds, with different mechanisms of action, which can be used for treatment of clinical cases of smallpox. As a result of this effort, three compounds (cidofovir, ST-246®, CMX001) that inhibit VARV replication in cell culture and in multiple animals utilizing surrogate orthopoxvirus models, have gained Investigational New Drug status from the FDA for the treatment of orthopoxvirus infections. Initial human studies are underway and phase I and II clinical studies have demonstrated an excellent safety profile for the two leading candidate compounds. Some animal data, especially from studies of ST-246®, showed that vaccination and simultaneous drug administration can be performed without attenuation of the immune response.

Three severe cases of vaccinia virus infection were successfully treated (compassionate use) with ST246. A 20-year old service member with progressive vaccinia (PV), a moribund two year old with eczema vaccinatum and a person with complications due to exposure to a live recombinant vaccinia virus vaccine. The proposed human therapeutic course would consist of two capsules (400 mg) once per day for 14 days.

- (2009) Human vaccinia infection after contact with a raccoon rabies vaccine bait – Pennsylvania, 2009. *MMWR*, **58** (43).
- (2009) Progressive vaccinia in a military smallpox vaccinee – United States, 2009. *MMWR*, **58** (19).
- (2007) Household transmission of vaccinia virus from contact with a military smallpox vaccinee – Illinois and Indiana, 2007. *MMWR*, **56** (19).

Vaccinia immune globulin (VIG) is approved and has efficacy against certain of the more common serious vaccine reactions. Whether VIG has an efficacy against actual smallpox disease (VARV) is a matter of discussion. VIG is only given intramuscularly and requires large volumes (e.g., 42 ml for a 70 kg person; 0.6 ml/kg). Repeated doses may be needed. VIG is in limited supply and is only available from the CDC under an FDA Investigation New Drug (IND) protocol.

References

1. Staib, C., Suezer, Y., Kisling, S., Kalinke, U. *et al.* (2006) Short term, but not post exposure, protection against lethal orthopoxvirus challenge after immunization with modified vaccinia virus Ankara. *J. Gen. Virol.*, **87**, 2917–2921.
2. Samuelsson, C., Hausmann, J., Lauterbach, H., Schmidt, M. *et al.* (2008) Survival of lethal poxvirus infection in mice depends on TLR9, and therapeutic vaccination provides protection. *J. Clin. Invest.*, **118**, 1776–1784.
3. Paran, N., Suezer, Y., Lustig, S., Israely, T. *et al.* (2009) Postexposure immunization with modified vaccinia virus Ankara or conventional Lister vaccine provides solid protection in a murine model of human smallpox. *J. Infect. Dis.*, **199**, 39–48.
4. Earl, PL., Americo, JL., Wyatt, LS., Espenshade, O., *et al.* (2008) Rapid protection in a monkeypox model by a single injection of a replication? Deficient vaccinia virus. *Proc. Natl. Acad. Sci. U.S.A.*, **105**, 10889–10894.

12
Arenaviruses: Hemorrhagic Fevers

Amy C. Shurtleff, Steven B. Bradfute, Sheli R. Radoshitzky, Peter B. Jahrling, Jens H. Kuhn, and Sina Bavari

Organism	Laboratory category	Transmissible by aerosol
Lassa virus, Machupo virus, Guanarito virus, Chapare virus, Lujo virus, most strains of Junín virus, and Sabiá virus	BSL-4	Yes

Organisms: are the causative agents of arenaviral hemorrhagic fevers.

Category

Lassa, Machupo, Guanarito, Chapare, Lujo, most strains of Junín, and Sabiá viruses are classified as BSL-4 arenaviruses. Lymphocytic choriomeningitis (LCMV) is the prototypical arenavirus. Certain strains of LCMV provide good models for the investigation of arenaviral hemorrhagic fever infection in nonhuman primates and guinea pigs, and they can be handled at BSL-2 or BSL-2/3, depending on the strain. Other arenaviruses such as Mopeia, Mobala, Pichindé, or Tacaribe are examples of arenaviruses which are close relatives of BSL-4 arenaviruses [1, 2]. These viruses are generally handled under BSL-2/3 conditions [2].

Recommended Respiratory Protection and Personal Protective Equipment

Arenaviruses are transmissible by direct contact with blood and body fluids, as well as indirectly by aerosolization of rodent excreta and airborne droplets. The viruses are chronically shed in excrement of the Natal praomys (*Praomys natalensis*) in endemic areas of western Africa, and this is likely the environmental source of aerosolized viruses [3, 4]. The viruses can be transmitted from patient to patient

BSL3 and BSL4 Agents: Epidemiology, Microbiology, and Practical Guidelines, First Edition.
Edited by Mandy C. Elschner, Sally J. Cutler, Manfred Weidmann, and Patrick Butaye.

through direct contact with blood or contaminated fluids, as well as via airborne droplets in a hospital environment [3, 5]. Through controlled experiments, it has been concluded that the biological half-life of aerosolized Lassa virus (LASV) at temperatures ranging from ambient room temperature to 38 °C, and in an environment of 30–80% relative humidity, was sufficiently long enough for airborne droplets to be transmitted between organisms [6]. In addition, aerosol transmission between experimentally infected animals has been observed in a laboratory setting [1]. Care must be taken to avoid generation of aerosols in the laboratory environment, for these viruses are highly infectious through aerosolization.

Based on the ease of transmission by aerosols, the lack of approved antiviral vaccines or therapeutics, and the high pathogenicity of the arenaviruses, BSL-4 laboratories are required for the handling and propagation of Lassa, Guanarito, Junín, Sabiá, Chapare, Lujo and Machupo viruses. BSL-4 laboratories require laboratory personnel to wear a one-piece positive pressure ventilated suit with a life support system in the form of house-supplied air or a similar air supply. These suits are manufactured by two sources: ILC Dover (Frederica, Del., USA) and Sperian (Roissy, France). The suits are made from heavy, durable, yet flexible chemically resistant vinyl or polyurethane-like material. They are affixed with replaceable, heavy duty rubber or nitrile work gloves. Personal respiratory protection such as high efficiency particulate air HEPA) filtered face masks or powered air purifying respirators (PAPRs) are not sufficient personal protective equipment (PPE). Personnel must wear sterilizable hospital scrub suits under these positive pressure suits, as well as latex primary gloves, earplugs, and socks. These pieces of clothing and PPE are removed after doffing the suit, and just before taking a personal shower before exiting the laboratory.

Arenaviruses can also be manipulated in a class III biological safety cabinet line, or "glove box," which protects the user from the virus contained entirely within a gas-tight, non-opening, negative pressure cabinet. Work in a class III cabinet is permitted without wearing the full one-piece positive pressure suit. Generally these class III hood lines are contained within a BSL-2/3 environment directly connected to a BSL-4 laboratory for loading, unloading, and decontamination, all of which is done on the BSL-4 side. Both the supply and exhaust air of the class III cabinet are HEPA filtered, and heavy duty rubber gloves are affixed to gas-tight ports in the side of the cabinet to allow a non-suited worker to manipulate the contents of the class III cabinet.

Best Disinfection

Arenaviruses are generally effectively inactivated through chemical and physical means. Solutions of 1% sodium hypochlorite (bleach), 2% glutaraldehyde, 10% neutral buffered formalin, and 5% Micro-Chem Plus (National Chemical Laboratories, Inc., Philadelphia, Pa.) have all been reported as effective chemical disinfectants [2].

Most viral hemorrhagic fever viruses can survive in a stored blood sample for about two weeks at room temperature. Heat has been shown to inactivate many of arenaviruses in laboratory media reagent solutions, or serum, or plasma [7, 8]. Heating serum at 60 °C for at least 1 h completely inactivated high titers of LASV [9]. Dilution of a LASV-containing sample at a 1 : 100 concentration into 3% acetic acid (pH 2.5), a diluent commonly used in hematological analysis, inactivated high titers of virus after only 15 min [9]. Other arenaviruses, LCMV, and Tacaribe viruses are sensitive to lipid solvents, such as ether and chloroform, as well as the detergents Triton X-100 and sodium deoxycholate [10], and it is reasonable to hypothesize that the BSL-4 arenaviruses would also be susceptible to these reagents. These arenaviruses are also rapidly inactivated by formalin, beta-propriolactone, hydrogen peroxide, and chloramine B [10].

Best Decontamination

All contaminated liquid and solid wastes from research materials are inactivated through chemical disinfection, autoclaving, or steam sterilization before removal from the containment suite [2]. Following primary decontamination, waste is removed from the facility and sent for incineration. Contaminated equipment is surface-decontaminated with 5% Micro-Chem Plus, removed from the biocontainment suite through an appropriate airlock chamber and exposed to formaldehyde gas. Infected animal tissue samples removed at necropsy must be fixed in 10% neutral buffered formalin for 21 days before they can be removed from biocontainment and taken to a laboratory for further processing. Cells grown on coverslips or in tissue culture plates must be fixed with a generous volume of 10% neutral buffered formalin for a full 24 h before they can be removed from biocontainment for further processing. Fresh virus-containing supernatants, infected cells, or animal tissue samples can be placed in Trizol (Invitrogen, Carlsbad, Calif.) or other phenolic compound at a volume of one part infectious sample to four parts Trizol and brought out of containment on the same day for RNA isolation processing [11].

Prevention

There are no FDA-approved vaccines available for the prevention of Lassa fever or any other arenaviral infection (but an efficacious and save vaccine to prevent Junín/Argentinian hemorrhagic fever, called Candid 1, is approved in the US as an Investigational New Drug). A multitude of studies are investigating vaccines based in different platforms for efficacy in animal models of Lassa virus infection. Recently Drs. Salvato and Lukashevich (Institute of Human Virology, University of Maryland, Baltimore, Md.) demonstrated the protective efficacy of a recombinant yellow fever 17D vaccine expressing LASV glycoproteins 1 and 2 [12]. They also described a reassortant vaccine made from the Mopeia and Lassa viruses, which provide sterilizing immunity in guinea pigs, which is also safe and effective in nonhuman primates [13]. Ongoing experimental work is evaluating virion-like particle as a vaccine platform for expression and immunogenicity of Lassa virus antigens [14]. Drs. Geisbert and Feldmann have engineered a vesicular stomatitis Indiana virus (VSIV) recombinant particle which expresses the LASV glycoprotein, and this replication competent vaccine protected nonhuman primates from a lethal LASV challenge [15]. Groups at USAMRIID and other locations are working on DNA vaccines expressing viral nucleoprotein or glycoproteins, with variable amounts of success [16].

Laboratory Accidents

Laboratory accidents, resulting in either inapparent infections or fully symptomatic cases with fatalities, have been documented for personnel handling arenaviruses. LCMV has caused fatal and non-fatal laboratory acquired cases, especially in workers handling infected animals which are suspected of generating aerosols [17–19]. Up to 10% of laboratory workers handling Junín virus, each of whom never reported symptoms of Junín virus infection, were shown to have anti-Junín antibodies which had functional and neutralizing characteristics similar to antibodies from 18 other Junín infection survivors [20]. Similarly, the handling of Pichindé virus, a South American BSL-2 arenavirus, has resulted in seroconversion in laboratory workers who manipulated cultures of very high concentrations [21]. There has been one case of Lassa fever infection in a laboratory worker who was handling cell cultures and infected mice [22]. In 1994, a laboratory investigator was infected with Sabiá virus while working at BSL-2/3, probably by the aerosol route due to a leak of virus from a centrifuge bottle into the chamber of a high-speed centrifuge [23]. The patient was isolated in a barrier facility in a hospital and fully recovered after a regimen of ribavirin administered following the Centers for Disease Control and Prevention protocol for exposure to other arenaviruses. These laboratory cases implicate the aerosol transmissibility of arenaviruses and underscore the need for these viruses to be handled in high biocontainment facilities.

Procedure Recommended in the Case of Laboratory Spill or Other Accident

At all times, a spill kit should be available immediately in the case of an accident in any biocontainment laboratory. The kit should contain nitrile gloves, some "Caution Biohazard – Do Not Enter" signs, autoclave tape, spray bottle with disinfectant such as bleach or Micro-Chem Plus, instructions for the disinfectant, absorbent towels such as wipes or paper towels, and any necessary PPE [24]. At BSL-4, wearing a positive pressure suit provides initial protection, but one must maintain a connection to the airline in the BSL-4 during the clean up process. Begin clean up by covering the spill with absorbent material to limit the spread of the liquid spill and aerosols, then spraying the chemical disinfectant onto the absorbent towels, working from around the perimeter of the area into the center. Allow sufficient contact time with the disinfectant to allow complete disinfection, then remove the soaked absorbent towels and repeat the process as necessary depending on the volume of the spill. Dispose soaked disinfectant towels in biohazard bags and proceed with autoclave waste disposal procedures. After clean up, the worker's supervisor and institutional officials responsible for safety and biosafety should be notified for appropriate accident and exposure reporting [2].

Treatment of Disease

The Centers for Disease Control and Prevention (CDC) recommends the treatment of symptomatic arenavirus infections, particularly Lassa virus, with ribavirin [25]. A loading dose of 30 mg/kg of body weight, followed by a dose of 15 mg/kg every 6 h for 4 days, and then by a dose of 7.5 mg/kg three times daily for 6 days is recommended [23, 25, 26]. There is no definitive experimental or clinical evidence that ribavirin effectively treats infection with each of the BSL-4 hemorrhagic fever arenaviruses, but there are published studies which report trends that indicate efficacy [27]. If exposure is suspected, the prophylactic dose regimen is oral ribavirin administration with 500 mg every 6 h for 7 days [25]. There are no other FDA-approved small molecule therapeutics available to treat arenavirus infection at this time.

References

1. Peters, C.J., Jahrling, P.B., Liu, C.T., Kenyon, R.H. *et al.* (1987) Experimental studies of arenaviral hemorrhagic fevers. *Curr. Top. Microbiol. Immunol.*, **134**, 5–68.
2. U.S. Department of Health and Human Services, Public Health Service Centers for Disease Control and Prevention, National Institutes of Health (2009) in *Biosafety in Microbiological and Biomedical Laboratories* (eds L.C. Chosewood and D.E. Wilson), U.S. Government Printing Office, *http://www.cdc.gov/biosafety/publications/bmbl5/BMBL.pdf* (accessed July 21 2009).
3. Monath, T.P. (1975) Lassa fever: review of epidemiology and epizootiology. *Bull. World Health Organ.*, **52** (4–6), 577–592.
4. Walker, D.H., Wulff, H., and Murphy, F.A. (1975) Experimental Lassa virus infection in the squirrel monkey. *Am. J. Pathol.*, **80** (2), 261–278.
5. Carey, D.E., Casals, J., Add, R.F. *et al.* (1972) Lassa fever. Epidemiological aspects of the 1970 epidemic, Jos, Nigeria. *Trans. R. Soc. Trop. Med. Hyg.*, **66** (3), 402–408.
6. Stephenson, E.H., Larson, E.W., and Dominik, J.W. (1984) Effect of environmental factors on aerosol-induced Lassa virus infection. *J. Med. Virol.*, **14** (4), 295–303.
7. Webb, P.A., Johnson, K.M., Mackenzi, R.B. *et al.* (1967) Some characteristics of Machupo virus, causative agent of Bolivian hemorrhagic fever. *Am. J. Trop. Med. Hyg.*, **16** (4), 531–538.
8. Allison, L., Salter, M., Mann, G. *et al.* (1985) Thermal inactivation of Pichinde virus. *J. Virol. Methods*, **11** (3), 259–264.
9. Mitchell, S.W. and McCormick, J.B. (1984) Physicochemical inactivation of Lassa, Ebola, and Marburg viruses and effect on clinical laboratory analyses. *J. Clin. Microbiol.*, **20** (3), 486–489.
10. Podoplekina, L.E., Shutova, N.A. *et al.* (1986) Influence of several chemical reagents on lymphocytic choriomeningitis and Tacaribe viruses. *Virology*, **37** (1), 43–48.
11. Blow, A., Dohm, D.J., Negley, D.L. *et al.* (2004) Virus inactivation by nucleic acid

extraction reagents. *J. Virol. Methods*, **119** (2), 195–198.

12. Bredenbeek, P.J., Molenkamp, R., Spaan, W.J., Deubel, V. *et al.* (2006) A recombinant Yellow Fever 17D vaccine expressing Lassa virus glycoproteins. *Virology*, **345** (2), 299–304.
13. Lukashevich, I.S., Carrion, R. Jr., Salvato, M.S., Mansfield, K. *et al.* (2008) Safety, immunogenicity, and efficacy of the ML29 reassortant vaccine for Lassa fever in small non-human primates. *Vaccine*, **26** (41), 5246–5254.
14. Branco, L.M., Grove, J.N., Geske, F.J., Boisen, M.L. *et al.* (2008) Lassa virus-like particles displaying all major immunological determinants as a vaccine candidate for Lassa hemorrhagic fever. *Virol. J.*, **7**, 279.
15. Geisbert, T.W., Jones, S., Fritz, E.A., Shurtleff, A.C. *et al.* (2005) Development of a new vaccine for the prevention of Lassa fever. *PLoS Med.*, **2** (6), e183.
16. Rodriguez-Carreno, M.P., Nelson, M.S., Botten, J., Smith-Nixon, K. *et al.* (2005) Evaluating the immunogenicity and protective efficacy of a DNA vaccine encoding Lassa virus nucleoprotein. *Virology*, **335** (1), 87–98.
17. Smadel, J.E., Green, R.H. *et al.* (1942) Lymphocytic choriomeningitis: two human fatalities following an unusual febrile illness. *Proc. Soc. Exp. Biol. Med.*, **49**, 683–686.
18. Scheid, W., Jochheim, K.A. *et al.* (1956) Laboratory infections with the virus of lymphocytic choriomeningitis. *Dtsch. Arch. Klin. Med.*, **203** (1), 88–109.
19. Baum, S.G., Lewis, A.M., Rowe, W.P., Heubner, R.J. *et al.* (1966) Epidemic nonmeningitic lymphocytic-choriomeningitis-virus infection. An outbreak in a population of laboratory personnel. *N. Engl. J. Med.*, **274** (17), 934–936.
20. Weissenbacher, M.C., Grela, M.E., Sabattini, M.M. *et al.* (1978) Inapparent infections with Junin virus among laboratory workers. *J. Infect. Dis.*, **137** (3), 309–313.
21. Buchmeier, M., Adam, E., and Rawls, W.E. (1974) Serological evidence of infection by Pichinde virus among laboratory workers. *Infect. Immun.*, **9** (5), 821–823.
22. Buckley, S.M. and Casals, J. (1970) Lassa fever, a new virus disease of man from West Africa. 3. Isolation and characterization of the virus. *Am. J. Trop. Med. Hyg.*, **19** (4), 680–691.
23. Barry, M., Russi, M., Armstrong, L. *et al.* (1995) Brief report: treatment of a laboratory-acquired Sabia virus infection. *N. Engl. J. Med.*, **333** (5), 294–296.
24. Headquarters Department of the Army (2009) Safety Standards for Microbiological and Biomedical Laboratories, Washington, DC, *http://www.fas.org/irp/doddir/army/pam385-69.pdf* (accessed July 21 2009).
25. CDC (1988) Management of patients with suspected viral hemorrhagic fever. *Morb. Mortal. Wkly. Rep.*, **37** (Suppl. 3), 1–16, *http://www.cdc.gov/mmwr/preview/mmwrhtml/00037085.htm* (accessed July 21 2009).
26. McCormick, J.B. (1986) Clinical, epidemiologic, and therapeutic aspects of Lassa fever. *Med. Microbiol. Immunol.*, **175** (2–3), 153–155.
27. Charrel, R.N. and de Lamballerie, X. (2003) Arenaviruses other than Lassa virus. *Antiviral. Res.*, **57** (1–2), 89–100.

13
Filoviruses: Hemorrhagic Fevers

Victoria Wahl-Jensen, Sheli R. Radoshitzky, Sina Bavari, Peter B. Jahrling, and Jens H. Kuhn

Organism	Laboratory category	Transmissible by aerosol
Marburg virus, Ravn virus, Bundibugyo virus, Ebola virus, Reston virus, Sudan virus, Taï Forest virus, Lloviu virus	BSL-4	Only in animal experiments

Filoviruses are transmitted by direct person to person contact (BDBV, EBOV, MARV, RAVV, SUDV, TAFV) and probably by animal to animal contact (LLOV, RESTV) during natural disease outbreaks. There is no evidence for aerosol transmission in natural settings. However, filoviruses can be transmitted to laboratory animals (rodents and nonhumans primates) via the aerosol route [1].

Recommended Respiratory Protection and Particular Personal Protective Equipment

Filoviruses are Risk Group 4 pathogens and therefore require maximum containment and the most vigilant personal protection equipment (PPE) and behavior in conjunction with rigorous standard operating procedures. Research with and handling of any materials suspected to contain infectious filoviruses must be done in BSL-4 containment, either in a class III cabinet line using standard PPE or in a full-body, hermetically sealed "space" suit connected to an air supply [2–5].

Disinfection/Decontamination

Filoviruses are enveloped RNA viruses, and are therefore most easily destroyed with detergents, heat, or irradiation. It is the current consensus that any effective

BSL3 and BSL4 Agents: Epidemiology, Microbiology, and Practical Guidelines, First Edition.
Edited by Mandy C. Elschner, Sally J. Cutler, Manfred Weidmann, and Patrick Butaye.

inactivation method works for all filoviruses even if it has been tested only with a particular filovirus. MARV can be inactivated with 1% sodium deoxycholate solutions, and with chloroform overnight at 4 °C when one part chloroform is mixed with three parts of cell-free MARV suspension. Other MARV inactivation methods include addition of acetone, diethyl ether, 1% formalin, methanol, sodium hypochlorite, Tego MGH, osmium tetraoxide, glutaraldehyde, 2% peracetic acid, phenol, SDS, Tween 20, chloramines B, β-proprionolactone, EGTA + 3% acetic acid, and Triton X-100. UV light (30 s at 10 cm distance) also inactivates MARV, as does photodynamic treatment (15 000 lux, 30 s at 50 cm) in the presence of 1 : 50 000 methylene blue at 1 : 100 000 thiopyronine [6–18].

Nanoemulsion ATB was shown to inactivate EBOV in cell culture supernatants within 20 min when diluted 1 : 100, or in tissues when diluted 1 : 10 [18]. EBOV could be inactivated within 30 min with heat (75 °C); likewise EBOV- and MARV-containing serum samples could be inactivated within 60–75 min at 60 °C or in combination with 0.2% SDS and 0.1% Tween 20 within 15 min. Blood smears can be inactivated by exposure to methanol for 30 min and subsequent exposure to 90–100 °C [19–23]. A novel method for filovirus inactivation is the use of the photoinduced alkylating probe 1,5-iodonaphthylazide (100 mmol/l INA for 30 min at 4 °C) in the presence of UV irradiation (310–360 nm) [24].

Prevention

There are currently no approved human or animal vaccines for the prevention or post exposure prophylaxis of filovirus infections. Experimentally, several vaccine platforms have been developed that can achieve protection from filovirus disease in nonhuman primates. However, all of these platforms only protect against challenge with homotypic filovirus – current experimental EBOV vaccines do not protect against MARV or vice versa. The most promising vaccine platforms are those based on DNA and/or recombinant replication-deficient adenoviruses, replication-competent vesicular stomatitis indiana viruses, and filovirus – like particles [25–29].

There are no case reports of filovirus-infected people being treated with an experimental vaccine or treatment regimen. DNA- and adenovirus-based vaccine platforms are currently evaluated in early-stage clinical trials.

Post Exposure Prophylaxis

There is no approved post exposure prophylaxis, but vaccines based on replicating recombinant vesicular stomatitis Indiana viruses and antisense molecules have protected nonhuman primates when administered within hours after exposure [30–33].

Laboratory Accidents

Laboratory accidents with filovirus usually occur during the inoculation stage within animal experiments. To date, six such infections have been described. Accidents with EBOV occurred in 1977 at MRE, United Kingdom [34, 35] (one nonlethal case) [36], in 1996 at Sergiev Posad-6, Russia (one lethal case) [37] and in 2004 at SRCVB "Vektor," Russia (one lethal case). Accidents with MARV occurred in 1988 [38] and 1990 [39], also at SRCVB "Vektor" (one lethal case, one nonlethal case).

Procedure Recommended in the Case of Laboratory Spill or Other Type of Accident

Spill

1) Spray paper towels with disinfectant and cover the area of the spill.
2) Gently apply additional disinfectant to the area, working from the margins toward the center with a smoothing flow of disinfectant to cover the entire area from a margin larger than the spill.
3) Let treated area stand for 20 min, then collect toweling and excess disinfectant in leakproof containers for same-day autoclaving. Mop floor or surface with disinfectant, and then rinse with water.
4) If spill contains radioactive materials, modify procedure as directed by specialists, who will test personnel and surfaces for residual radioactivity.
5) If spill contains broken glass, do not use hands to pick up glass; always use whiskbroom and dust pan.
6) Notify suite supervisor, personnel supervisor, and appropriate authorities in the building.

Accident with Personnel Involvement

Follow the protocol put in place by the Occupational Medical Services (OMS), which usually includes immediate exit from the suite with subsequent immediate disinfection of the wound; inform all supervisors and OMS and then follow their instructions.

Treatment of Disease

There is currently no approved specific treatment for filovirus disease. Care of patients is therefore supportive, using oral and intravenous nutritional support, electrolytes, analgesics, antibiotics, and antimycotics to treat secondary infections, and using plasma expanders, and blood or platelet substitutions to control blood loss and thrombocytopenia. There are no statistical data that prove the actual effectiveness of these measures. Experimentally, certain pharmaceuticals have been used to protect nonhuman primates from filovirus infection, disease, or death. The most promising are antisense platforms [40, 41].

References

1. Kuhn, J.H. (2008) *Filoviruses. A Compendium of 40 Years of Epidemiological, Clinical, and Laboratory Studies*, Springer Wien, NewYork, Vienna, Austria.
2. Henkel, R.D., Sandberg, R.L., and Hilliard, J.K. (2002) in *Anthology of Biosafety. V. BSL-4 Laboratories* (ed. Y. Richmond Jonathan), American Biological Safety Association, Mundelein, Ill, pp. 237–251.
3. Marklund, L.A. (2003) Patient care in a biological safety level-4 (BSL-4) environment. *Crit. Care Nurs. Clin. North Am. (Philadelphia)*, **15** (2), 245–255.
4. Stuart, D., Hilliard, J., Henkel, R., Kelley, J. *et al.* (1999) in *Anthology of Biosafety. I. Perspectives on Laboratory Design* (ed. Y. Richmond Jonathan), Chapter 10, American Biological Safety Association, Mundelein, Ill, pp. 149–160.
5. U. S. Department of Health and Human Services, Centers for Disease Control and Prevention, National Institutes of Health (2007) *Biosafety in Microbiological and Biomedical Laboratories (BMBL)*, 5th edn, HHS Publication No. (CDC) 93-8395, U.S. Government Printing Office, Washington, DC.
6. Bowen, E.T.W., Simpson, D.I.H., Bright, W.F., Zlotnik, I. *et al.* (1969) Vervet monkey disease: studies on some physical and chemical properties of the causative agent. *Br. J. Exp. Pathol.*, **50** (4), 400–407.
7. Grolla, A., Lucht, A., Dick, D., Strong, J.E. *et al.* (2005) Laboratory diagnosis of Ebola and Marburg hemorrhagic fever. *Bull. Soc. Pathol. Exot.*, **98** (3), 205–209.
8. Kiley, M.P., Bowen, E.T.W., Eddy, G.A., Isaäcson, M. *et al.* (1982) Filoviridae: a taxonomic home for Marburg and Ebola viruses? *Intervirology*, **18** (1–2), 24–32.
9. Kunz, Ch. and Hofmann, H. (1971) in *Marburg Virus Disease* (eds G.A. Martini and R. Siegert), Springer-Verlag, Berlin, pp. 109–111.
10. Malherbe, H. and Strickland-Cholmley, M. (1971) in *Marburg Virus Disease* (eds G.A. Martini and R. Siegert), Springer-Verlag, Berlin, pp. 188–194.
11. May, G. and Herzberg, K. (1969) Comparison of an agent of a communicable disease of monkeys (Cercopithecus aethiops sic: Cercopithecus aethiops) with a vesicularstomatitis virus. *Zentralbl. Bakteriol. Parasitenkd. Infektionskr. Hyg.*, **211** (2), 133–147.
12. Mitchell, S.W. and McCormick, J.B. (1984) Physicochemical inactivation of Lassa, Ebola, and Marburg viruses and effect on clinical laboratory analyses. *J. Clin. Microbiol.*, **20** (3), 486–489.
13. Pereira, M.S., a reply from Lloyd, G., Bowen, E.T.W., Slade, J.H.R., and Simpson, D.I.H. (1982) Inactivating Lassa and Marburg/Ebola viruses. *The Lancet*, **320/ii** (8290), 155.
14. Peters, C.J., Jahrling, P.B., Ksiazek, T.G., Johnson, E.D. *et al.* (1992) *Filovirus Contamination of Cell Cultures*, Developments in Biological Standardization, Vol. **76**, S. Karger, Basel, Switzerland, pp. 267–274.
15. Slenczka, W.G. (1999) in *Marburg and Ebola Viruses*, Current Topics in Microbiology and Immunology, Vol. **235** (ed. H.-D. Klenk), Springer, Heidelberg, pp. 49–75.
16. van der Groen, G. and Elliot, L.H. (1982) Use of betapropionolactone inactivated Ebola, Marburg and Lassa intracellular antigens in immunofluorescent antibody assay. *Ann. Soc. Belg. Méd. Trop.*, **62** (1), 49–54.
17. Muntyanov, V.P., Kryuk, V.D., and Belanov, Y.F. (1996) Disinfecting action of chloramine B on Marburg virus. *Vopr. Virusol.*, **41** (1), 42–43.
18. Chepurnov, A.A., Chuyev, Y.P., Pyankov, O.V., and Yefimova, I.V. (1995) Effects of some physical and chemical factors on inactivations of Ebola virus. *Vopr. Virusol.*, **40** (2), 74–76.
19. Bray, M., Davis, K., Geisbert, T., Schmaljohn, C. *et al.* (1998) A mouse model for evaluation of prophylaxis and therapy of Ebola Hemorrhagic fever. *J. Infect. Dis.*, **178** (3), 651–661.
20. Grolla, A., Lucht, A., Dick, D., Strong, J.E. *et al.* (2005) Laboratory diagnosis of Ebola and Marburg hemorrhagic fever. *Bull. Soc. Pathol. Exot.*, **98** (3), 205–209.

21. Hambling, M.H., Freeman, R., Tovey, L.D., and Stevenson, J. (1978) Decreasing any viral infectivity in blood-smears for malarial parasite examination. *The Lancet*, **311/i** (8057), 222.
22. Mitchell, S.W. and McCormick, J.B. (1984) Physicochemical inactivation of Lassa, Ebola, and Marburg viruses and effect on clinical laboratory analyses. *J. Clin. Microbiol.*, **20** (3), 486–489.
23. World Health Organization - Division of Emerging and Other Communicable Diseases Surveillance and Control (1997) WHO recommended guidelines for epidemic preparedness and response: Ebola Haemorrhagic Fever (EHF), Geneva, Switzerland.
24. Warfield, K.L., Swenson, D.L., Olinger, G.G., Kalina, W.V. *et al.* (2007) Ebola virus inactivation with preservation of antigenic and structural integrity by a photoinducible alkylating agent. *J. Infect. Dis.*, **196** (Suppl. 2), 276–283.
25. Jones, S.M., Feldmann, H., Ströher, U., Geisbert, J. *et al.* (2005) Live attenuated recombinant vaccine protects nonhuman primates against Ebola and Marburg viruses. *Nat. Med.*, **11** (7), 786–790.
26. Sullivan, N.J., Sanchez, A., Rollin, P.E., Yang, Z.Y. *et al.* (2000) Development of a preventive vaccine for Ebola virus infection in primates. *Nature*, **408** (6812), 605–609.
27. Sullivan, N.J., Geisbert, T.W., Geisbert, J.B., Xu, L. *et al.* (2003) Accelerated vaccination for Ebola virus haemorrhagic fever in non-human primates. *Nature*, **424** (6949), 681–684.
28. Sullivan, N.J., Geisbert, T.W., Geisbert, J.B., Shedlock, D.J. *et al.* (2006) Immune protection of nonhuman primates against Ebola virus with single low-dose Adenovirus vectors encoding modified GPs. *PLoS Med.*, **3** (6), 865–873.
29. Warfield, K.L., Swenson, D.L., Olinger, G.G., Kalina, W.V. *et al.* (2007) Ebola virus-like particle-based vaccine protects nonhuman primates against lethal Ebola virus challenge. *J. Infect. Dis.*, **196** (Suppl. 2), 430–437.
30. Daddario-DiCaprio, K.M., Geisbert, T.W., Ströher, U., Geisbert, J. *et al.* (2006) Postexposure protection against Marburg haemorrhagic fever with recombinant vesicular stomatitis virus vectors in non-human primates: an efficacy assessment. *The Lancet*, **367** (9520), 1399–1404.
31. Geisbert, T.W., Daddario-DiCaprio, K.M., Williams, K.J., Geisbert, J.B. *et al.* (2008) Recombinant vesicular stomatitis virus vector mediates postexposure protection against Sudan Ebola hemorrhagic fever in nonhuman primates. *J. Virol.*, **82** (11), 5664–5668.
32. Geisbert, T.W., Lee, A.C., Robbins, M., Geisbert, J.B. *et al.* (2010) Postexposure protection of non-human primates against a lethal Ebola virus challenge with RNA interference: a proof-of-concept study. *Lancet*, **375** (9729), 1896–1905.
33. Warren, T.K., Warfield, K.L., Wells, J., Swenson, D.L. *et al.* (2010) Advanced antisense therapies for postexposure protection against lethal filovirus infections. *Nat. Med.*, **16** (9), 991–994.
34. Bowen, E.T.W., Lloyd, G., Platt, G., McArdell, L.B. *et al.* (1978) in *Ebola Virus Haemorrhagic Fever – Proceedings of an International Colloquium on Ebola Virus Infection and Other Haemorrhagic Fevers, Antwerp, Belgium, 6–8 December, 1977* (ed. S.R. Pattyn), Elsevier/North-Holland Biomedical Press, Amsterdam, pp. 95–102.
35. Emond, R.T.D., Evans, B., Bowen, E.T.W., and Lloyd, G. (1977) A case of Ebola virus infection. *Br. Med. J.*, **2** (6086), 541–544.
36. Borisevich, I.V., Markin, V.A., Firsova, I.V., Yevseyev, A.A. *et al.* (2006) Hemorrhagic (Marburg, Ebola, Lassa, and Bolivian) fevers: epidemiology, clinical pictures, and treatment. *Vopr. Virusol.*, **51** (5), 8–16.
37. Akinfeyeva, L.A., Aksyonova, O.I., Vasilyevich, I.V., Ginko, Z.I. *et al.* (2005) A case of Ebola hemorrhagic fever. *Infektsionnye Bolezni*, **3** (1), 85–88.
38. Alibek, K. and Handelman, S. (1999) *Biohazard*, Random House, New York.
39. Nikiforov, V.V., Turovskii, Y.I., Kalinin, P.P., Akinfeyeva, L.A. *et al.* (1994) A case of Marburg virus laboratory infection. *Zh. Mikrobiol. Epidemiol. Immunobiol.*, **3**, 104–106.

40. Geisbert, T.W., Lee, A.C., Robbins, M., Geisbert, J.B. *et al.* (2010) Post-exposure protection of non-human primates against a lethal Ebola virus challenge with RNA interference: a proof-of-concept study. *Lancet*, **375** (9729), 1896–1905.

41. Warren, T.K., Warfield, K.L., Wells, J., Swenson, D.L. *et al.* (2010) Advanced antisense therapies for postexposure protection against lethal filovirus infections. *Nat. Med.*, **16** (9), 991–994.

14
Bunyavirus

BSL3 and BSL4 Agents: Epidemiology, Microbiology, and Practical Guidelines, First Edition.
Edited by Mandy C. Elschner, Sally J. Cutler, Manfred Weidmann, and Patrick Butaye.

14.1
Crimean Congo Hemorrhagic Fever Virus

Ali Mirazimi

Organism	Laboratory category	Transmissible by aerosol
Crimean Congo hemorrhagic fever virus	BSL-4	Yes

Recommended Respiratory Protection

FFP3 masks or positive ventilated masks attached to FFP3 filters are recommended to avoid an infection via the respiratory route. There are several suppliers on the market (e.g., 3M, Dräger). This protection is recommended because of the BSL4 categorization. Several instances of nosocomial infection indicate the infectivity of fomites. The level of infectivity however has never been assessed. Several fact sheets by governmental organizations mention that in some Russian cases aerosol transmission was suspected and that aerosols from rodent excreta may be infectious. The original literature however is difficult to trace.

Recommended Personal Protective Equipment

Standard BSL-3/BSL-4 PPE recommended.

Best Disinfection

The enveloped virus is easily disinfected with 70% EtOH or detergents.

Best Decontamination

The enveloped viruses are easily decontaminated with 70% EtOH or detergents.

BSL3 and BSL4 Agents: Epidemiology, Microbiology, and Practical Guidelines, First Edition.
Edited by Mandy C. Elschner, Sally J. Cutler, Manfred Weidmann, and Patrick Butaye.

Prevention

An inactivated suckling mouse brain-derived vaccine was developed and is used in Bulgaria.

- Papa, A., Papadimitriou, E., and Christova, I. (2010) The Bulgarian vaccine Crimean-Congo haemorrhagic fever virus strain. *Scand. J. Infect. Dis.*, **43** (3), 225–229.

Case Reports

There are several clinical case reports available.

- Jamil, B., Hasan, R.S., Sarwari, A.R., Burton, J. *et al.* (2005) Crimean-Congo hemorrhagic fever: experience at a tertiary care hospital in Karachi, Pakistan. *Trans. R. Soc. Trop. Med. Hyg.*, **99** (8), 577–584.

Laboratory Accidents

Laboratory accidents have not been reported but there are many cases of nosocomial transmission in hospitals. One publication reports accidents that occurred while taking samples from patients.

- Tutuncu, E., Gurbuz, E., Ozturk, Y., Kuscu, B. *et al.* (2009) Crimean Congo haemorrhagic fever, precautions and ribavirin prophylaxis: a case report. *Scand. J. Infect. Dis.*, **41**, 378–380.

Procedure Recommended in the Case of Laboratory Spill or Other Type of Accident

Standard BSL-4 procedures.

Treatment of Disease

There are anecdotal reports of the use of ribavirin for the treatment of CCHF. Results are conflicting and the use of ribavirin is under an ongoing debate. Apart from that only a nonspecific treatment is available.

- Fisher-Hoch, S.P., Khan, J.A., Rehman, S., Mirza, S. *et al.* (1995) Crimean Congo-haemorrhagic fever treated with oral ribavirin. *Lancet*, **346** (8973), 472–475.

Clinical Guidelines

A lot of experience has been collected in Turkey where more than 4000 cases have occurred since 2004.

- Ergönül, O. (2006) Crimean-Congo haemorrhagic fever. *Lancet Infect. Dis.*, **6** (4), 203–214.

14.2 Rift Valley Fever Virus: a Promiscuous Vector Borne Virus

Manfred Weidmann, F. Xavier Abad, and Janusz T. Paweska

Organism	Laboratory category	Transmissible by aerosol
Rift valley fever virus	BSL-3	Yes

Recommended Respiratory Protection

Use respiratory protection when handling or processing carcasses, infected organs, tissues, or blood. In field conditions, a FFP3 or N-95 mask in combination with face shields should be worn to avoid direct contact of aerosols with the external mucosa.

During laboratory work, when processing virus stocks or heavily infected tissues from experimental animal infection, or when performing aerosol-generating procedures, such as centrifugation, a positive pressure mask equipped with battery-powered blowers (also called a powered air-purifying respirator (PAPR) or a FFP3 mask are recommended (*http://pathport.vbi.vt.edu/pathinfo/pathogens/Rift_Valley_Fever_virus.html*).

Recommended Personal Protective Equipment

In the field and in the laboratory the precautions include wearing protective clothing including, at the minimum, goggles or face shield, waterproof apron or disposable protective gown or waterproof apron, and resistant gloves, but also a wrap around laboratory gown or sleeves, and boots. In the laboratory environment a second pair of gloves is also highly recommended (*http://pathport.vbi.vt.edu/pathinfo/pathogens/Rift_Valley_Fever_virus.html*).

Best Disinfection

Rift valley fever virus (RVFV) is an enveloped virus and not expected to persist in the environment for prolonged periods. It is also not a very resistant virus to chemical or

BSL3 and BSL4 Agents: Epidemiology, Microbiology, and Practical Guidelines, First Edition.
Edited by Mandy C. Elschner, Sally J. Cutler, Manfred Weidmann, and Patrick Butaye.

physical inactivation. The virus is destroyed by strong sunlight/ultraviolet radiation [1] but UV radiation is difficult to standardize. The RVFV is very susceptible to acid pH and readily inactivated below pH 6.2; it is most stable within pH 7–8 [1–3]. Commonly used disinfectants, when properly used, will render viral suspensions inactive.

The best choice is to perform a first step disinfection with Virkon 1% (contact time 10 min) and a final disinfection (which allows the elimination of Virkon residues) with ethanol 70% for 10 min more. All paper and material used in the disinfection has to be autoclaved before disposal. Virkon can be replaced by PERAsafe, a peracetic-based disinfectant which is less aggressive to metals.

Best Decontamination

For contaminated areas (animal boxes, necropsy rooms, electronic devices, etc.), fumigation with formaldehyde [1, 2] and hydrogen peroxide vapors are very effective in inactivating enveloped viruses [4]. Low concentrations of formalin are able to totally inactivate liquid RVFV suspensions [5]. For the decontamination of thermostable materials use dry heat for 2 h at 160 °C or autoclave at 120 °C and 1 atm for 20 min.

To decontaminate and safely handle sera from infected animals, mix the sera and PBS with 1% Tween (final concentration 0.5%) and inactivate at 56 °C for 1 h to keep the immunological properties of the material unchanged [6].

RVFV is highly sensitive to typical lipid solvents (ether, chloroform) [5].

With respect to the use of molecular techniques, most the extraction reagents (e.g., TRIzol® LS reagent or AVL buffer) appear to inactivate RVFV completely among other arboviruses, supporting the reliance upon either to render clinical or environmental samples non-infectious, which has implications for the handling and processing of samples outside biocontainment [7].

RVFV is sensible to heat: viremic blood became non-infective after 40 min of incubation at 56 °C in phosphate buffer, pH 7.2 [8]. A knock down of 2 $\log_{10}$ of infectious titer at 42 °C in 48 h has been reported [9]. Heating for 1 h at 60 °C completely destroyed the infectivity of both the sucrose/acetone-extracted antigen of RVFV and RVFV-infected mouse brain [10]. Pasteurization is able to inactivate the virus [2].

Prevention

A live veterinary vaccine is effective, highly immunogenic, and confers lifelong immunity, but it tends to be teratogenic and induces abortions in up to 15% of pregnant animals in the susceptible breeds of sheep and goats.

Known Laboratory Accidents

The potential for infection of humans by contact or aerosols transmission was first recognized in veterinarians performing necropsies; contact with infected animal tissues and infectious aerosols is dangerous; many infections have been documented in herders, slaughterhouse workers, and veterinarians [11]. Most of these infections resulted from exposure to blood and other tissues, including the aborted fetal tissues of sick animals. In the first half of the past century, 47 laboratory infections were reported; before modern containment and vaccination became available virtually every laboratory that began work with the virus suffered infections suggestive of aerosol transmission [12, 13]. It was reported that workers were infected when scraping the walls of an animal room used three months earlier for RVF studies [14]. Well documented secondary infections among contacts and medical personnel not parenterally exposed have occurred (*http://pathport.vbi.vt.edu/pathinfo/pathogens/Rift_Valley_Fever_virus.html*).

Procedure Recommended in the Case of Laboratory Spill or Other Type of Accident

Use standardized procedures when dealing with spills inside biosafety cabinets, on the working bench, or on the floor. In all cases, the chain of activity is to delimit the area, pouring an appropriate disinfectant (Virkon 1%, or household bleach diluted 1 : 10, or PERAsafe) over the spill, from the outside edge inwards, leaving this to act for at least 15 min, and further collection of all adsorbent materials for later disposal in a biological waste container. Currently, a second disinfection is performed, using ethanol 70% to eliminate the first disinfectant residues. Finally, report the incident to the biosafety officer.

Treatment of Disease

Concerning therapeutic treatments, the administration of antibodies, interferon, interferon inducer, or rivabirin as assayed in experimentally infected RVFV mice, rats, or monkeys has been shown to be efficient in protecting against the disease, but these treatments have never been tested to treat RVFV-infected patients.

References

1. FAO (2001) Manual on Procedures for Disease Eradication by Stamping Out. Part 3: Decontamination Procedures. Produced by: Agriculture and Consumer Protection, *http://www.fao.org/DOCREP/004/Y0660E/Y0660E00.HTM#TOC* (accessed August 13 2011).
2. Department of Primary Industries and Energy (1996) Australian Veterinary Emergency Plan, AUSVETPLAN, Disease Strategy Rift Valley Fever, 2.0 edn, Electronic version 2.0, DPIE, Canberra, ISBN: 0 642 24506 1.
3. The Center for Food Security and Public Health and Institute for International

Cooperation in Animal Biologies, Iowa State University (2006) Rift Valley Fever Infectious Enzootic Hepatitis of Sheep and Cattle.

4. Heckert, R.A., Best, M., Jordan, L.T., Dulac, G.C. *et al.* (1997) Efficacy of vaporized hydrogen peroxide against exotic animal viruses. *Appl. Environ. Microbiol.*, **63**, 3916–3918.
5. Swanepoel, R. and Coetzer, J.A.W. (1994) in *Infectious Diseases of Livestock*, vol. **I** (eds J.A.W. Coetzer, G.R. Thomson, and R.C. Tustin), Oxford University Press, Cape Town, pp. 688–717.
6. Van Vuren, P.J. and Paweska, J.T. (2009) Laboratory safe detection of nucleocapsid protein of Rift Valley fever virus in human and animal specimens by a sandwich ELISA. *J. Virol. Methods*, **157**, 15–24.
7. Blow, J.A., Dohm, D.J., Negley, D.L., and Mores, C.N. (2004) Virus inactivation by nucleic acid extraction reagents. *J. Virol. Methods*, **119**, 195–198.
8. Finlay, G.M. (1932) The infectivity of Rift Valley fever for monkeys. *Tran. R. Soc. Trop. Med. Hyg.*, **26**, 161–168.
9. Terasaki, K., Murakami, S., Lokugamage, K.G., and Makino, S. (2011) Mechanism of tripartite RNA genome packaging in Rift Valley fever virus. *Proc. Natl. Acad. Sci.*, **108**, 804–809.
10. Saluzzo, J.F., Leguenno, B., and Van der Groen, G. (1988) Use of heat inactivated viral haemorrhagic fever antigens in serological assays. *J. Virol. Methods*, **22**, 165–172.
11. WHO (2010) RVFV Fact Sheet n° 207.
12. Francis, T. and Magill, T.P. Jr. (1935) Rift valley fever: a report of three cases of laboratory infection and the experimental transmission of the disease to ferrets. *J. Exp. Med.*, **62**, 433–448.
13. Smithburn, K.C., Haddow, A.J., and Mahaffy, A.F. (1949) Rift valley fever: accidental infections among laboratory workers. *J. Immunol.*, **62**, 213–227.
14. Brès, P. (1981) Prevention of the spread of Rift Valley fever from the African continent. *Contrib. Epidemiol. Biostat.*, **3**, 178–190.

14.3
Hantaviruses: the Most Widely Distributed Zoonotic Viruses on Earth

Jonas Klingström

Organism	Laboratory category	Transmissible by aerosol
Hantavirus	BSL-3Animal experiments with the HPS-causing Andes virus and Sin Nombre virus are classified in BSL-4	Yes

Recommended Respiratory Protection

For Hantaviruses categorized as BSL-3 no special protection is recommended, other than that routinely used in BSL-3. However in animal laboratories, animal breeding facilities, and in the field aerosols from rodent feces and urine can cause infections. FFP3 masks or positively ventilated masks attached to FFP3 filters are recommended to avoid infection via the respiratory route. There are several suppliers on the market (e.g., 3M, Dräger).

Recommended Personal Protective Equipment

Standard BSL-3/BSL-4 PPE recommended. Goggles, gloves, and gowns are also recommended in animal facilities.

Best Disinfection

The enveloped viruses are easily disinfected with 70% EtOH or detergents.

Best Decontamination

The enveloped viruses are easily decontaminated with 70% EtOH or detergents [1].

BSL3 and BSL4 Agents: Epidemiology, Microbiology, and Practical Guidelines, First Edition.
Edited by Mandy C. Elschner, Sally J. Cutler, Manfred Weidmann, and Patrick Butaye.

Prevention

A mouse brain-derived, formalin-inactivated vaccine Hantavax® is used in Korea/China but has no FDA approval [2, 3].

Post Exposure Prophylaxis Available?

None.

Laboratory Accidents

There are more than 100 recorded animal laboratory infections:

- Umenai, T., Lee, H.W., Lee, P.W., Saito, T. *et al.* (1979) Korean haemorrhagic fever in staff in an animal laboratory. *Lancet*, **1**, 1314–1316.
- Pedrosa, P.B. and Cardoso, T.A. (2011) Viral infections in workers in hospital and research laboratory settings: a comparative review of infection modes and respective biosafety aspects. *Int. J. Infect. Dis.*, **15**, e366–e376.

Procedure Recommended in the Case of Laboratory Spill or Other Type of Accident

Use standard BSL-3/BSL-4 procedures. All individuals inside P3 must evacuate the laboratory at once. In the case when an individual has been in physical contact with infectious material, decontaminate them first (70% EtOH). Decontamination of the spill is then, later, done by individuals dressed in protective gear, including respiratory protection.

Treatment of Disease

There are reports of ribavirine use against HFRS, but it has not been clearly shown that this has any positive effects. Only symptomatic treatment is available.

Clinical Guidelines

- Braun, N., Haap, M., Overkamp, D., Kimmel, M. *et al.* (2010) Characterization and outcome following Puumala virus infection: a retrospective analysis of 75 cases. *Nephrol. Dial. Transplant.*, **225** (9), 2997–3003.
- Tulumovic, D., Imamovic, G., Mesic, E., Hukic, M. *et al.* (2010) Comparison of the effects of Puumala and Dobrava viruses on early and long-term renal outcomes in patients with haemorrhagic fever with renal syndrome. *Nephrology*, **15** (3), 340–343.

- Simpson, S.Q., Spikes, L., Patel, S., and Faruqi, I. (2010) Hantavirus pulmonary syndrome. *Infect. Dis. Clin. North Am.*, **24** (1), 159–173.

References

1. Kraus, A.A., Priemer, C., Heider H., Krüger D.H. *et al.* (2005) Inactivation of Hantaan virus-containing samples for subsequent investigations outside biosafety level 3 facilities. *Intervirology*, **48**, 255–261.
2. Choi, Y., Ahn, C.J., Seong, K.M., Jung, M.Y. *et al.* (2003) Inactivated Hantaan virus vaccine derived from suspension culture of Vero cells. *Vaccine*, **21** (17–18), 1867–1873.
3. Schmaljohn, C. (2009) Vaccines for hantaviruses. *Vaccine*, **27** (Suppl. 4), D61–D64.

Index

BSL3 and BSL4 Agents: Epidemiology, Microbiology, and Practical Guidelines, First Edition.
Edited by Mandy C. Elschner, Sally J. Cutler, Manfred Weidmann, and Patrick Butaye.